Developments in Economic Geology, 22

mineral economics of africa

Further titles in this series

1. *I.L. ELLIOTT and W.K. FLETCHER (Editors)*
GEOCHEMICAL EXPLORATION 1974
2. *P.M.D. BRADSHAW (Editor)*
CONCEPTUAL MODELS IN EXPLORATION GEOCHEMISTRY
The Canadian Cordillera and Canadian Shield
3. *G.J.S. GOVETT and M.H. GOVETT (Editors)*
WORLD MINERAL SUPPLIES
Assessment and Perspective
4. *R.T. SHUEY*
SEMICONDUCTING ORE MINERALS
5. *J.S. SUMNER*
PRINCIPLES OF INDUCED POLARIZATION FOR GEOPHYSICAL EXPLORATION
6. *R.A. RICH, H.D. HOLLAND and U. PETERSEN*
HYDROTHERMAL URANIUM DEPOSITS
7. *J.G. MORSE (Editor)*
NUCLEAR METHODS IN MINERAL EXPLORATION AND PRODUCTION
8. *M. KUŽVART and M. BÖHMER*
PROSPECTING AND EXPLORATION FOR MINERAL DEPOSITS
9. *C.R.M. BUTT and I.G.P. WILDING (Editors)*
GEOCHEMICAL EXPLORATION 1976
10. *G.B. FETTWEIS*
WORLD COAL RESOURCES
Methods of Assessment and Results
11. *R.G. TAYLOR*
GEOLOGY OF TIN DEPOSITS
12. *H.K. GUPTA*
GEOTHERMAL RESOURCES
An Energy Altenative
13. *C.R.M. BUTT and R.E. SMITH (Editors)*
CONCEPTUAL MODELS IN EXPLORATION GEOCHEMISTRY, 4
Australia
14. *G. BÁRDOSSY*
KARSTIC BAUXITES
15. *A.W. ROSE and H. GUNDLACH (Editors)*
GEOCHEMICAL EXPLORATION 1980
16. *R.W. BOYLE*
GEOCHEMICAL PROSPECTING FOR THORIUM AND URANIUM DEPOSITS
17. *G.R. PARSLOW (Editor)*
GEOCHEMICAL EXPLORATION 1982
18. *M. KUŽVART*
INDUSTRIAL MINERALS AND ROCKS
19. *P. LAZNICKA*
EMPIRICAL METALLOGENY
Depositional Environments, Lithologic Associations and Metallic Ores
20. *O. RUDAWSKY*
MINERAL ECONOMICS
Development and Management of Natural Resources
21. *M. KUŽVART and M. BÖHMER*
PROSPECTING AND EXPLORATION OF MINERAL DEPOSITS (second completely revised edition)
23. *B.L. GULSON*
LEAD ISOTOPES IN MINERAL EXPLORATION

DEVELOPMENTS IN ECONOMIC GEOLOGY, 22

mineral economics of africa

N. de Kun

Rue Marie Depage 16, Brussels, 1180, Belgium

ELSEVIER, Amsterdam — Oxford — New York — Tokyo 1987

ELSEVIER SCIENCE PUBLISHERS B.V.
Sara Burgerhartstraat 25
P.O. Box 211, 1000 AE Amsterdam, The Netherlands

Distributors for the United States and Canada:

ELSEVIER SCIENCE PUBLISHING COMPANY INC.
52, Vanderbilt Avenue
New York, NY 10017, U.S.A.

Library of Congress Cataloging-in-Publication Data

De Kun, N. (Nicolas), 1923-
Mineral economics of Africa.

(Developments in economic geology ; 22)
Bibliography: p.
Includes index.
1. Mines and mineral resources--Africa. 2. Mineral industries--Africa. 3. Natural resources--Africa.
I. Title. II. Series.
TN115.D38 1987 553'.096 87-6769
ISBN 0-444-42795-3

ISBN 0-444-42795-3 (Vol. 22)
ISBN 0-444-41250-6 (Series)

Printed in The Netherlands

A OVERVIEW

This is a book about Africa. It is, therefore, a book about the development of water, fertilizers, energy, minerals and processing which are synonymous with prosperity and growth on the continent. On a land which faces its worst crisis in a century. Dedicated to a New Deal for Africa and to a better future, this volume sets some targets and describes industry and resources as they stand today. The 'Mineral Resources of Africa' (Elsevier 1964) provides more detail, most of which is still amply relevant. The data base has deteriorated since then as the World Bank pointed out and considerable effort has been made to gather as many as possible. If any praise is due for their collection, it must go to those who so generously provided documentation and background information.

A 1 INTRODUCTION

Acknowledgements

All I can do is to reiterate how much I appreciated the help of Dr. Paul G. Teleki, of the U.S. Geological Survey, Dr. Ebraham Shekarchi of the U.S. Bureau of Mines, Mr. Henri Martin of the European Economic Community, and Dr. Pierre Goossens of Bugéco.

M. Jean-Claude Lambeau and Mme Tielemans generously opened their files and libraries at the Royal Museum of Central Africa. Without the unfailing help of M. Daniel Leprince of Pétrofina, M. Philippe Parmentier and Dr. Jos Jonkers of the EEC and the efficient departments of the Bibliothèque Royale this volume would have been incomplete. EEC figures Eurostat of Luxemburg so helpfully produced were most valuable. Mr. John W. Stamper, Mr. John R. Lewis, Dr. G. Rabchevsky, and Mr George J. Coakley of the U.S. Bureau of Mines and Mr. John H. De Young Jr. of the U.S. Geological Survey provided indispensable data they and their associates gathered with the greatest patience and accuracy.

My special thanks go to M. Dominique Vastel of Cogéma, M. J. Meuret of Sominki, Dr. Eugene C. Shannon of the Liberian Geological Survey, Señor Alberto del Hoyo de Monteverde of the Junta de Canarias and Señor Alfonso Alvarez Miranda of Fosfatos de Bou-Craa. Dr. P. J. Hugo of the Minerals Bureau of South Africa kindly provided key publications ; so did Dr. D. J. Winterbank of the Geological Survey.

I must record a debt of gratitude to Drs. I. de Magnée and G. Panou of Brussels University and to Drs. H. Schmidt, U. Krauss and F. Barthel of the Bundesanstalt für Geowissenschaften und Rohstoffe.

Mr. H. von Wistinghausen of the Agence Internationale de l'Energie at OECD, M. F. del Marmol of the Belgian and Mr. M. T. Mutandingwe of the Zimbabwe diplomatic services, and Dr. Joseph Vandenberg of ECA were most helpful. Count Jacques de Villelume of the Moroccan polytechnic and Ir. R. Vanbellingen of the Comité belge des Grands Barrages assisted in every possible way.

To Dr. Enzo Fano of the United Nations, Dr. Dumitrescu and M. Z. M. Amevoh of Unesco I cannot express my appreciation enough.

The Columbia University Seminar on water contributed insights to many chapters of this volume ; its chairman Dr. G. Halasi-Kun and my colleagues there continue to earn my deepest gratitude.

Mrs. Isabelle Want most efficiently typed this manuscript, spiked with African and Arabic names and 140 tables.

TABLE OF CONTENTS

	Page
A Overview	5
A1 Introduction	5
Table of contents	7
Abbreviations	11
Legend	13
List of symbols classified according to groups	14
A2 The African share of world resources	19
A3 Dependence on Africa	34
A4 Distribution of resources within Africa	37
A5 History and development of the industry	57
1 The Maghreb	68
11 Morocco and Western Sahara	68
12 Algeria	79
13 Tunisia	88
14 Libya	93
2 Northeast Africa	101
21 Egypt	101
22 Sudan	109
23 Ethiopia	115
24 Djibouti	119
25 Somalia	120
3 The Sahel	123
31 The Gambia	123
32 Senegal	125
33 Mauritania	129
34 Mali	133
35 Burkina Faso	138
36 Niger	140
37 Chad	144
4 West Africa	146
40 Guinea-Bissau	146
41 Guinea	148
42 Sierra Leone	153
43 Liberia	157
44 Ivory Coast	160
45 Ghana	163
46 Togo	167
47 Benin	170
48 Nigeria	171
49 Cameroon	182

5 Central Africa 186

51 Central African Republic 186
52 Gabon 188
53 Congo 193
54 Angola 197
55 Zaīre 203
56 Zambia 214

6 East Africa 223

61 Burundi 223
62 Rwanda 225
63 Uganda 229
64 Kenya 232
65 Tanzania 236
66 Malaŵi 241
67 Moçambique 243

7 Southern Africa 249

71 Zimbabwe 249
72 Botswana 259
73 Namibia 266
74 Swaziland 272
75 Lesotho 274

8 South Africa 276

81 Organization and performance 276
82 Structure of the industry 279
83 Gold and diamonds 283
84 Energy 288
85 The metals of the Bushveld 293
86 Other metals 297
87 Building materials 302
88 Refractories and ceramics 305
89 Food and processing 307

9 Islands and the Sea 311

91 Malagasy Republic 311
92 Smaller Islands 318
93 Atlantic Islands 321
94 The Sea 324

List of acronyms 329

Further reading 337

Index 340

Definitions

Except for the traditionalist industries of oil and gas, tables and text follow the metric system, whose Greek building blocs are :

h hecto	k kilo	M mega	G giga		T tera
h hundred	k thousand	M million	B billion	and	T trillion

They combine, as shown in the following list of abbreviations, with the usual grams (g), meters (m), tons (all metric, t) and watts (w) ; b is shorthand for barrels, ct for carats (metric), d for day, ft for feet, l for liter, s or sec. for seconds and y for year.

Following the list of Abbreviations and Conversion factors, the symbols of the figures have been itemized, since the same are used on all maps except for figures 1 to 8. Lastly, the reader will find a list of Acronyms of companies and organizations involved in Africa, at the end of the book.

Unless otherwise specified, production data concern periods of one year rounded and averaged for the last five years. They are only indicative and should be read in conjunction with the text, which elucidates changes and trends. As to yearly statistics and comments, the reader is referred to Codigaz, the Minerals Yearbook , Mining Annual Review, the Oil and Gas Journal, World Oil, &.

So called granite and marble in the columns are trade terms. Additionally, production data on semi-precious stones are all too often unavailable ; these lines, therefore, reflect typical outputs of recent years.

Shipments of the past (signaled with a $^{+}$) are shown where sales foundered, so are projects (^{P}) predicting growth. Illicit exports plague countries with gold and diamonds (and even some other commodities) ; they are included in production data, which are, therefore, rough estimates. So are recoveries of the minor metals, usually refined abroad. With trade patterns shifting continuously, these data only illustrate trends.

In contrast, the word reserves is used in the strictest sense. 'Reserves' which will never be worked or not in the foreseeable future, because they have become uneconomical, have been therefore eliminated. Coal reserves are recoverable, so are certain EUR data on oil, but all others are in situ estimates out of which anything between 50 and 95% will be recovered. For purposes of comparison phosphate (and apatite) reserves have been reduced to their P_2O_5 content and iron ore to iron.

On the one hand, the reserve base of the U.S. Bureau of Mines includes both measured and indicated ; economic, marginally economic and some currently subeconomic resources. The South African Minerals Bureau equates this base with reserves. On the other hand, the U.S. Bureau of Mines uses the word resources in a broad sense unless classified in the United Nations categories of reliable estimates.

R1E, R1M and R1S for economically exploitable, marginally economic and sub-economic known deposits and R2E for economically exploitable extensions. Lastly, hydropower estimates in the columns reserves and resources define potentials.

Abbreviations

Am ammonia
AmNi ammonium nitrate

b barrels
boe barrels oil equivalent
by barrels per year
B billion = 1000 million
Bb billion barrels
Bcf billion cubic feet
Bcfy billion cubic feet per year
Bm^3y billion cubic meter per year
BPL bone phosphate of lime
Bt billion tons (gigaton)
BTU British thermal unit

c capacity
cal calory
cal/g calory per gram
cap capacity
carb carbonate
cc concentrates
cf cubic feet
cf/b cubic feet per barrel
cfd cubic feet per day
cm centimeter
coltan columbite-tantalite
ct metric carat = 0.2 grams
ct/m^3 carats per cubic meter
ct/Mo carats per month
cty carats per year
(C) center
Casa' Casablanca or Dar el Beïda
(CE) centereast, &

d day
d in cement tabulations : dry process
DAP diammoniumphosphate ($46\%\ P_2O_5$, $18\%\ N_2$)

E East
EUR estimated ultimate recoverable reserves
ECU European currency unit

f few, little

g/l gram per liter
g/m^3 gram per cubic meter
g/t gram per ton
G giga = billion
GDP gross domestic product
Germany Federal Republic or West Germany

GNP Gross National Product
Gw gigawatt = one million kilowatts
Gwh gigawatts/hour

H hecto, hundred
ha hectare = 10,000 m^2
Hm^3 hundred m^3

ind. indicated

k kilo = thousand
kb thousand barrels
kct 1000 carats
kg kilogram
kg/t kilograms per ton
km kilometer
km^2 square kilometer
kt kiloton = thousand tons
kty kilotons per year
kVA kilovolt ampère

l liter
l/d liter per day
l/t liter per ton
LDC less developed country
LNP liquified natural gas
LPG liquified petroleum gas

m meter
md millidarcy
mg/l milligram per liter
mm millimeter
m^3 cubic meter
m^3d cubic meter per day
m^3h cubic meter per hour
m^3Mo cubic meters per month
m^3s cubic meter per second
m^3t cubic meter per ton
M mega = million
MAP monoammoniumphosphate ($52-55\%\ P_2O_5$, $11\%\ N_2$)
Mb million barrels
Mboe million barrels oil equivalent
Mcf million cubic feet
Mcfd million cubic feet per day
Mct million carats
Mcty million carats per year
MJ megaJoule
Ml million liters
Mm^3 million cubic meters
Mm^3y million cubic meters per year
Mo month

MRA de Kun, N., 1964, Mineral Resources of Africa, Elsevier pp 766, Libr. Congr. N° 64-14180
Mt million tons = megaton
Mty million tons per year
µ micron = 0.001 mm

(N) north
NAc nitric acid
NPK nitrogen-phosphat-potash fertilizer

o.b. overburden
o e oil equivalent
OFS Orange Free State (South Africa)

P planned, project
p.a. per annum (per year)
PAc P acid, phosphoric acid (52-55% P_2O_5)
Pc cement process :
d dry
w wet
sd semidry
PGM Platinum Group Metals
PK phosphate-potash fertilizer
ppm parts per million
Pot potential
prod production
produc production

R Rand
RCA Central African Republic
reinj reinjected gas
RE Rare earth metals or lantanides, including cerium and yttrium
Rsrcs resources
Resv reserves
rv reserves
rvb reserve base
R1E economically exploitable resources
R1M marginally economic resources
R1S subeconomic resources
R2E economically exploitable extensions of known resources

s,sec. second
s,sd semidry process in cement tabulations
smelt(er)
(S) south
SAc S acid = sulfuric acid
SDR special drawing rights of the IMF
Soc. société or company
SP superphosphate (16-22% P_2O_5)
SPac superphosphoricacid (68-72% P_2O_5)
\$ U.S. dollars
\$...B... billion dollars
\$...M... million dollars
Σ sum, total

t metric ton or tonne
th tons per hour
ty tons per year
toe tons oil equivalent
td tons per day
tMo tons per month
T tera = trillion
Tcf trillion cubic feet
Tcfy trillion cubic feet per year
Tm^3 trillion cubic meters
TSP triple superphosphate (44-48% P_2O_5)
Tv Transvaal (South Africa)
Twh terawatthour = trillion kilowatthours

u.g. underground
U.K. United Kingdom of Great Britain and Northern Ireland
USA United States of America

w watt
w wet process in cement tabulations
(W) West

y per year

Conversion factors

1°API = (141.5 : specific gravity at 60°F) - 131.5
7.64 barrels Algerian oil = 1t
7.7 barrels Libyan oil = 1 t
1 bd = 50 ty
1 boe = 6 M BTU
1%BPL = 0.4576% P_2O_5
1 BTU = 252 cal.
1 BTU/lb = 0.556 cal/g
1000 cf = $28m^3$
1 cfd = 10 m^3y

50 CFA francs = 1 French franc
100 g/t = 3.26 troy ounces per long ton
1 ls (l/sec) = 15.8 gallons per minute
1 toe = 39.68 M BTU
= 10 M kcal
= 4 Gwh (at 34% thermal efficiency)
= 1.43 t coal

LEGEND

Alphabetic list of symbols

Alum
Aluminum
Andalusite
Antimony
Arsenic
Asbestos
Attapulgite

Barite
Bauxite
Bentonite
Beryl
Bismuth
Bituminous shales

Cadmium
Cement
Cesium
Chromite
Clay
Coal
Columbium
Copper
Copper and nickel
Corundum

Diamonds
Diatomite

Feldspar
Ferrochromium
Ferromanganese
Ferrosilicon
Fertilizer plant
Flint clay
Fluorspar
Fuller's earth

Garnet
Gas
Geothermal steam
Germanium
Gold
Graphite
Groundwater
Gypsum

Heavy oil
Hydropower plant
Hydropower site
Iron ore
Kaolin
Kyanite

Lead, zinc & silver
Lignite
Limestone
Lithium

Magnesite
Manganese nodules
Marble
Meerschaum
Mercury
Mica
Molybdenum

Nickel
Niobium

Oil

Peat
Perlite
Petroleum refinery
Phosphate
Platinum group
Potash
Pozzuolana
Pumice

Rare earths

Salt
Sand, silica
Semi precious stones
Silicon plant
Silver with lead
with gold
Soda and brines
Steel
Stone
Strontium
Sulfur

Talc
Tantalum

Thorium
Tiger's eye
Tin
Tin & niobium
Titanium minerals
plant
Tungsten

Uranium

Vanadium
Vermiculite

Wollastonite
Wonderstone

Yttrium

Zinc
with lead
Zirconium

List of symbols classified according to groups

BASE METALS GROUP

Barite B
Cadmium ↄ
Copper
Fluorite F
Germanium G
Lead, zinc and silver
Mercury Σ
Strontium Sr
Zinc Z

BUILDING MATERIALS

Asbestos
Cement ±
Gypsum ∪
Limestone L
Marble ⊏
Pumice q
Stone □
Vermiculite >
Wonderstone

CLAYS

Bentonite
Clay c
Flint clay 6
Fuller's earth
Kaolin

ENERGY

Bitumen ⊕
Coal
Gas ◎
Geothermal steam g
Heavy oil ⊕
Hydropower plant site
Lignite
Oil
Peat π
Uranium U

IRON-BAUXITE

Asbestos
Bauxite ≈
Iron
Manganese nodules
Vanadium ∀

METAMORPHICS

Andalusite group K
Corundum Я
Garnet ▼
Graphite ▦
Wollastonite ʍ

PRECIOUS, &

Antimony A
Arsenic
Diamonds △
Gold and silver •
Semi-precious stones ▽
Tiger's eye ◁

SEDIMENTS and WATER

Alum
Diatomite 0
Groundwater ⁝⁝
Phosphates
Potash P
Salt ∩
Silica ◇
Soda, brines S
Sulfur

TIN-TITANIUM

Beryl b
Bismuth β
Cesium ↄ
Feldspar
Lithium i
Mica m
Molybdenum
Niobium x
Rare earths R
Tantalum x
Thorium R
Tin +
Titanium t
Tungsten W
Zircon

ULTRABASICS

Chromite v
Magnesite M
Nickel >
Platinum group ^
Talc t

▌ signals processing

e.g. bauxite ≈
aluminum
limestone L
cement ±

Poverty amidst wealth

Africa dominates the world markets of the strategic metals : chromium, cobalt, manganese and platinum. Better than any other region, it can provide these substances which are scarce in North America, the European Economic Community and Japan, but in adequate supply in the Comecon area. Similarly, the continent is the leading supplier of diamonds, gold and uranium.

Needed and encouraged by the rest of the world, the newly independent countries stood, therefore, twenty years ago, when the "Mineral Resources of Africa" were published, on the threshold of progress and prosperity based on abundant natural resources. The food supply was satisfactory even if diets lacked balance and exports of commodities boomed : sales of gas and bauxite started and soared, and oil, coal and phosphate shipments trebled.

Based on these resources and a gross investment of 200 billion dollars, great strides were thus made in education and nation building ; sophisticated elites appeared and acquired wealth. But, with some notable exceptions, the vision of wealth and prosperity remained just that - a vision. By the seventies robust growth came to a halt, in many countries food became scarce, the infrastructure deteriorated, harbors were clogged or empty and many roads became inpracticable even in the dry season. Except for the first steps taken in the Arab North and the white-ruled South industrialization was only noticeable by its absence ; plants remained underutilized because of inadequate planning and coordination, and poor markets ; since spare parts could not be imported, equipment was cannibalized. The ensuing recession did more harm to the LDC's than to the industrialized world and many countries sank into poverty. Even the richest were debt ridden and to quote the secretary of the Organisation for African Unity, the continent is now in dire need.

Why ?

Firstly, some phenomena affected the whole developing world :

- With spreading synthetics, the share of solid minerals within the world product declined in some cases quite strongly.
- Commodities markets weakened during the last decade and prices fell by 10% in SDR terms.
- Sluggish demand triggered overproduction since governments tried to maintain (or even increase) their revenue based on lower unit value because they were unable or unwilling to reduce expenditures.
- The terms of trade turned against the exporters of raw commodities and, as the

World Bank pointed out, they were at their lowest since the recession of 1927.

- In response to the revolution of expectations, urbanization gathered speed in the tropics. More people joined thus the monetary economy creating thereby a demand for services that were hitherto unavailable or unattainable.

The causes of the crisis in Africa

The second set of phenomena is more specific for Africa :

- Albeit slowly, the rate of population growth declined in Asia and levelled out in Latin America. It is, however, still growing over and above 3% per year in Africa.
- At the same time, food production increased dramatically in Asia and even in Latin America. Indeed, the new breeds of wheat rice and corn were particularly well suited for these.
- In contrast, centralization of decision making and the fixing of inadequate food prices, encouraged the rush to the towns. The flight from the country side further reduced some governments commitments to farmers.
- To compound these problems, two lengthy periods of drought affected one African country out of three. On-going civil war destroyed the infrastructure and further diminished the water and food supply in three more.
- With the concentration of administration and spending in the urban sprawls, the erosion of soils gathered momentum in many countries. Likewise, drilling programs for water and the construction of wells had to be curtailed because of lack of funds or spares.
- While half a dozen Far Eastern countries and Brazil emerged as major manufacturers, the continent continued to rely on the export of its traditional, low priced, raw materials.
- Even if South Africa did not follow this trend, the ratios of beneficiation is on the whole the lowest in the world. Indeed, the value added by processing on non fuel commodities averages only 10% in Africa. As a comparison, processed materials are worth in the United States 900% of raw materials (energy excluded).
- Even these few plants are all too often not fully operational in Africa. To quote the World Bank : "Much industrial capacity stands idle, the victim of falling domestic incomes, poor investment choices, a failure to develop export opportunities and inadequate foreign exchange for materials and spare parts."

It is not surprising, therefore, that the GDP grew by only 4% from 1975 to 1980 in the Sub-Saharan area, which extends from the desert to the Limpopo and real incomes per head may not reach again their level of 1980 before 1990. To the north, the Maghreb and Egypt initially fared much better, but oil exporters had the biggest setbacks in recent years ; revenues of Libya and Nigeria were, for instance, halved. The ensuing cutbacks affected ambitious programs of processing. Although not as acute, the decline of South African commodities from their peaks of 1978-1980 was quite sharp.

Although Latin debt is much larger, African countries owe some $200-250 billion and according to the IMF the debt service of the fourty sub-Saharan states rose from $5 billion in 1983 to $11 billion in 1985. Further steep jumps are to be expected. Even wealthy South Africa had to curtail interest payments on half of its debt of $20 billion or more, and would wish to roll over several loans. In sub-Saharan Africa, access to new credit all but dried up and new private investment only caters to projects which have good prospects of profitability in countries which offer a good investment climate. Sadly enough, capital flows were halved in five years' time, and aid was sometimes counterproductive. Yet investments are still required and as Robert S. McNamara, former president of the World Bank pointed out, reversing the decline in sub-Saharan Africa is "the overriding development task that faces the world today".

The following ten points itemize the most urgent needs.

Remedial action

I. Obviously, the top priority is better management of existing resources and facilities. Even marginal improvements in decisions relating to the operations of agencies and enterprises would have substantial benefits in terms of resource savings and other efficiencies. Moreover, significant upgrading of performance has a multiplier effect on whole economies, particularly on African economies, which are in distress.

If management training and incentives are particularly important, enhancing managerial ethics follow suit. While the managerial culture is well established in South Africa, the training and education of Africans to share responsibilities in its mineral industries and in Namibia has acquired more urgency.

II. Better management would ultimately improve policies and the selection of priorities. Although agricultural commodities are beyond the scope of this volume, such immediate targets as the protection of soils against erosion and rapid water resources development are not. Yet, the latter only receives limited aid transfers and is relatively inexpensive ; it certainly deserves additional allocations. So does the upgrading of the quality of drinking water, which is all too often substandard.

III. Improvements of existing infrastructure, which has deteriorated and in some cases broken down would require more funds, better maintenance of facilities also demands more and unrelenting attention.

IV. With debt service often ranging from a third to a half of income from exports, African economies cannot be rehabilitated without more revenue from mineral commodities, which match a half of the export earnings of the richer countries of Africa and as much as 90% for some oil producers.

With such notable exceptions as the Ivory Coast, Kenya and Malaŵi, the poor countries are in effect, poor because they do not ship minerals or do not earn enough from these. Albeit water and food have top priority, commodities will thus retain crucial importance at least for one or two generations.

V. Rational policies would undoubtedly stimulate demand on the continent, which ranks last, since it hardly consumes 3% of the world's energy and metals and barely uses per capita a quarter of the fertilizer utilized in the rest of the LDCs.

VI. Cooperative organizations have proliferated. Yet the only common currency is the CFA franc backed by the Bank of France. The rand plays a similar role in four countries surrounding South Africa. SACU is the only customs union, but the French speakers have lowered tariffs among themselves in West Africa.

Further translating laudable intentions into effective cooperation would ultimately enhance economic independence as targeted in the Lagos resolutions. On a broader scale, only a better understanding between all developing countries would give them more weight in negotiating improvements in the terms of trade.

VII. Similarly, if Africa is to shed its image as the eternal supplier of raw materials, it has to strive for closer cooperation since only broader markets would justify processing.

VIII. Needless to say that only favorable legislation, applied with efficiency, will attract funds to the petroleum, mineral and processing industry. Ideally, the political and social risk should also be minimal. Should this not be the case additional incentives have to be created in order to find capital. The same holds for all three areas of Africa from Caïro to the Cape.

IX. Small is beautiful. This adage also holds for the mineral industries and small and medium sized operations should be ecouraged if they do not enhance illicit exports since these workings provide much needed employment.

X. In the last resort, only initiative and perseverance will improve the situation, which requires a detailed review in each country. Meanwhile, preliminary goals have been set under each heading.

A2 THE AFRICAN SHARE OF WORLD RESOURCES

The Continent plays a key role as a source of precious minerals, fuels, ferro-alloys, light metals, and fertilizers ; it ranks
FIRST for the andalusite group, bauxite, chromite, cobalt, diamonds , germanium, ferrochromium, fluorspar, gold, flake graphite, hafnium, LNG, manganese ore, phosphate, the platinum group, uranium and vanadium ;
SECOND for cesium, ilmenite, manganese metal, sodium carbonate, tantalum, vermiculite and zircon ;
THIRD for asbestos, antimony, beryllium, garnet, gas, helium, hydropower, indium, rutile, selenium and tin and
FOURTH for coal, ferromanganese, iron ore, lead, niobium (columbium) and oil.

The reader will find further details concerning these in the following tables arranged according primary usage or enduse. In some cases even the second most important application is listed, thus : (2). Each table specifies the rank of Africa, the percentage share of reserves and production, and the first or second ranking continent or region, that is North America : U.S.A., Canada, Greenland ; Latin America : all other American countries except for Cuba ; Comecon : U.S.S.R., its European allies, Cuba, Mongolia, Afghanistan and Indochina (commodities where Comecon data are unavailable are shown thus : ¯) ; Europe : the rest of Europe ; West and South Asia : the Near East and the Indian sub-continent from Turkey to Burma, and the Far East : China, east Asia and the Pacific.

Energy, nuclear industry and aerospace

Table 1 Energy, Nuclear Industry, Aerospace.

Africa	Rank	Reserves	Production	Area	Rank	Reserves	Production
Oil	4	8	9	W + S Asia	1	60	22
Gas	3	6	2	Comecon	1	45	30
Heavy oil	3	15		L.America	1	65	
Oil shale	3	3		N.America	1	70	f
Peat	5		1	Comecon	1	40	95
Coal	4	11	5	Comecon	1	25	25
Hydropower	3	15	2	L.America	1	30	10
Uranium¯	1	30	30	Far East	2	30	15
Beryllium	3	10	2	N.America	1	30	10
Hafnium	1	30	20	Far East	2	25	50
Indium (2)	3	20	20	N.America	1	30	30
Cobalt (2)	1	55	60	Far East	2	15	10
Rutile	3	8	40	L.America	1	65	2
Niobium¯	4	5	10	L.America	1	80	70
Thorium (2)	6	1	1	N.America	1	40	60

Even though lignite resources hardly reach 1% of the world total, and peat bogs even less, the Continent as a whole has an excellent combination of fuels. Yet, most African countries import oil and coal. By the end of the century known oil reserves will be more than half depleted in Africa, but vast amounts of gas will be left to compete with hydropower. If only easy to harness sites are considered the African share of hydropotential exceeds indeed a third of the world total.

Even earlier, South Africa may move to the second position as a shipper of low cost black coal, even though its reserves are outranked by Comecon, the Far East and North America.

Little is known, however, about geothermal potential and only few of the vast Rift valley reserves have been developed to date.

Uranium resources of market economy countries are almost evenly divided between Africa, Australia and North America, but sales of African producers and the discovery of low cost-high grade resources in Canada may jeopardise growth when the uranium glut will be over ten years from now. As to thorium, the reader is referred to the 'electric industry' and table 5.

The Soviet Union produces half of the world's titanium and the Comecon group and North America hold most beryllium resources. Yet, since the West builds few reactors demand for the metal is flat. Alloy applications are growing and as shown by the Bureau of Mines, usage of other nuclear and aerospace materials is expected to grow rapidly in the U.S. at least until 1990 : hafnium, niobium (columbium) and titanium by 5% , cobalt, indium and rare earth metals by 3% per annum.

Hafnium is used in naval reactors, and the nuclear industry is also a growth application for zircon, the only known source of hafnium. Indium alloys are also used in control rods and half of U.S. imports came from a Belgian plant which refines African metals. Cobalt alloys used in high temperature jets are also extracted from Zaïre concentrates, niobium is a component of superalloys and the principal raw material of titanium metal is rutile produced in Australia and Africa.

Transportation and oil industry

Africa plays a dominant role for half of the commodities concerned. South Africa holds most of the platinum group reserves, resources and capacity and rivals the U.S.S.R., the principal palladium producer. Demand is expected to grow rapidly along with stricter emission control legislation in Europe. Platinum is also used as a catalyst in the petroleum industry and by the year 2000 at least 40% of the palladium will be utilized in feed stock.

Twenty per cent of the aluminum and manganese produced is built into cars, and 15% of antimony and three quarters of all lead is mined for usage in batteries. Demand for the two latter is more erratic and projections range from -1% to + 2% growth per year. The oil industry is the principal user of the rare earth metals. Among these, lanthanum and praeseodymium mostly serve as petroleum catalysts for the recovery of chloride. Lanthanum oxide is also the cheapest among ceric oxides and demand for some of these metals may increase rapidly. Although ilmenite dredging supplies monazite, bastnäsite resources of China far outrank Africa. Likewise, Métallurgie Hoboken recovers African rhenium in Belgium, but North America has vast resources of this catalyst which enhances octane rating. Consumption is to increase by 3-5% per year and spectacular price increases have enhanced the usage of substitute iridium, mainly recovered in South African platinum and gold mines.

Table 2 Transportation and Oil.

Africa	Rank	Reserves	Production	Area	Rank	Reserves	Production
Platinum	1	85	43	Comecon	2	13	53
Bauxite(2)	1	35	20	L. America	2	25	25
Lead	4	5	7	N. America	1	35	20
Antimony (2)	3	7	20	Far East	1	55	30
Barite	4	15	5	N. America	1	30	15
Rare Earths	4	2	.2	Far East	1	80	40
Rhenium	4	2	2	N. America	1	55	60
Bentonite	6	3	2	Far East	1	60	10
Gold	1	60	50	Comecon	2	15	15
Silver (3)	5	4	2	N. America	1	30	30
Diamonds gems	1	60	60	Far East		20	10
Diamonds industrial	1	50	40	Far East	2	40	40
Garnet	3	20	5	N. America	1	30	70
Corundum	1	40	45	Comecon	2	35	45

In contrast, Africa plays only a minor role as a producer of barite which increases the weight of drilling muds which incorporate bentonite. To conclude the United States and Japan dominate the bromine and iodine trade. None of these are recovered in Africa.

Precious minerals and abrasives

Africa has been the major source of gold for millennia and reserves are abundant as shown in table 2. To these can be added 60,000 tons or 2.2 B ounces already extracted, which are mainly in vaults where their value increased 20 times in 20 years time. Wage increases in South Africa would have further spectacular effects on prices and/or output since demand is expected to grow by 2% per year in the U.S. until 1990 and by a total of 50-100% by the year 2000 from its 1978/80 base*.

*Further references to short and long term growth concern the same time frames of the U.S. Bureau of Mines and 20 years price increases refer to 1960-1980.

More and more platinum is used in Japanese jewelry. On the other hand silver has other major applications in photography.

Most of the world's gem diamonds have come from Africa. While demand for large stones has declined since the recession, reserves of smalls, which compete with synthetics are particularly abundant. Furthermore Africa is a major or significant source of aquamarine, emeralds, garnets, ruby and other semi-precious stones.

Africa also has the largest resources of industrial diamonds and boart but production is growing rapidly in Australia and new discoveries might further enhance its position.

Prices have doubled following the last meagre years and demand is growing at a rate of 4% p.a.

Reserves of garnet are abundant in Africa and resources rival those of North America. Natural garnet also competes with Australian abrasives and the synthetic variety, but prices have trebled in 20 years and U.S. consumption is growing at 2% per year.

The continent is the major source of corundum. Prices have increased fourfold in 20 years, enhancing, thereby, substitution. Consequently demand may decline. Lastly, while little pumice is used in abrasives in Africa, some bauxite has been incorporated in these.

Steel industry

Africa is the major source of alloying elements, ferroalloys and fluxes- and a net importer of steel. This group has also suffered most from the recession of the early eighties, but demand for steel has bounced back since then in the industrialized countries.

Prices have also firmed up. Demand should also increase in the nineties especially in less advanced economies. Nonetheless, the African share of iron ore markets has been declining and low cost Brazilian shipments will further reduce its competitiveness, although resources remain abundant. The continent boasts much of the world's land based manganese reserves and resources and there are important concentrations of nodules in the south east Atlantic. The African market share has declined during the recession. Then demand increased dramatically during the recovery, but it is expected to level out since proportionately less manganese is used per unit of steel. Meanwhile manganese ore prices have doubled in 20 years in money terms ; they have declined, however, in real terms by 20%.

Africa is also the leading source of cobalt, but this metal is scarce in the U.S. and U.S.S.R. It is, therefore, the strategic metal par excellence. Even though real prices declined by 20% over the last ten years, a tenfold increase in inflated prices put a break on demand, which bounced back and is expected to increase by at least 3% per year. Meanwhile Zaïre curtails shipments and Zambia recovery in order to maintain prices. The usage for nickel is to grow slightly slower in accordance with the demand for stainless steel. Prices have declined 20% in constant dollars over 20 years, some marginal mines, therefore, temporarily shut down, creating thereby a tight situation as an after effect. At any rate, Africa plays only a minor role since most nickel reserves are only accessories of the platinum deposits.

Table 3 Steel industry and alloys.

Africa	Rank	Reserves	Production	Area	Rank	Reserves	Production
Iron ore	4	10	5	Comecon	1	30	40
Manganese	1	80	15	Comecon	2	15	45
Cobalt	1	55	60	Far East	2	15	10
Chromium (2)	1	90	35	W + S Asia	2	3	1
Nickel	4	7	6	Far East	1	40	30
Vanadium	1	50	40	Comecon	2	30	35
Tungsten	7	0.5	1	Far East	1	40	40
Molybdenum	7	0.5	0.5	N. America	1	60	60
Niobium (2)	4	5	10	L. America	1	80	70
Graphite flake	1	45	10	Europe	2	20	25
Fluorspar(2)	1	35	10	Comecon	2	20	35
Tellurium	3	15	20	L. America	1	30	7

Contrary to conventional knowledge, the chemical industry was, during the recession, the leading consumer of chromium, the most 'African' of major metals. Demand for stainless and heat resisting steels is now booming - so are, as a matter of fact - prices. South Africa is also the key producer of vanadium, whose inflated prices doubled in twenty years' time. Demand declined in the early eighties, it should increase by more than 3% per year until 1990 and probably double by 2000.

Other ferroalloys, fluorite and graphite

Africa has been the major source of columbite-tantalite, which lost its primacy as a source of niobium (columbium) to Brazilian pyrochlore and pandaïte. Resources of these exist in Africa, which now produces some pyrochlore too, but Brazil is the only developing country which built a ferrocolumbium plant. Further, African recovery of byproduct columbite might decline sharply along with tin prices. The same holds for tungsten, which is scheduled for spectacular growth rates in the U.S., following a sixfold price increase in the last twenty years, threefold for molybdenum, fourfold for bastnäsite, and twofold for monazite, another rare earth mineral. Among these especially cerium, lanthanum, and associated yttrium are alloy bases. Further, demand for samarium-cobalt magnets is expected to grow rapidly.

Some 40% of natural graphite is used in foundry molds, steel works and crucibles. Almost half of the world's reserves and 40% of the resources are in the Malagasy Republic. Its market share has declined, however. Meanwhile prices have increased sevenfold in 20 years and usage in foundries is expected to level out.

A third of all fluorite recovered is used as a flux in foundries. Africa holds the largest fluorspar resources and demand for metallurgical fluorite is expected to grow at least as fast as for the chemical grade.

Africa is also a major source of tellurium which increases the machinability of low carbon steels. Prices have increased seven fold. Consequently, demand is expected to level out or to grow by only 1% p.a. In contrast, the usage of low melting point indium alloys in sprinklers and fire alarms might increase by 3% per year. While this metal has been discussed already under the 'nuclear' heading, cadmium also used in plating alloys will be reviewed with the pigments.

Packaging, light metals and paper

Africa is the principal source of chemical grade fluorite and of bauxite, but its marketshare is much lower. U.S. demand is expected to increase at a rate of 6% per year until 1990 and world bauxite needs might grow two to fourfold by the year 2000.

Fluorite is extensively used in aluminum smelting. Africa holds the largest resources and is a major exporter, but deposits are widespread. Consequently, shipments from Africa declined during the recession. In longer perspective, prices increased 2.5 times in 20 years and they have firmed up again. Demand is increasing at a yearly rate of 2.5% in the U.S. and is expected to treble worldwide by the year 2000.

A third of all lithium produced is used in aluminum smelting, but the African share has declined spectacularly. The same holds for tin, whose prices increased eightfold in twenty years. Simultaneously African production declined to one eighth and many hard rock mines risk closure as a result of the recent drop in prices.

The sea is, of course, an inexhaustible reservoir of magnesium, but no brines are extracted in Africa.

Paper and pulp manufacturers use a third of the sodium sulfate and half of the kaolin produced. Understandably Africa makes only a minor contribution in this area.

Table 4 Packaging, light metals, food, fertilizers and paper.

Africa	Rank	Reserves	Production	Area	Rank	Reserves	Production
Bauxite	1	35	20	L. America	2	25	25
Tin	3	5	5	Far East	1	75	45
Fluorspar	1	35	10	Comecon	1	20	35
Magnesium	7		-	Far East	1	50	5
Lithium (2)	3	10	5	L. America	1	90	25
Phosphate	1	65	25	N. America	2	15	35
Potash	6	1		N. America	1	50	35
Sulfur	7	1	1	Comecon	1	30	40
Salt (2)	7		1	N. America	1		45
Na sulfate (2)	5	5	1	Comecon	1	20	45
Kaolin	6	5	1	N. America	1	30	30

Food, fertilizers and soil

By most criteria, Africa ranks last as a food producer. Only Australia has less soil for its area, but even so more per capita. In contrast, Africa holds two thirds of the world's phosphate reserves and a third of its vermiculite. It ranks, however, low for potash, salt and sulfur.

Phosphate usage trebled during the last decade and considering the need for food, demand should again treble by the year 2000 even if Africa does not follow suit. On the other hand, escalating costs erode the competitiveness of American exporters.

Agriculture absorbs a quarter of the vermiculite winnowed. Since this application has been declining, the reader is referred to 'construction'.

Measured potash reserves are insignificant in Africa. There is, nonetheless, a considerable potential. Prices have quadrupled and demand is expected to double or treble.

The chemical industry is the principal customer for salt worldwide, but in Africa salt is mainly used for seasoning. Although reserves of rocksalt are low thick beds of salt lie in oil basins. Meanwhile, only little marine salt is ponded. While international salt prices have doubled, African prices depend on local conditions. Worldwide demand is expected to double and considering population trends, this should ultimately affect Africa too.

Fertilizers are the largest consumers of sulfur. Prices have quadrupled and demand should double or quadruple too by the year 2000. Little native sulfur is produced in Africa though and the continent ranks low as a source of pyrite or byproduct sulfur.

Electronics and instruments

The second most important application of gold is in electronics. Africa also holds important reserves of cesium, a metal used in thermionics and magnetohydrodynamics with a yearly growth rate of 3% or more, as in the case of tantalum used in capacitors and personal computers, prices of which have increased 15 times. They have declined, however, with the recession. So will the African contribution if tin mines close because of sinking prices. Although more tantalum is extracted from Asian tin slags, the major primary resources remain in Africa.

Small workings in Africa are an important source of sheet mica, but India is the major producer. Madagascar is the major source of high priced phlogopite. Prices have increased by 50% but demand is expected to decline further. Similarly, Africa has lost its preeminence as a supplier of piezoelectric quartz and Brazil dominates lascas exports, which are expected to increase at a rate of 21% to the U.S. at least until 1990...

Likewise, the African contribution to gallium production is expected to decline with the startup of a U.S. plant. Since prices have sunk by four fifths, demand has doubled in a few years time and is increasing in the U.S. at a rate of 10% per year and even faster in Japan. Gallium is used in semiconductors, bubble memories, solid state physics and solar energy and Africa holds large reserves in two zinc-germanium deposits. Most of the lowgrade resources are contained in its bauxites.

Table 5 Electronics, instruments and electric industry.

Africa	Rank	Reserves	Production	Area	Rank	Reserves	Production
Gold (2)	1	60	50	Comecon	2	15	15
Tantalum	2	30	15	Far East	1	60	65
Quartz crystal	3			L. America	1		
Yttrium	3			W & S Asia	1		
Indium	3	20	20	N. America	1	30	30
Selenium	3	10	8	N. America	1	35	50
Thallium	3	10	5	N. America	1	25	25
Copper (2)	3	15	15	L. America	1	25	20
Manganese (2)	1	75	25	Comecon	2	20	45
Palladium	1	75	40	Comecon	2	20	57
Mercury	5	3	5	Europe	1	60	30
Thorium	6	1	1	N. America	1	40	60
Ruthenium	1	80	45	Comecon	2	20	50
Iridium	1	90	60	Comecon	2	10	40

Indium has already been discussed with the nuclear industry and yttrium is used in microwave and laser production along with the rare earths (steel heading). Among these, terbium is specified for X-rays. On the other hand, selenium will be described with photography and strontium, used in TV sets with glass (Table 6).

Thallium was mainly used in gamma radiation equipment (and pharmacology). It is recovered as a residue of zinc flue dusts. Prices remained unchanged and many African zincs are low in thallium.

Electric industry

A third of all copper, a quarter of the manganese, a fifth of all tin and 10% of aluminum constitute electric wire and material. However, these major commodities have broader applications in the building, steel and packaging industries where they are discussed. Africa is the leading source of most of these and the prime supplier of palladium, ruthenium and iridium. Yet from the latter nineties on the petrochemical industry and feedstock will require even more palladium. Its prices oscillate less than those of platinum and they have increased sixfold. Since they are lower than those of platinum demand is growing and should attain 50 to 100% more before the end of the century.

In contrast, ruthenium is the cheapest of PGMs; supply is growing, as a result of increased capacity on the UG2 reef which has proportionately more palladium and ruthenium. Meanwhile, prices have declined and demand is more difficult to predict - it might level out or double. In contrast, iridium prices ranged between $320 and $600 over two decades with a long term growth trend. As pointed out under the oil heading demand might increase long term threefold.

Algeria is a minor producer of mercury ; larger deposits are on the northern side of the Mediterranee. Mercury prices have doubled. Usage of this pollutant has declined, but may pick up in battery applications.

On the other hand, Africa is a major producer of sheet mica, used in equal amounts in electricity supplies and electronics where this item has been discussed. The major application of beryllium is in the nuclear industry, q.v. followed by electric apparatus. Similarly, thorium is widespread as an accessory of rare earth minerals, which even caused oversupply. In the past, most African monazite has been, therefore, stockpiled . However, american usage is already growing at a rate of 10% p.a. and might increase 2 to 9 times by 2000. Thorium is mainly used in incadescent lamps and indium, in electrical and electronic components and plating. A Belgian plant which refines African concentrates is a key supplier of the latter.

Photography, optics, pigments, coating

The most important metal used in this field is silver, since sterlingware and electricity require less. On the other hand, Africa is a major source of other rarer metals used in this industry.

Silver prices have peaked with speculation in 1980, increasing 25 fold in two decades and the metal still costs 10 times more than in 1960. Demand is increasing at a yearly rate of 2% and total needs may grow 30 to 230% by the year 2000. Yet the share of photography may decline from one half to one third. North America holds a third of world reserves and Latin America provides a third of the supply. In Africa silver is only a byproduct of lead and gold mining.

Most delineated cesium reserves are in Canada and Southern Africa. The metal is also used in photoelectrics and rubidium, a byproduct of lithium and cesium mining finds applications in vacuumtubes and photosensors. Demand is dependent on the prices of cesium to which level rubidium has now declined. The U.S. Bureau of Mines forecasts only modest growth at least short term.

The continent holds more than half of the world's germanium resources which was discovered in African minerals. The major usage of germanium is in infrared fiber optics and photovoltaics with an expected yearly growth of 6%. By the year 2000 usage may, therefore, double. Prices have also doubled by now.

Selenium is also a byproduct of zinc mining. Along with Canada and Japan, Africa has important reserves. Prices have first doubled, then reverted to their 1960 level and are now increasing again. The metal is mainly used in photoelectronics and xerography and usage may increase by 50% to 350%.

Neodymium (a rare earth) oxide is used in television glass screens, and europium oxide in color television, above all in Japan.

Africa holds a fifth of the world's ilmenite resources out of which titanium oxide is usually made and - exceptionally - its market share is even higher ; so is, as a matter of fact that of the Far East. Pigment prices have doubled, demand grows by 2% yearly and may increase by 20-50% at the end of the century.

It is difficult to assess the resources and production of iron oxide pigment. North America, Europe and India are major producers. Reserves of scrap mica are also moot, but North America dominates these markets, and India the resources. Similarly, North America ranks first for cadmium resources, a byproduct of zinc smelting. Prices have increased by 60% and demand is growing 2% per annum. Usage of wollastonite grows now much faster, however. This filler is now worked in Africa ; other deposits are in India, North America, Finland, etc.

Ceramics, glass, refractories

A quarter of all graphite is used as a refractory and the Comecon group, South and East Asia outrank Africa as a supplier of amorphous and lump graphite. U.S. demand is increasing at a rate of 3% per year.

The refractory alumino-silicates andalusite, kyanite and sillimanite compete with synthetic mullite. Africa holds half of the reserves of andalusite, but large kyanite resources are available in the U.S. where consumption increases at a yearly rate of 2% only . While worldwide usage might double prices are declining after a fourfold increase between 1960 and 1980.

Table 6 Photography, optics, pigments, coating, ceramics, glass and refractories.

Africa	Rank	Reserves	Production	Area	Rank	Reserves	Production
Silver	5	4	3	N. America	1	30	20
Cesium	2	30		N. America	1	65	
Germanium	1	55	30	N. America	2	20	15
Gallium	1	40	5	Far East	2	30	5
Sheet mica	3	10	4	W & S Asia	1	60	70
Rubidium	2	20		N. America	1	70	
Selenium	3	10	8	N. America	1	35	50
Ilmenite	2	15	20	N. America	1	20	20
Cadmium	4	8	7	N. America	1	40	20
Iron pigment	6		1	W & S Asia	1		30
Scrap mica	7		1	N. America	1		55
Graphite, amorphous	4	20	3	Comecon	1		50
Andalusite[a]	1	40	35	N. America	2	20	25
Magnesite	7		2	Far East	1	50	40
Talc	6	1		N. America	1	45	15
Chromite	1	90	35	W & S Asia	2	3	1
Sodium carbonate	1		2	N. America	2		70
Zircon	2	30	25	Far East	1	35	55
Rare Earths (3)	4	2	.2	N. America	1	80	40
Lithium	3	10	5	L. America	1	90	25
Feldspar	7		2	Europe	1		45
Strontium	4		3	Europe	1	50	30
Meerschaum	2		45	W & S Asia	1		45

[a]includes kyanite and sillimanite

African zircon resources match those of Australia, the principal producer, but reserves are lower. Zircon is used in foundry sands and molds and prices have doubled. Demand is growing in the United States at a yearly rate of 4% and may reach 60 to 300% more by the year 2000. South Africa is the only source of baddeleyite (ZrO_2), a ceramic colorant which experiences even faster growth. On the other hand , hafnium, a ceramic metal included in zircon has been discussed with the nuclear industry.

Although African chromite resources are vast, little refractory ore is produced for 20% of all applications.

The Far East is the principal source of magnesite. Furthermore, seawater contains 0.13 weight % Mg. Demand is growing by 2%, but African production is insignificant by world standards. The situation of talc is somewhat similar, with the U.S. holding 40% of the reserve base. Crude talc prices have trebled. Usage advances at 3% per annum and might double or treble in two decades. Further, pipes are manufactured out of meerschaum or sepiolite and African resources rival those of Turkey.

Exports of refractory flint clay continue to decline, few comparisons are available on fire clay and Africa contributes little to kaolin supplies. Europe is the principal source of widespread feldspar, a mineral flux and wollastonite, a ceramic filler, has been mentioned under the heading 'paint'.

Glass and ceramics manufacturers are the principal users of lithium in the market economy countries, where Africa has lost ground to Chile. Meanwhile the U.S.A. and the U.S.S.R. produced most concentrates. American demand grows by 6% per year and consumption may be twice to four times as large in 2000 as in 1980.

The largest sodium carbonate deposit is in the United States, but soda occurs widely in Africa. One half of all soda ash recovered serves the glass industry. Prices increased 2-5 times and 1% yearly increment in current demand might result in an ultimate growth of 30 to 230% by the end of the century.

Strontium serves in the manufacture of the glass of color TV tubes. Major resources are in Spain and Mexico is an important producer. Celestite and strontianite have been worked from time to time in the Maghreb. Prices have doubled and demand might increase by up to 50% over twenty years in the U.S. and conceivably faster in Japan.

Some arsenic is used in the manufacture of glass and will be discussed under the chemical heading. The United States are the leading producers of boron, and reserves are largest in the U.S. and Turkey, African resources have never been systematically explored.

Europium, a rare earth, and yttrium oxide are also used as fluorescents, phosphorescents and in TV tubes, neodymium in TV glass screens, cerium oxides in glass and ceramics and lanthanum in optical glass. Large resources of these exist in Africa, but few have been inventoried.

Chemical industry

Africa is the dominant source of chromium and platinum, an important source of rubidium, but the continent has an only minor importance for other commodities in this industry.

Ninety per cent of all chromite reserves and resources are located in Africa and the continent also was the principal producer of chemical ore, now rivalled by Comecon. Prices have doubled and, following a turnaround, demand may almost double. The chemical, medical and dental industries use some 30% of the platinum group metals, which have been discussed under the heading oil.

Rubidium, reviewed in the photography section, also serves tracer technology and medical diagnostics.

Bismuth suppliers, mainly in the Far East, expect only moderate growth. The same holds for arsenic, a Latin American commodity. In contrast, diatomite usage shows increments of 3% p.a., even though prices have trebled. North America and Europe are the principal beneficiaries. Usage of fuller's earth, another absorbent, also grows fast. Similarly, the United States dominate helium sales, but Algeria hopes that needs may grow even faster than 7% per year in Europe, which is also the second consumer of lime, following Comecon. Lime prices have increased fourfold in the western world and U.S. demand swells by 3.7% yearly, that is twice as fast as needs for salt.

Table 7 Chemical, Medical, Building and Insulation Industry.

Africa	Rank	Reserves	Production	Area	Rank	Reserves	Production
Chromium	1	90	35	W & S Asia	2	3	1
PGM	1	80	42	Comecon	2	15	55
Bismuth	6	.5	-	Far East	1	50	55
Arsenic	5	5	6	L. America	1	50	25
Salt	7		2	N. America	1		20
Lime	7		3	Comecon	1		40
Diatomite	6	0.3	0.5	N. America	1	25	35
Fuller's ea.	3	5	7	N. America	1	85	80
Rubidium	2	20		N. America	1	75	
Helium	3	10		N. America	1	70	60
Copper	3	15	15	L. America	1	25	20
Zinc	6	5	4	N. America	1	40	20
Asbestos	3	10	10	N. America	1	40	20
Vermiculite	2	45	35	N. America	1	50	55
Gypsum	6	10	2	Europe	1	30	30
Stone crushed	6		2	N. America	1		30
", dimension	5		4	Europe	1		55
Clays	5		5	N. America	1		25
Mica, scrap	7		1	N. America	1		55
Perlite	7	.3		W & S Asia	1	60	10
Pumice	7		.2	Europe	1		70

Construction and insulation

Africa is a major source of copper, and the second source of asbestos and vermiculite. It is not a major supplier of zinc and of quarry materials because its industrial base is still narrow.

A third of all copper is used in buildings. Demand has bottomed out, so have prices. Needs may grow by up to 2% per year long term, but the share of high cost African produces has declined. The Americas hold half of the land based reserves. In the next millennium seabed nodules should also become economical.

Zinc is also an "American" metal. Additionally, the Soviet Union is also an important producer. There are also several deposits in Africa. Prices have almost trebled in two decades ; they are growing again, but slower than inflation and after bleak years, the market has turned around.

The demand for cross fibre (chrysotile) asbestos has also decreased, even though prices increased 2.5 times to 1980, when they declined sharply because of health concerns. The Soviet Union is a more important user of this mineral than the United States; Canada has the largest reserves, followed by Africa.

Following considerable growth, demand has declined to a third for crocidolite, mainly used in cement pipe, but amosite production was only halved. South Africa is the only source of this asbestos used in light weight materials such as felted insulation blankets.

Exfoliated vermiculite is a component of cements and serves insulation. African resources are second only to North America. Prices have increased four fold and sluggish demand may ultimately double.

There are vast deposits of gypsum and anhydrite in the African oil basins. None of these are, however, commercial since there are more accessible deposits in Europe, North America. Prices have increased 2.3 times ; yet demand should increase by only 40% in two decades.

The usage of crushed and dimension stone, sand and gravel, and ball clay is tied to a large industrial base - and to urbanization. World prices have increased 2.5 times.

Scrap mica is mostly used in plasters and construction. Deposits are widespread, prices have doubled, but demand grows slower than the population. Perlite, a glassy volcanic rock expands up to 20 fold. The Comecon and Turkey have the largest reserves and the United States one third of world consumption. Prices remained flat for years and then edged forward with demand growing at 3% p. a. in the United States. Italy is the major source of pumice and pozzolana which also occur in the Atlantic islands. Prices of light weight volcanic cinders have doubled and demand may increase by 50 to 100%.

Processed materials

Copper, lead and zinc smelters have been built in Africa during the colonial period, but the continent has only a minor market share of such products as aluminum, steel, cement and fertilizers, mainly because of demand and capital, and trade barriers, not withstanding its leading position as a supplier of raw materials as shown below :

Table 8 Comparison of resource and market shares of processed commodities.

Share of....	resources %	markets	%	Resources	%	Markets	%
Phosphates	65	fertilizer	6	Oil	8	petroleum	2
		ammonia	6	Gas	6	LNG	50
Limestone	large	cement	6	Coal	11	coke	1
Bauxite	35	aluminum	3	Iron ore	10	steel	2
Hydropower	15			Chromite	90	ferrochromium	40
Vanadium	50	slag, oxides	30			ferrochromium -silicon	30
Ilmenite	15	titanium slag,&	18				
Antimony	7	pentoxide	30	Manganese ore	80	ferromanganese	8
Gold, bullion	60	coins	80			manganese metal	80*
Gems, raw	60	cut	2	Silica	large	silicon	6

*export share in the market economy countries

Considering a growth rate of 6 to 7%, aluminum excess capacity may be resorbed soon and Africa, with only 3% of world production, has both vast bauxite and hydropower resources which can be easily developed near the coasts. In effect, more than 2 kw of power are required to produce 1 t of aluminum per year and several pojects have been under consideration for years.

In contrast excess steel capacity is not expected to disappear worldwide before the end of the century and as a net importer, Africa is expected to at least hold its small share of 2%. Ferroalloys compensate for this deficiency. With 10% of shipments, South Africa is the world's third producer of ferroalloys following the U.S.S.R. and Japan ranks first for ferrochromium. The African contribution is shown in table 8a.

Table 8a Percentage share of ferroalloy production.

Weight	Type	Africa	Comecon	Far East	Europe	Other areas
20	Ferrochromium	33	30	15	15	7
2	Ferrochromium-silicon	8	6	66	12	14
40	Ferromanganese	8	40	20	20	8
8	Silicomanganese	4	4	25	35	32
25	Ferrosilicon	3	30	15	30	22
5	Silicon	4	15	5	45	31

South Africa is also the world's principal exporter of manganese metal.

Similarly, large proportions of vanadium, ilmenite and antimony are processed into slags and oxides, but the net tier, such semi-manufactures as pigments and metal lag.

At the opposite end, krugerrands still dominate the coin trade and the Maghreb is a major shipper of LNG. Meanwhile, only a fraction of wet gas is marketed and oil refining capacity barely reaches 3% of the world total.

Market shares are, of course, not static. While tables 1 to 7 indicate the current situation, Africa has registered spectacular growths for some commodities over twenty years :
gas from 0 to 2%, coal 0.5 to 5, bauxite 6 to 20, PGM 27 to 45, nickel 1 to 6, ilmenite 1 to 20, rutile 0.5 to 40, zircon 10 to 20, vanadium 30 to 40 and oil from 5 to 9% following a high of 14% in 1970 ; counterbalanced by declines : lead, zinc and tin 10 to 3%, silver 4 to 2, gem and industrial diamonds 95 to 60 and 40%, asbestos 15 to 10, manganese ore from 20 to 12, and beryl, lithium, niobium and tantalum 45, 66, 65 and 80% to 2, 5, 15 and 15% respectively.

A 3 DEPENDENCE ON AFRICA

Although Comecon exploration and construction teams are active in Africa, most projects are limited to Angola, Ethiopia, Mali, Moçambique and Morocco. Chromite and nickel reserves of the bloc would barely last 30 to 50 years at least at current grades. Similarly, the Soviet Union imports bauxite from Guinea since it has only 1% of world reserves. Its relationship - partnership and competition - with the diamond cartel is old standing and a combination of South Africa and Comecon would dominate world resources (and production) of PGM 96% (97), vanadium 80 (75), manganese 80 (60), ferromanganese (48), chromite 77 (60), ferrochromium (62), gold 70 (65), gem diamonds 80 (70), industrials 60 (50) and conceivably more than half of uranium.

The Near East exports crude oil to Africa and imports some phosphate and fertilizer. So does the Far East, which has become a low cost competitor for copper. Taïwan imports uranium from Namibia. More important, Japan is the leading customer for palladium and ruthenium and takes half of the platinum. Ten per cent of its uranium imports came from Niger, and an undisclosed amount from South Africa, which also supplies 4% of its iron ore, 40% of its manganese chromite and 70% of ferrochrome and most vanadium, yet with 4% of imports, Zaïre shipped more copper.

Considering similarities in the geological makeup of Africa, Siberia and Latin America, these regions essentially compete for markets. With more initiative, the Latins have thus been able to improve their position, e.g. in copper, iron ore and rare metals at the expense of Africa.

Africa imports capital, knowhow and equipment mainly from the U.S. and the EEC, which is its principal trading partner.

It is difficult, on the other hand, to assess imports of gold into the EEC since it moves partly through Switzerland. Having excluded it, the share of oil and gas has declined among other mineral imports from 82 to 77% in five years' time, a period during which all these imports have increased by only 8% in value. Consequently, they declined in constant ECU's.

Simultaneously, copper moved from a 4.2% share to 4.5%, coal from 3.4 to 5.2% and iron ore from 2.6 to 3.5%. In contrast, all other commodities combined only represent 10% of mineral imports.

Contrary to some assumptions, the African ACPs only play a minor role as a source of materials and their share only reaches 15% of mineral imports, or if fuels are excluded, hardly more than 10%. The Maghreb, thus, outclasses them as an exporter. Their contribution to tin imports plummeted to 6% and while their market share declined from 30 to 20% for iron ore, it doubled for bauxite.

The provenance of the commodities can be schematized as follows (in %), including :

	oil	gas	fertilizers	manganese	chromite	bauxite	iron ore	copper	industrial diamonds	graphite
Maghreb	25	35	50					5		
African ACPs	15		15	45		40	20	25	35	25
South Africa				35	70		2	5	5	

Africa and the United States

Yet, the dependence on Zaïrian cobalt, South African chrome, PGM, manganese and some less critical materials can be better demonstrated in the United States. Except for copper, nickel and refractory bauxites, African materials dominate, in effect, the strategic stockpile. Admittedly, their contribution to the supply of the rare metals beryllium, cesium, niobium (columbium) and tantalum has declined sharply and low grade cobalt resources appeared south of the Keyes in Cuba... The rapid rise of production costs also hampers the development of alternative supplies. Zimbabwean chromite could only partly replace, for instance, South African ore after long delays, and the building up of a million tons of ferrochromium capacity would take years. Chromite remains thus high on the critical list of the International Strategic Minerals Inventory. Likewise, the ratio of concentration among a few producers is striking. Thirty commodities listed in table 9 indicate that South Africa is the key source of 20 of these and Zaïre of 6, with two to one products supplied by Botswana, Gabon, Guinea, the Malagasy Republic, Namibia, and Zimbabwe.

Finally, let us add that in many cases import reliance of the primary commodity is 100% with the balance recovered from scrap.

Table 9 U.S. import reliance, the stockpile and imports from Africa.

	Reliance %	Stockpile stock	goal	% of import from South Africa	rest of Africa
Steel industry					
Cobalt	95	kt 20	40		Zaïre 40, Zambia 10
Chromite metallurgical	100	Mt 3	1.8		
Ferro-chromium high C	85	kt 170	360		
Ferro-chromium low C		kt 70	50	40	Zimbabwe 10
Ferrochromiumsilicon		kt 80	50		
Manganese ore	100	Mt 2	2.5	30	Gabon 30
Manganese metal	99	kt 10	-		
Ferromanganese	80	kt 600	400	40	France 30 (some ex Gabon)
Vanadium	40	kt 0.5	8	45	
Graphite	100	-			Malagasy R. 10
Electricity, electronics, photo, optics					
Phlogopite mica	100	t 15	200		Malagasy R. 100
Iridium	100	t 1	3	80 direct and	
Rhodium	90	-	-	80 indirect	
Cesium	100	-	-		reserves : Zimbabwe Namibia
Germanium	40	t -	30		Zaïre 60
Indium	60	-	-		Zaïre 25 viaBelgium
Selenium	50	-	-		Zaïre 7
Transport and aeronautics					
Rutile	80	kt 35	100	10	Sierra Leone 15
Platinum	90	t 12	30	80 direct and	
Ruthenium	95	-	-	80 indirect	
Chemical, medical and packaging					
Palladium	90	t 35	90	70	
Chromite chemical	100	kt 200	600	70	
Fluorspar, acid		Mt 0.8	1.2	60	
Bauxite		Mt 17	28		Guinea 35
Gems, abrasives, insulation					
Diamonds gems	100	-	-	20	rest of Africa 70
Diamonds industrials	100	Mct 14	8	30	rest of Africa 60
Corundum	100	-	-	in 1979	
Amosite fiber	100	Mt 30	15	100	
Crocidolite fiber	100	kt 30	-	70	

A 4 DISTRIBUTION OF RESOURCES WITHIN AFRICA

Between 1978 and 1985 fuels represented 70 to 80% of all mineral raw materials sales, the precious 12-17% and all other minerals only 13-8%, depending essentially on oil shipments, the price of oil and gold, and the dollar rate of the rand... Simultaneously, product value grew from $50 to $90 billion and then dropped $60B. A glance at pages 10 and 11 of the Mineral Resources of Africa also shows that fuels increased fifty times and the precious tenfold while all other commodities only quadrupled in two decades in inflated dollars - which explains the problems of the mining industry. Then, in 1986, oil prices slipped...

Table 10 Value of African minerals produced in million dollars.

Year	1963	1978	1985	Year	1963	1978	1985
Oil	800	31000	45000	Manganese	60	320	180
Gas	-	1500	4000	Vanadium	4	140	100
Coal	100	1200	3000	Nickel	-	190	130
Uranium	100	600	300				
Energy	1000	35300	52300	Ferrous metals	220	1800	2000
Gold	800	4600	8000	Copper	550	1500	1700
Diamonds	200	1500	1700	Phosphate	150	940	700
Platinum group	30	500	800	Bauxite	30	310	250
Silver	3	70	200	Zinc	50	80	160
Precious	1030	6670	10700	Asbestos	50	230	140
				Lead	50	100	130
Iron ore	100	780	700	Tin	40	130	100
Cobalt	40	190	500	Other minerals	920	3290	3200
Chromite	20	180	400				

Fluorite, barite, salt, clays, titanium minerals and soda contributed together $400 million. Surprisingly enough semi-precious stones, led by emeralds might range up to $250 million. While all other minerals add up to less, limestone (and dolomite), stone (with aggregate and sand) may respectively reach $300 to $500 million each, depending on local prices. In other words, the value of ferrous metals, non-ferrous metals, and industrial and building materials combined is roughly comparable.

As to processed materials, petroleum products, fertilizers, cement and ferroalloys represent respectively $25, 2, 1.5 and 1 billion. If aluminum, steel, lime, & are included, they match half of the total of raw materials. Thus if we exclude the fuels, the ratio of raw to processed materials is 11 to 1. As a comparison, the same ratio is 1 to 9 in the United States - which proves the importance of processing.

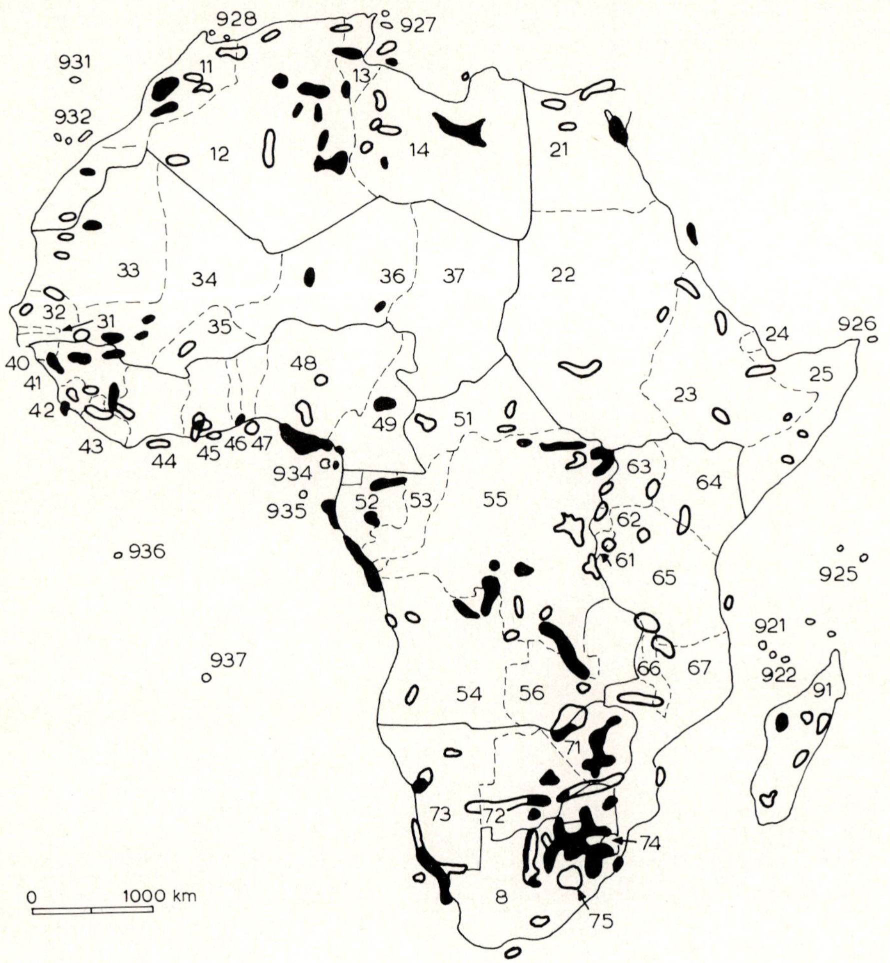

Fig. 1. The principal (solid areas) and the important (open areas) deposits.

1 The Maghreb 11 Morocco and Western Sahara, 12 Algeria, 13 Tunisia, 14 Libya
2 Northeast Africa 21 Egypt, 22 Sudan, 23 Ethiopia, 24 Djibouti, 25 Somalia
3 The Sahel 31 The Gambia, 32 Senegal, 33 Mauritania, 34 Mali, 35 Burkina Faso, 36 Niger, 37 Chad 4 West Africa 40 Guinea Bissau, 41 Guinea, 42 Sierra Leone, 43 Liberia, 44 Ivory Coast, 45 Ghana, 46 Togo, 47 Benin, 48 Nigeria, 49 Cameroon
5 Central Africa 51 Central African Republic, 52 Gabon, 53 Congo, 54 Angola, 55 Zaïre, 56 Zambia 6 East Africa 61 Burundi, 62 Rwanda, 63 Uganda, 64 Kenya, 65 Tanzania, 66 Malaŵi, 67 Moçambique 7 Southern Africa 71 Zimbabwe, 72 Botswana, 73 Namibia, 74 Swaziland, 75 Lesotho 8 South Africa 9 The Islands 91 Malagasy Republic, 921 Comoros, 923 Mayotte, 924 Réunion, 925 Seychelles, 926 Suqutra, 927 Pantelleria and Pelagians, 928 the Presidios, Chafarinas, Alhucemas 931 Madeira, 932 Canary Islands, 933 Cape Verde Islands, 934 Equatorial Guinea, 935 Sao Thomé and Principe, 936 Ascension, 937 St. Helena, 938 Tristan da Cunha, 939 Gough.

Distribution according to subregions

In order to facilitate this review we have divided Africa into nine subregions :

I Four countries of the Maghreb, from Morocco to Libya have a quarter of the continent's GDP and some 6% of mineral employment.

II A fifth of the continent's population lives in the Nile Valley and the Horn, that is North-East Africa.

III The seven countries of the Sahel from Senegal to Chad fight the desert and the dry steppe with only 2% of the total GDP.

IV Nigeria and nine other countries on the West African coast between Guinea and Cameroon concentrate more than a quarter of the population and GDP.

V Population density is much lower in the rain forests where six countries centered on Zaïre exhibit great differences of income. With 15% of mineral employment, Central Africa ranks second.

VI The seven East African countries include from Burundi to Moçambique a sixth of the population but only 5% of the GDP.

VII With only 2% of the population the five countries of Southern Africa from Zimbabwe to Namibia have higher standards of living.

VIII The Republic of South Africa has a quarter of the continent's GDP, including the homelands, and 70% of all mineral employment.

IX Finally, the Malagasy Republic and 20 other island groups shown on figure 69 comprise 2% of the area and population and 3% of GDP.

The countries are numbered on figure 1 and in the detailed description. The unequal distribution of mineral wealth is shown below in per cent according to sub-regions and categories. The first entry shows the weight of oil during peak years and the following line demonstrates that South Africa relies on the gold trade.

Production	Maghreb	West	South	Central	Northeast	Southern	Sahel	East[i]
All minerals	39	24	18	7	9	2	0.6	0.1
Energy excluded	7	4	66	13	0.8	8	1	0.2
Precious also excluded	17	8	28	27	3	10	6	1

[i]includes the Islands

Distribution according to countries and commodities

Five countries hold 77% of the mineral wealth each with $15 to $5 billion : Algeria, Libya, Nigeria, South Africa and Egypt.

Eight countries share 18%, with $2 to 1 B each : Angola, Cameroon, Gabon, Tunisia, Congo, Zaïre, Zambia and Morocco.

Eight countries total 4.7% of the minerals produced, with \$800 to \$200 million each : Namibia, Botswana, Zimbabwe, Ghana, Liberia, Guinea, Niger, Mauritania. The following eleven produce \$100 to \$10M each, i.e. only 0.3% of the total : Canary Islands, Togo, Sierra Leone, Senegal, the Central African and Malagasy Republics, Swaziland, Kenya, Rwanda, Tanzania and Ethiopia, with the remaining twenty countries contributing less than 0.02%.

The concentration of mineral wealth into small areas is even more striking than it was in 1964 (MRA p. 13) : the Niger delta, the bay of Sirte, parts of the northern Sahara and South Africa which cover 1% of the continent contribute 80% and following the deduction of four more areas 98% of the continent provides less than 1% of the mineral product. Similarly, Africa is well known for the unequal distribution of income among social classes and of the per capita mineral product which has peaked in Libya with \$6000 (250), Gabon \$1800 (\$30), Namibia \$800 (130), South Africa \$800 (80), Botswana \$600 (1) and Tunisia \$300 (6).- The bracketed figures show the same data for 1963 and the progress made by the 'rich' who have the principal reserves as shown below.

The Maghreb Barite production, Bentonite, Diatomite, Fertilizer, Gas, Helium, Lead production, LNG, LPG, Mercury, Oil, Petroleum products, Phosphates.

Northeast Africa Gypsum, Meerschaum, Nitrogenous fertilizer, Potash, Salt, Wollastonite.

Sahel Attapulgite, Molybdenum.

West Africa Alumina, Aluminum capacity, Bauxite, Gallium, Rutile production.

Central Africa Beryl, Cadmium, Cobalt, Copper, Industrial diamonds, Emeralds, Gallium production, Germanium, Hydropower, Indium, Lithium, Niobium, Oil shale, Rhenium, Selenium, Tantalum, Thallium, Tellurium, Tin and Zinc production.

East Africa Bismuth, Piezoelectric quartz, Rare earths, Soda and Tungsten production, Thorium, Zeolites.

Southern Africa Chrysotyl Asbestos, Arsenic, Bentonite, Beryl and Lithium production, Cesium, metallurgical Chromite resources, Corundum, Gem Diamonds, Garnet, Magnesite, Rubidium, Scandium, Soda, Tin.

Islands : Graphite, Heavy oil, Sheet mica, Pumice.

While there is little boron, bromium and iodine on the continent, South Africa ranks first for all fourty other mineral commodities and processed materials.

Water

Water and soil are the most important geological resources. It is unfortunate that because of past and present climates, Africa should be so poorly endowed with these substances. Happily, the same factors and its particular structure have provided the continent with a vast hydropotential. Water is thus scarce, but waterpower is plentiful. As a wit said, Africa is either too dry or too wet.

Nonetheless, groundwater is plentiful in the arid 40% of its surface where the Late Continental and the Intermediate Continental are the principal aquifers. They drain the Nile and runoff from the Atlas, but most groundwater in the Sahara is 10,000 to 40,000 years old. Since Kgalagadi sands hold less water in the southern desert, even draining the Zambezi and its spillways over thousands of kilometers to South Africa and Namibia have been considered.

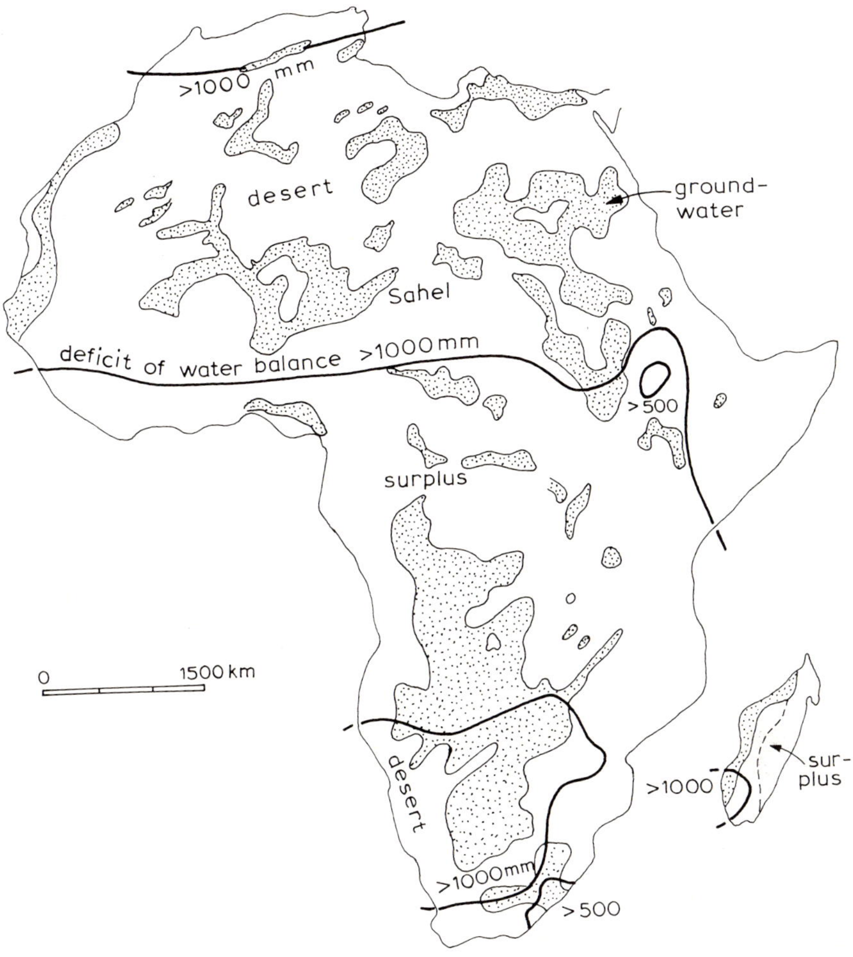

Fig. 2. (a) Precipitation and groundwater.

Unfortunately, the coastal strip of sediments is narrow. Water is often brackish even dozens of kilometers inland and salt invades the Nile delta as a result of the construction of the High Dam.

A third problem area is the Basement which covers half of the continent ; its hard rocks only provide localized, secondary aquifers. It rains normally, though, in many of the Precambrian areas, but when the rains failed half of the continent suffered. Even if the value of water cannot be easily quantified, its social cost is, thus, obvious.

Soil

Africa has fed itself until independence. Because of urbanization, underpriced farm products, overpriced currencies, the growing birthrate and life expectancy and other social and political factors, it imported 2 Mt grain in 1960, 5 in 1970, 15 in 1980 and the figure will reach 30 Mt in 1990. Concomitantly its share of food aid will rise from 6% to two thirds of the world total. Consequently, the erosion of millions of tons of soil in Ethiopia, Lesotho and other highlands has been called a major tragedy and the conservation and upgrading of soils has acquired great importance.

While climate influces soil formation in the deserts and in the tropics more than in temperate areas, in Africa the dynamics of soil evolution are such that even minor variations of the bedrock determine major pedologic changes. Indeed, high temperatures and heavy rains facilitate the leaching of bases, the supply of which is particularly short among the acid granites, gneisses and schists of Precambrian Africa, as well as in the quartz sands of the Sahara and the Kgalagadi. Thus, relatively good soils are restricted to the basic volcanics of the East African and Ethiopian Highlands and the Cameroon, and to the alluvia of the major arteries - the Nile, Niger, Congo and Zambezi - and to smaller rivers. Twelve groups of soil provinces can be schematically distinguished :

(1) In North Africa, a 30-300 km wide band of largely chestnut soils covers the Atlases.

(2) In the Sahara, a 1,500 km wide band of desert extends between latitudes 32° N and 17°N, and is bordered in the northwest and south by 250 km wide belts of light-brown steppe soils.

(3) The Sahel, a 300 km wide band undulating between latitudes 10°N and 15°N, consists of a northern, chestnut and a southern, black soil zone.

(4) In West Africa, a 600 km wide band of red-brown soils extends between latitudes 12°N and 6°N as far as Ethiopia.

(5) in eastern Ethiopia, an irregular zone of chestnut soils extends to Mombasa (Kenya).

(6) In the Horn of Africa, light-brown steppe soils cover Somalia and adjacent Ogaden.

(7) The equatorial area of red tropical soil occupies the coastal strip of the Gulf of Guinea, Gabon and the Zaïre basin between latitudes 4°N and 3°S.

(8) The Plateaux, covered by red-brown soils, connect Angola, the southern part of the Congo basin, and eastern Tanganyika as far as latitude 12°S and eastern Moçambique.

(9) The Zambezi Basin, an 800 km wide area of slightly leached soils, extends from southern Angola to western Moçambique and thence southward to Durban.

(10) The Southern Steppes of light-brown soils cover southwestern Angola, Namibia and Botswana.

(11) A belt of chestnut soils surrounds (10) in a 400 km wide ring which lies mostly in South Africa, cf. 1.

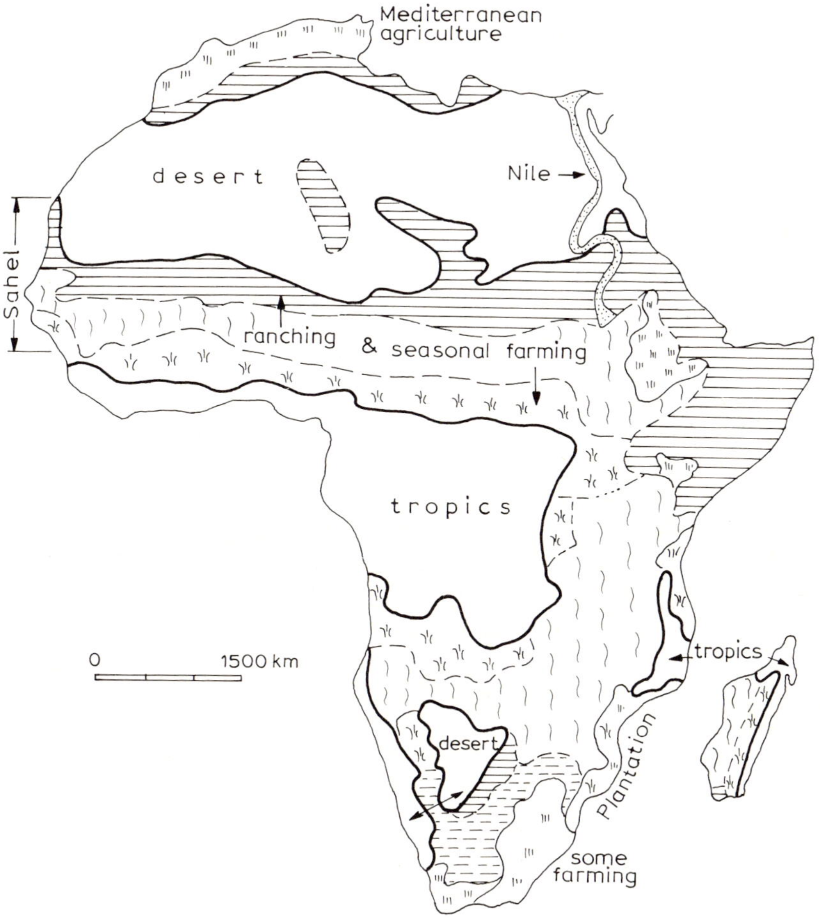

(b) Soil and subsistence.

(12) In Madagascar, laterite plateaus are surrounded by various smaller zones. A comparison of soil and climate maps thus shows a distinct zoneography, with soils distributed along east-west trending belts of latitudes, with the exception of the Horn and Southern and South Africa, which are influenced by other structural and meteorological patterns: A glance at the geological map of Africa will reveal a considerable control of the formation of soils.

Energy

As shown below reserves and production are unevenly distributed with most gas (and LNG) in Algeria and two thirds of oil in Libya and Nigeria. These countries control three quarters of refining capacity together with Egypt and South Africa. The Republic also holds half of the uranium (Fig. 5), most coal, 2 billion tons of anthracite, followed by Morocco and Swaziland and most subcoking coal together with Zimbabwe.

Table 11 Energy reserves, resources::, capacity (cap) and production.

Hydropower potential		Uranium		Coal			Gas		Oil	
		resvs.	prod.	resvs.	prod.	Country	resvs.	marketed	resvs.	prod.
Gw	Gw	kt	kt	Mt	kt		Tcf	Bcf	Bb	Mb
5	0.3	30		50	2	Algeria	110	2-3000	7	200
						Libya	20	150	20	350
10	12			90		Nigeria	50	20	17	400
4	2.7	5		30	250^{p}	Egypt	10	120	3	300
		::				Tunisia	4	15	2	40
10	23			10		Angola	4	8	2	80
23				500::		Cameroon	5		0.5	45
18		20	1			Gabon	0.6	5	0.6	70
9						Congo	2.5		0.8	50
130	120	3		80	300	Zaïre	1		0.2	15
						All others	10		2	15
		190	5	59B	150M	South Africa	1			
5	4			250	3M	Zimbabwe				
		120	4	3B::		Namibia	2			
				4B	400	Botswana				
0.7						Swaziland				
12		::		400	500	Moçambique				
64		5		100		Malagasy R.				
21	10			30	10	Tanzania				
4	12	8		300	500	Zambia				
2	0.2	160	3	4	100	Niger				
1	2.4	::		50	50	Morocco				

The first column reproduces World Energy Conference estimates and the second column World Bank Data.

Yet, out of 200 million tons of lignite mostly in Nigeria, but also in Niger, Angola and Madagascar only 10 million tons will be mined one day in Ethiopia.

In contrast, 10 Mt peat have acquired importance in Senegal, followed by 0.5 Mt in Burundi and probably more in Rwanda. The Malagasy have likewise 20 Bb of heavy oil, followed by the Ivory Coast and other West African oil producers, Mali and Niger. Similarly, shales of Zaïre contain 10 Bb of bituminous oil. Although Morocco holds only 0.5% of these reserves, it built a pilot recovery plant ; its phosphates, further, boast most of the 30 Mt of low grade uranium resources.

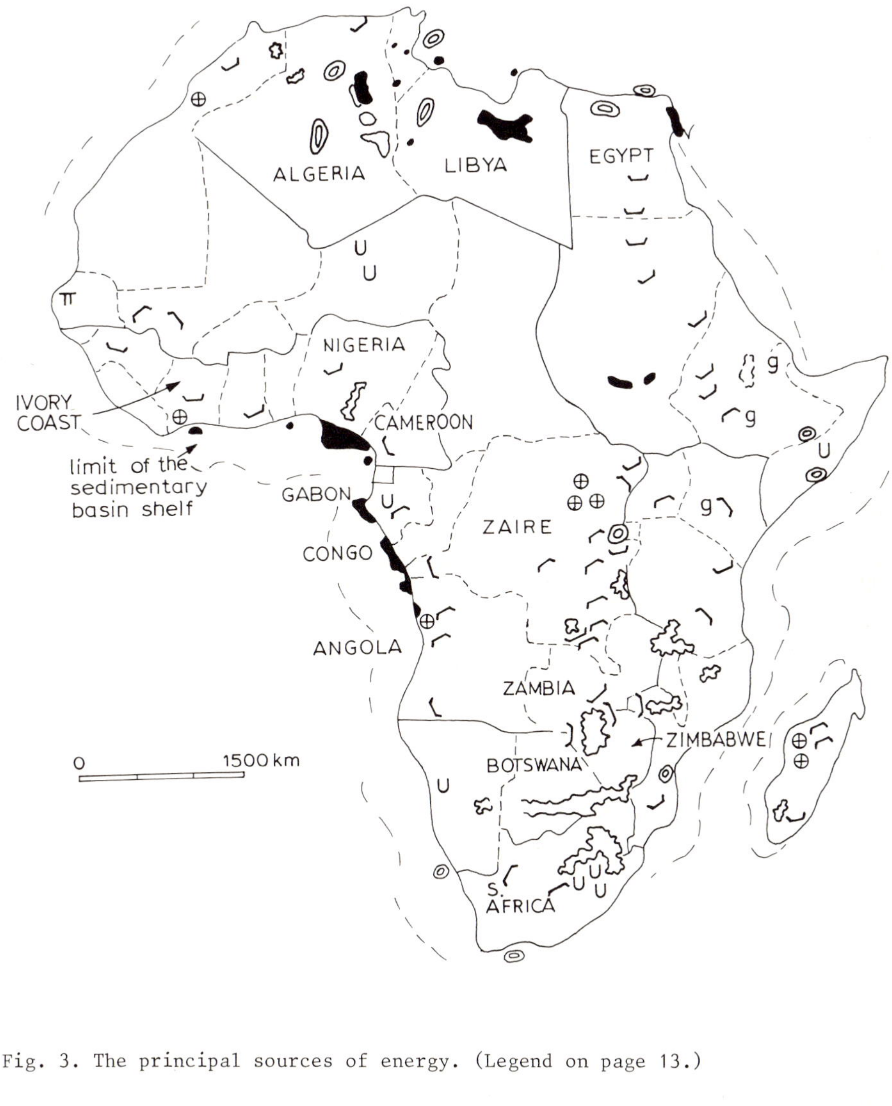

Fig. 3. The principal sources of energy. (Legend on page 13.)

Most geothermal reservoirs are in the nine Rift countries ; they may generate e.g. 500 Mw in three Kenya fields alone. Following this example and Zaïre, steam will be recovered soon in Djibouti and some day in Ethiopia.

In addition to the countries listed in table 10, the following have hydropotentials of 3 Mw or more : Sudan 16 (2.7) Mw, Uganda 12 (2.2) Central African Republic 11 (?), Guinea 6 (?), Senegal 4 (0.5), Mali 3.5, Chad, Sierra Leone (1.3) and Botswana 3 Mw each, Ivory Coast 0.8 (3) Mw. (Bracketed figures are World Bank data).

Gold, diamonds, ultrabasics

Table 12a Resources (1) and Production (2) of gold and diamonds.

Reserves		S. Africa	Namibia	Botswana	Zimbabwe	Angola	Zaïre	Tanzania	RCA	Ghana	S.Leone	Guinea	Mali
Gold	t	K20		10	600		150	40	f	150	f	2	50
Diamonds	Mct	370	20	210		30	230	5	5	22	25	5	
Production													
Gold	t	700			15		9	2^P	0.2	15	0.3	f	0.7
Gems	Mct	4.5	1	4.5		1.4	6	0.2	0.5	0.2	0.6	0.4	
Industrials	Mct	6	0.05	7		0.3	9	0.1	0.1	0.2	0.2	0.05	

South Africa dominates gold and antimony reserves (Fig. 4) and production and the diamond trade. The Republic shares with Namibia most high priced gems, with Botswana, Zaïre and Angola low priced stones and industrials. Depressed markets for large diamonds, depth and political considerations delay the development of additional reserves in Namibia. Moreover, Morocco supplies a tenth of the antimony.

Likewise, South Africa is preeminent in platinum group metals, chromite and vermiculite. The ratio of palladium to platinum is higher in Zimbabwe, where chrysotile, asbestos and magnesite are also abundant. So are resources of high grade chromite, which is costlier to mine.

Botswana also produces nickel and the Malagasy Republic might do so. Lateritic nickel resources are also widespread, for instance, in the Ivory Coast, Burundi and Tanzania. Egyptian production of talc matches the weight of South African output, but the value of pyrophyllite (wonderstone) is higher. Finally, Kenya occasionally produces vermiculite.

Table 12b Commodities related to ultrabasics.

Commodity		Reserves: S.Africa	Botswana	Zimbabwe	Malagasy	Sudan	Burundi	Ivory C.	Production: S.Africa	Botswana	Zimbabwe	Malagasy	Sudan
PGM	kt	50					f		t 100		kg 150		
Nickel	kt	M 6	450	220	100		1.2	3	kt 20	18	10	t50	
Chromite	Mt	B 3.2		750	40	1.5			3		0.5	kt50	t20
Chrysotile	Mt	Mt 3		5		f			kt200		150		Swaz.30

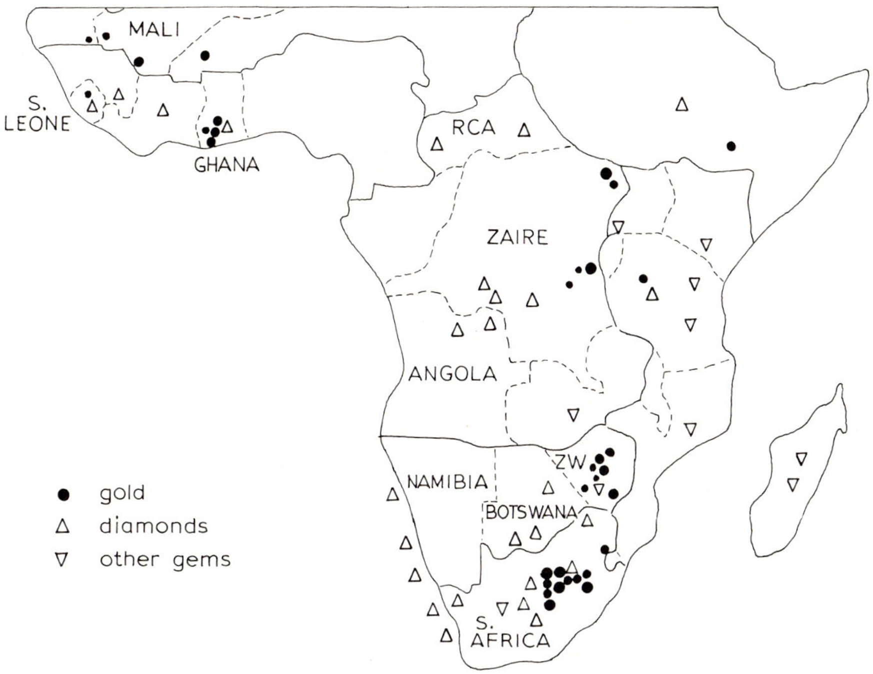

Fig. 4. Precious minerals and metals.

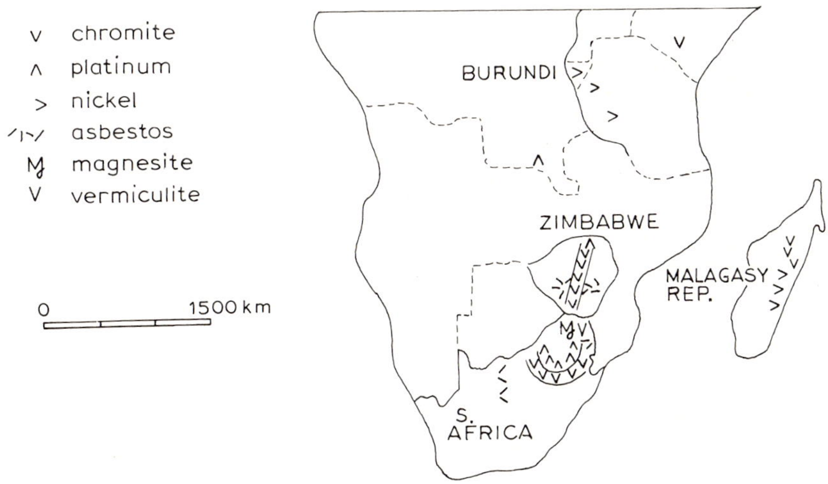

Fig. 5. Ultrabasic group.

While vast, often remote, resources of iron ore are abundant, South Africa remains the major supplier. Guinea is to replace the high grade ores of Liberia which are being depleted and Senegal and Gabon also wish to develop their deposits. The Republic also has unrivaled amphibole asbestos, vanadium and manganese resources, but Gabon is an important producer of the latter.

Table 13 Iron content of ore, bauxite, manganese ore, vanadium and amphibole asbestos.

Iron	Reserves metal Mt	Production Ore Mt	Reserves, resources^{rc}	Mt	Bauxite	Resvs. Bt	Prod. Mt
S.Africa	6000	15	Guinea	400	Guinea	7	13
Liberia	230	10	Ivory C.	500	Cameroon	0.8	
Mauritania	500	5	Senegal	420	Mali	0.6	
Algeria	50	2	Gabon	700	Ghana	0.5	
Egypt	170	1	Angola	80	S.Leone	0.1	0.6
S.Leone	30	0.5	Libya^{rc}	450	G.Bissau	0.2	
Zimbabwe	100	0.5	Zaïre^{rc}	7000	8 others	0.3	
Manganese ore	**Mt**	**Mt**	**South Africa**			**Mt**	**kt**
South Africa	12700	4	vanadium			8	20
Gabon	400	2	amphibole asbestos			5	90
Ghana	20	0.2					
Morocco	4	0.1					

While minor resources of vanadium also exist in Burkina Faso, Burundi and Moçambique, much of the world's bauxite spreads south of the Sahel between latitudes 6° and 11°N to which might be added some 200 Mt in the Malagasy Republic, Malawi and Zaïre.

Base and minor metals, silver, sulfur, barite and fluorite

Zaïre holds half of the copper, most cobalt and follows with Morocco the lead of South Africa for the other commodities. Like fluorspar and pyrite, base metals are widespread and Burkina Faso, Congo, Egypt, Nigeria and Sudan plan moderate expansions of output, but zinc and most silver recovery will cease in Zambia before 1990. Generally, silver is a byproduct of leadmining and since gold contains 8 to 12% silver, the Witwatersrand includes two thirds of African reserves, as well as most economically recoverable pyrite. Production depends on requirements for sulfuric acid in uranium precipitation and for leaching in the Copperbelt. Industrial extraction of native sulfur has ceased, but the oil refineries supply the byproduct, for instance some 20,000 ty in Libya and 4,000 ty in Egypt.

Although arsenic is widespread in the Maghreb, only Namibia recovers more than 1,000 ty. In contrast, Algeria still produces some 350 ty of mercury out of 5,000 t of reserves as well as 3,000 ty of strontium minerals which also occur in Tunisia and in the carbonatites. Lastly, Gabon planned to produce barite out of 1.2 Mt of reserves.

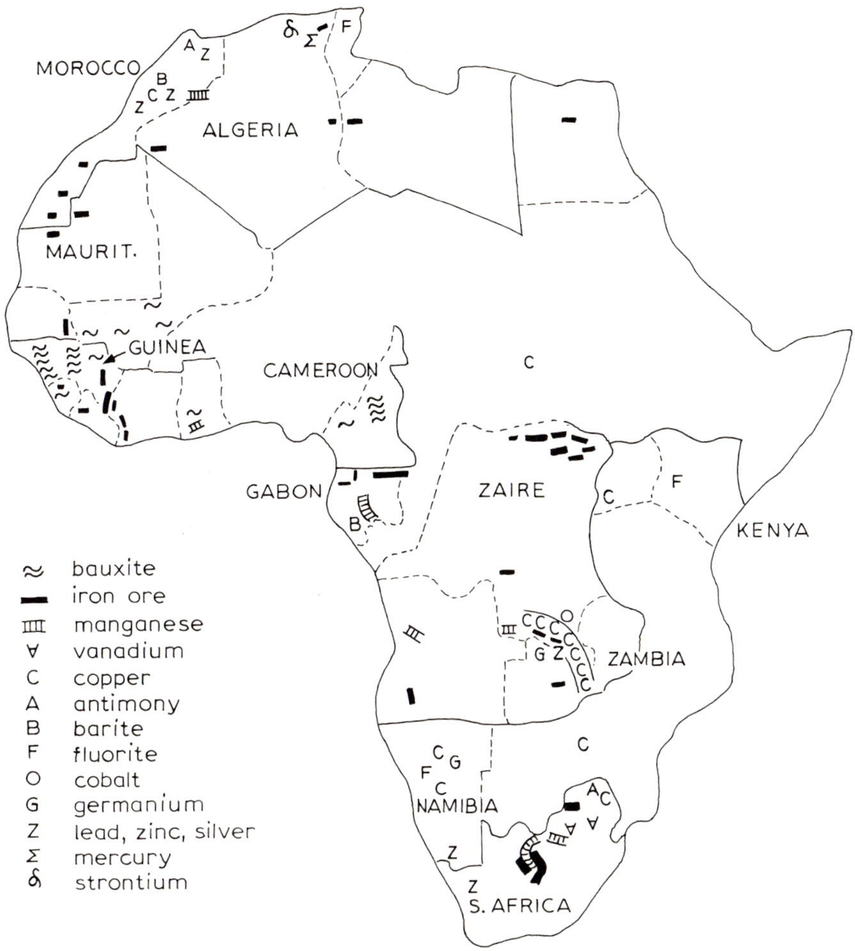

Fig. 6. The iron and ultrabasic groups.

Data are missing on South African <u>cadmium</u> recovery (if there is any) and they are even more sketchy on the minor metals which accompany zinc and copper notwithstanding the considerable share of the world supply.

Table 14 Base metals, silver, cadmium, barite and fluorite.

RESERVES	Copper Mt	Cobalt kt	Lead kt	Zinc kt	Silver t	Cadmium kt	Fluorite Mt	Barite Mt	
Zaïre	25	M 1.2		M 1	k 1.5	20			
Zambia	14	350	70	200	150	0.2			
South Africa	6	30	M5	M12	k 10	40	90	30	
Namibia	0.6		350	300	500	1	10		
Botswana	0.4	8							
Zimbabwe	0.2	8			300				
Morocco	0.4	5	M2	150	K 2	0.6	6	5	
Algeria			100	300	30	1		2	
Tunisia			200	300	50	1.5		4	
Mauritania	0.4				100				
Kenya							15		
PRODUCTION	kt	kt	kt	kt	t	t	kt	kt	Sulfur kt
Zaïre	480	10		70	100	300			130
Zambia	550	4	15	40	30	20			100
South Africa	200	0.3	100	60	200		350	6	500
Namibia	50		40	30	100	70			90
Botswana	20	0.2							
Zimbabwe	0.2	0.1			30		0.5		5
Morocco	20		90	7	100		60	100	40
Algeria			5	10	3			100	15
Tunisia			7	5	4		40	20	
Kenya							80		

Table 15 Minor metals production and reserves (magnitudes).

		Germanium	Selenium	Tellurium	Indium	Gallium	Rhenium
Production Zaïre		45 t	100 t	2 t	10 t	300 kg	100 kg
Reserves	Zaïre	3 kt	6 kt	2 kt	200 t	50 t	3 t
	Zambia		7 kt	2 kt			
	Namibia	4 kt	1 kt			3 t	

Namibia produces germanium again and Zambia supplies 20 t selenium. In addition to these, South Africa has been credited with a 150 t indium reserve base and much of the world's gallium is, of course, a trace element in Guinea bauxites.

Tin - rare metal - titanium group

Tin, rare metals and most semiprecious stones are found in areas surrounding granite and the large grained pegmatites which accompany them. They are also eroded into placers and titanium, minerals are dredged from beach sands. Thirdly, rare metals and earths and thorium occur in carbonatite-alkaline complexes mostly in East Africa along with 'limestone', 'phosphate', fluorite, barite and vermiculite.

Table 16 Tin and titanium group reserves, resources^{rc} and production.

Reserves & Resources	Tin	Tungsten	Niobium	Tantalum	Beryl	Lithium	Mica	Feldspar	RE oxides	Th	Ilmenite	Rutile	Zirc.
	kilotons										million tons		
Nigeria	10		0.5	0.1									
Rwanda	20	12	0.3	0.2	10								
Zaïre	32	1.7	180	0.4	5	180^{rc}							
Zimbabwe	20	1			5	30	50						
Namibia	60	2											
South Africa	30	20			15				320			7	6.8
Moçambique					5						12^{rc}	2^{rc}	
Malagasy R.									55	9	4^{rc}		0.2
Sierra Leone												4	1.5
Egypt											1.8		0.4
Production	kt	tons									kilotons		
Nigeria	1		50	10				6^{+}					
Rwanda	1.5	400	15	10	8	2^{+}							
Zaïre	3.2	70	30	20	100^{+}								
Zimbabwe	1.5	30	10	10	50	150	K1^{+}	1					
Namibia		150^{+}	p	p		20					Ti. slag		
South Africa	3				40	scrap	4	50			400	60	160
Moçambique					20^{+}	5^{+}	500^{+}	1^{+}					
Malagasy R.					10		300						
Sierra Leone												85	
Egypt								K7					

Abundant, but high cost tin resources are available above all in Nigeria and Zaïre, but the recent price drop would force the closure of many mines of this group. More and more tailings are, therefore, retreated and tinslags have become a key resource of tantalum. There are also 10,000 tons of lithium in Mali, 10,000 and 25,000 tons of cesium along with 0.1% rubidium in Namibia and Zimbabwe, an important emerald producer which competes with illicit exports from Zambia. Lapidaries collect other semi-precious stones in Tanzania, Kenya, South Africa and Madagascar where recovery of highpriced phlogopite mica has plummeted out of vast reserves. Some monazite and uranothorianite is left further south and there are also 330,000 and 33,000 tons of rare earth and thorium oxides in Malaŵi. Further, Zaïre produces some monazite.

Although some molybdenum shows in placers too, the uranium plants can produce 400 ty in Niger and scandium capacity is one ton each in Niger and Gabon and 5 ty in Namibia.

Black sands darken the beaches from the Canary Islands to Aden via the Cape and in the Nile Delta. In Senegal alone, 5 Mt ilmenite and 1.2 Mt zircon are still left. Meanwhile, most rutile and ilmenite are dredged in Sierra Leone and South Africa, which produces titanium slag. Some of its zirconium comes though from an alkaline complex.

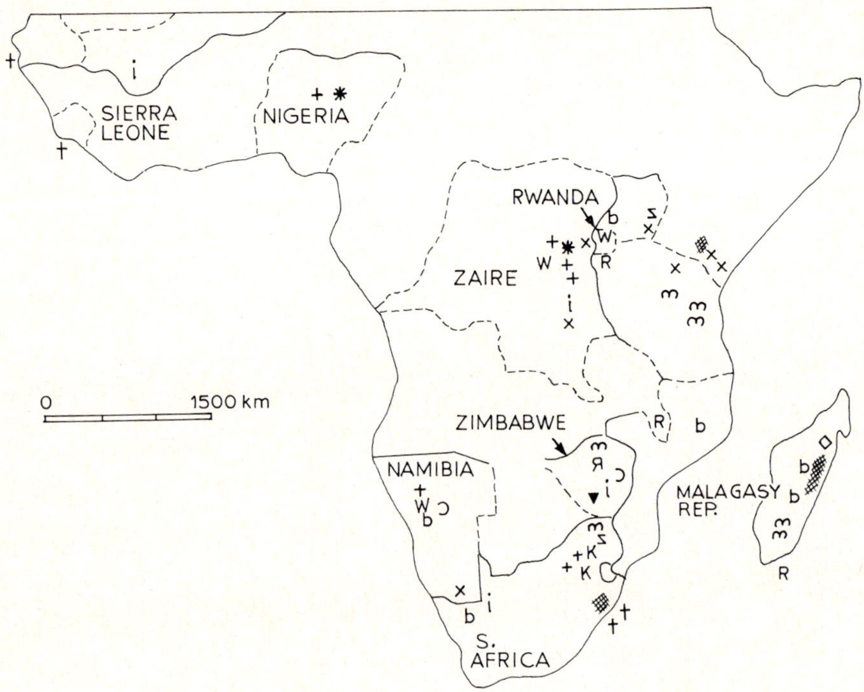

Fig. 7. Tin and titanium groups, metamorphics. (Legend on page 13.)

Metamorphics, sediments and building materials

The Republic also ranks first for most commodities tied to metamorphic rocks. Andalusite production peaked with 200,000 tons in 1980 and sillimanite recovery plummeted to a 1,000 tons out of 50 and 2 Mt reserves.

Some 20 to 50 ty of corundum are still sold; capacity attains 10,000 ty of garnet out of 300,000 tons of reserves and mining of amorphous graphite is starting out of 20 Mt of reserves. Malagasy reserves of highpriced crystalline flake stood at 5 Mt and resources exceeded 30 Mt. Production declined, however, to 15,000 ty. Zimbabwe's garnet and corundum reserves rank first with 600,000 t and 3 Mt respectively, to which we can add 6 Mt of resources and a production of 9,000 ty. Finally, Sudan is on the verge of shipping wollastonite out of 450,000 tons of reserves.

Morocco holds, of course, 90% of phosphate reserves and 85% of resources. Tunisia and South Africa beneficiate low grade concentrates. South Africa and

Egypt are the leaders in most of the other categories. Sierra Leone, Namibia and Ethiopia produce 220,000, 200,000 and 130,000 ty of salt. Although some is ponded, in most countries, production has grown slower than population. Potash accompanies other salts next to the Gabon and Congo oilfields. Congo has 20 Mt of reserves near the flooded Hollé mine and Ethiopia has 210 Mt of resources in the Danakil depression. Yet, the Tunisian project of 40,000 ty appears to be nearest to execution. So is, as a matter of fact, the Botswana soda and brine scheme which would draw on 610 Mt of reserves. Kenya has meanwhile expanded capacity out of 50 Mt of reserves. Although 150 Mt of Tanzanian reserves are larger, the project has been scaled down to 60,000 ty. Additionally, Uganda and Chad have 20 and 7 Mt of reserves and 5 to 20,000 ty are recovered from the Lake.

Table 17 Phosphate reserves and production (P_2O_5) of sediments and processing planned^{P}, past^{+}.

Phosphate		S.Africa	Zimbabwe	Togo	Nigeria	Senegal	Egypt	Libya	Tunisia	Algeria	Morocco	Canary
Reserves	Mt	180	2	90		50	70		80	80	B 4.3	
Resources	Mt	B2.3	6	360	f	80	200		350	90	B19.3	
Production	kt	M 1	40	800		800	200		M1.8	300	M 5	
Fertilizer prod	kt	500	20				60		600	50	300	
Fertilizer cap	kt	M 1	40	330^{P}		260	130		M 1.2	160	M 1→1.5	
N in fertilizer	kt	650	70	Kenya	150^{P}		700	250		150		
Salt	kt	700		60	3	170	M 1	10	400	150	80	
Soda	kt			300			40					
Diatomite	kt	1		2			0.3			5		
Bentonite	kt	40	80				3			35	10	
Attapulgite,&	kt	5				150				5	30	
Kaolin	kt	130	3				100			15		
Gypsum	kt	300					800	180	80	250		
Lime	kt	M 2		30			100	250	500	40		
Cement	Mt	8	0.4	1.5	4	0.4	6	5	3.5	6	4	1.2
Steel	kt	M 8	500	10	400^{P}		600	10	150	600	6	
Ferroalloys	Mt	1.5	0.2				K 60^{P}					
Aluminum	kt	170					160					
Refining cap.	Mb	140		30	100	10	200^{P}	120	15	180	30	60
Petroleum	Mb	130		20	60	6	150	40	10	150	25	60
Hydropower	Twh	2	6	3	4	P	9		f	1	3	
	Twh	Zambia 12, Zaïre 5, Ghana 5, Cameroon 2, Ivory Coast 2, Sudan 1.2, Ethiopia 1.										

Algeria has 4 Mt of diatomite reserves and large bentonite deposits. Senegal expands the production of fuller's earth and Togo also plans to compete with the sinter on the market for cat litter. Production of refractory flint clay continues to decline to 100,000 ty in South Africa, but 30 Mt of reserves are left. Few countries record the tonnage of fire clay and ball clay, South Africa and Zimbabwe recover 2,000 and 1,000 tons of iron oxide pigment and Egypt reports ochre.

Vast gypsum and limestone beds lie near to the oil basins and also in Somalia and Mauritania. Ash is abundant in the East African volcanics and Cameroon and the Cape Verde Islands quarry 50-100,000 ty pumice and the Canary Islands produce about a million tons of pozzolana.

Processing

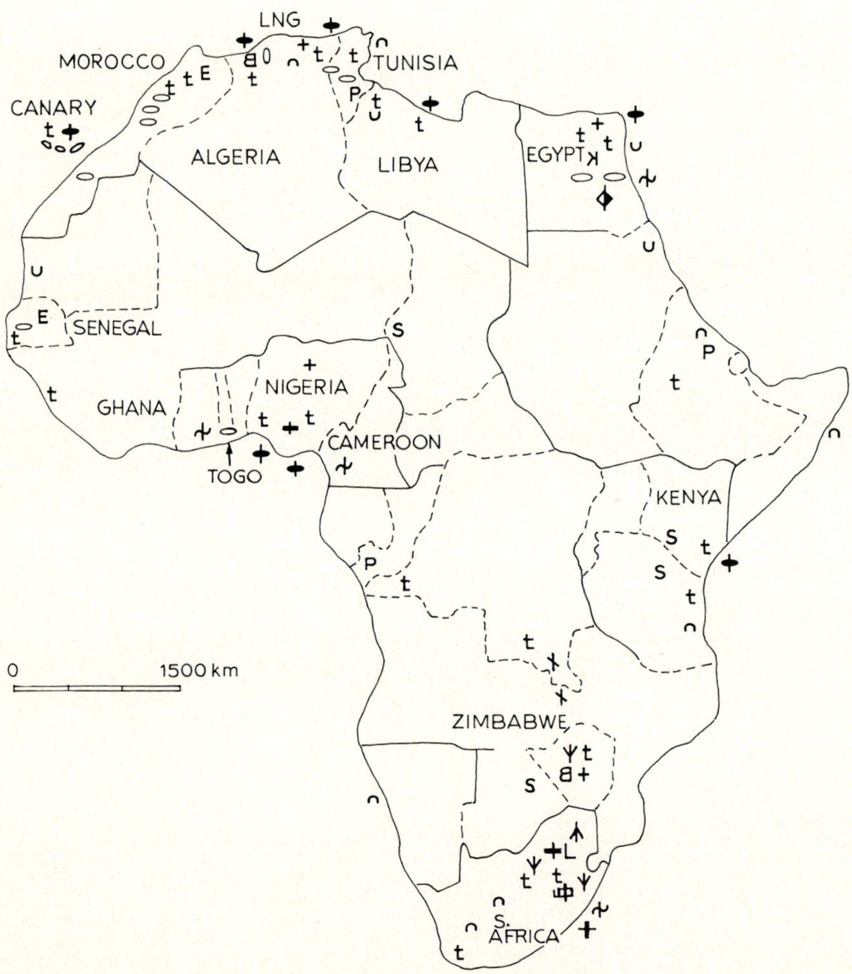

Fig. 8. Processing and sedimentary resources. (Legend on page 13.)

In addition to ten major producers of <u>cement</u>, Tanzania, Cameroon, Zaïre and Zimbabwe ship 400 to 500,000 tons each and Ghana 300,000 tons. All in all twenty five countries have kilns, but many operate at part of their capacity. South African industry uses most <u>lime</u> and Zambia and Zaïre supply 200,000 and 160,000 ty for their copper plants.

Although Morocco builds up <u>fertilizer</u> capacity, Tunisia still produces more phosphatic and Egypt more nitrogenous fertilizer. A dozen countries have plants, some of which operate under capacity, notwithstanding dire needs. Tunisia and Senegal will offer a full line of fertilizers just as South Africa does. The major gas producers, Libya, Algeria and Nigeria all plan urea conversion and five million tons of additional fertilizer capacity are in various stages of design and build-up, mainly in the Maghreb.

Algeria, Egypt, South Africa, the Canary Islands and Nigeria refine 80% of the <u>petroleum</u>, Algeria is a major exporter and Kenya serves the hinterland. More than twenty countries refine oil, many plants are underutilized, yet additional capacity is being built up, mostly in Algeria, Libya and Nigeria. Algeria is also the dominant producer of <u>LPG</u> and <u>LNG</u>, followed by Libya for the latter. Market conditions and Soviet competition hamper the entry of Nigeria and Cameroon into this field. The Arab mediterranean follows the steps of South Africa and <u>plastics</u> industries take shape in Egypt. Reduced oil revenues have slowed down, however, their build-up in Libya and Algeria.

South Africa is the world leader in <u>oil-from-coal</u> technology and this feedstock serves synthetics, rubber, and <u>explosives</u>. The Republic produces 5 Mt <u>coke</u>, only followed by Zimbabwe (200,000 ty). South Africa has also built the first uranium processing facility and <u>nuclear power</u> plant and Egypt would wish to follow suit considering rapidly growing needs.

Although South Africa has only limited <u>hydropotential</u>, the traditional source of energy on the continent, it owns half of the continent's generating capacity, some 90% of these 25 Gw are coal fired and have grown sixfold in 25 years. With more than a third of the continent's potential, Zaïre still generates less than 10%, leaving first place to Zambia, a country that also sells power to Zimbabwe. These three Egypt, Ghana and Nigeria generate 60% of the hydropower, but only ten countries do not have plants.

<u>Aluminum</u> smelting is generally tied to hydropower too. During the drought, most of Ghana's 250,000 ty of potlines lay idle, therefore. Conversely, South Africa and Cameroon would wish to double their capacity of 170,000 and 80,000 ty and Zaïre would prefer to use Inga power. Meanwhile, all producers import alumina on the very continent which has the world's largest bauxite reserves, and Guinea, the leader, remains a minor producer of alumina.

Nickel is also won in Zimbabwe, but platinum <u>refining</u> and gold <u>coinage</u> are

South Africa's preserve. In contrast, base metals and cobalt smelter and refineries date from colonial days. Some wire is also drawn in Zambia, Zaïre and South Africa, which also has three quarters of the continent's steel capacity. Algeria, Egypt, Nigeria and Libya use or plan to use gas in the direct reduction of iron ore, which is mostly imported. The same countries and Nigeria roll steel and produce shapes.

Zimbabwe produces 200,000 tons of ferrochromium and Egypt plans to supply 60,000 ty of ferrosilicon. Yet, with a million tons of ferrochromium, 300,000 ty of ferromanganese, 2-300,000 tons of other alloys, 40,000 ty of silicon and 30,000 ty of manganese metal, South Africa operates on a different scale.

To these should be added 50-70,000 ty of vanadium slag, ferrovanadium and pentoxide, 4-500,000 ty of titanium slag, dioxide and pigments, 10,000 ty of antimony in oxide, zirconium oxides and a variety of semi-manufactures, which are not produced yet in the rest of Africa.

A 5 HISTORY AND DEVELOPMENT OF THE INDUSTRY

The beginnings

The search for water and salt is older than the human race, which had its cradle in the African Rift some three million years ago and the quarrying of stone for weapons, called 'tools' spans more than half a million years. Men already recovered iron oxide from Bomvu Ridge in today's Swaziland 30,000 years ago and artists decorated with mineral pigments rocks of South and East Africa, the Atlas and above all the Sahara, from the Xth millennium on. Then, at the dawn of civilization, metal mining started in Africa. Indeed, when underdeveloped nomads still roamed in Europe and North America, Africans already extensively used metals. By the Vth millennium B.C. copper made its appearance in Egypt. Prior to 3100 B.C., slaves of the Egyptian state already mined gold in the Southeastern Desert and Berenice ; and in 1950, forced labour still worked the Ethiopian placers. In the VIIIth century B.C., Ethiopian invasions forced the Ptolemeans to abandon the mines of the Wadi Olaki. Here, Agatharcides and Diodoros described sophisticated mining and smelting, including sluice boxes and cupellation, methods still used in our times. The Pharaohs used cobalt, and gemstones even earlier, and the Nok civilization of Nigeria sculptured heads from stone in the Vth century B.C. Further south, in Ife and Benin smiths alloyed brass and bronze from the Vth century A.D. on.

To the west, the Berbers and then the Romans worked lead, limestone and salt and used bitumen. Another North African, Hannon, the Carthaginian navigator, visited the coast and bought gold from the Bambutos. Although the Romans returned there, it was not until 1714 that a French explorer saw the placers of the Upper Falémé in Senegal. Long before this, in the Xth century A.D., Persians established their gold trading posts in Monomatopa, south of the Zambezi, and Arab geographers described the metal mining of western Africa.

The fame of the golden throne of the King of Ghana reached Leon the African, in the XIth century A.D. ; El-Bekri by this time had already visited the iron mines of Mauritania, which ,rediscovered by pilots, recommenced operations in 1963. In 1354, Ibn Batuta, another great Arab explorer, reached Azelik and In Gall, near the current uranium mines of Niger and found that copper mining had gone on there for many years. Indeed, already in the IVth century, iron appeared in Egyptian tombs and western African artifacts. In central Africa, the age of iron appears to have come only a few centuries ago, and small scale smelting continues even now. Gold mining continued in the Red Sea Hills and in Ghana until 1873, when British soldiers looted Ashanti. Obuasi reopened as an industrial operation in 1899. However, geologists have evaluated that the total pre-industrial production was less than a recent year's output.

Central Africa

After the establishment of the Lunda Empire, the copper of Shaba reached Europe. Indeed, the XVIth century archives of the Oud West Indische Compagnie of The Hague record that copper transported from the interior to the estuary of the Zaïre River was regularly shipped to Europe. In the hinterland of the coast in the XVIIth century Catholic Kingdom of the Congo, specialists already worked the deposits of the Mavoio-Niari area. Similarly, in the XVIIIth century, each caravan travelling the Bihé trail, a route the Benguela railway now follows, the path to Ambriz or the Zaïre transported 5-7 tons of copper, along with ivory. The slaves spent 3-8 months en route before they reached the coastal strip controlled by the Portuguese. Ladislas Magyar, a naval officer who married the daughter of the King of Bihé, vividly described the horrors of these caravans. In 1853, the Hungarian explorer also visited another area where Africans smelted copper, namely, Tsumeb in today's Namibia. In 1688, south of the Orange River, Dutch explorers reported on the copper of O'okiep and by 1837 a British company planned the exploitation of these. But to return to Shaba, in the second half of the XIXth century, the exploitation of copper took place each season after the sorgho harvest, under the control of the Bayeke warriors and Babemba chiefs. However, historians believe that during that period not more than 1,000 tons of copper were smelted on both sides of the Copperbelt. Then, at the turn of the century, Africans led the British and Belgians to the ancient workings, and when the railway reached the Copperbelt in 1911, the world's largest copper complex commenced operations.

South and North Africa

Let us now turn southward to the Cape. Unlike gold, Africans did not appreciate diamonds until after their discovery at Kimberley in 1867. By 1887, 2 years before one of the great recessions of our economic system, cumulative production already was in excess of 3 million carats ! But miners soon turned their attention again to gold. While for centuries Africans washed gold, not until 1872 did Australian and American prospectors find gold in South Africa. The Witwatersrand was not discovered until 1884, and the goldfield of the Orange Free State had to wait for its exploitation until after World War II. This activity triggered the mining of coal, a fuel Africans already used. In 1888, the Portuguese granted a concession at Tete. In South Africa, from its beginnings in 1859, its output was raised to 70,000 tons in 1891 and to a million tons in 1894. Yet, it was not until 1904, when the railway reached Wankie and the bridge on the Zambezi, that this colliery could sell coal. The Malagasy already had noticed another form of carbon, and for centuries used graphite as a pigment. However, graphite was not sold for export until 1907. Meanwhile, in what was then Southern Rhodesia, industrial gold mining commenced in 1896, in northeastern Congo (Zaïre) in 1905 and south of Lake Victoria in 1921.

If we now turn our attention to the Maghreb, another focus of industrial development, we witness the discovery of the pyrites of Guelma (in Algeria) in 1848, the beginnings of a century of mercury mining at Ras el Ma in 1860, and the opening of the Aïn Mokra iron mine in 1865. In 1873, Philippe Thomas, a geologist, discovered the first phosphates of Algeria and, in 1886, those of Gafsa in Tunisia. In 1917, one of his associates, Major Bursaux, discovered Khouribga (Morocco) and, also in 1917, General Lyautey founded the Office Chérifien des Phosphates, which was to 'outproduce' Tunisia. Meanwhile, in 1910 already, R. Chudeau explored the phosphates of the Thiès Plateau, but not until after World War II did production start in Senegal. As to coal, lagging behind South Africa, the deposits of Kenadza and Jerada were not discovered until the first decade of this century. Their exploitation started respectively in 1918 and 1931. Kenadza is already closed, but Jerada has found a new lease on life.

After the railway reached Selukwe, Southern Rhodesia (the future Zimbabwe) exported 9,000 tons of chromite in 1906. Although chromite had been discovered in South Africa in 1868, the first shipment did not leave until 1929. Union Minière du Haut Katanga commenced the exploitation of cobalt in the same year. Platinum, which neighbours chromite in the Bushveld, was first recovered in 1924. Czarist Russians also discovered this noble metal in Ethiopia. Yet, output always remained modest. In contrast, Shaba palladium-platinum production reached the mark of 400 kg in 1936. The British worked tin in northern Nigeria in 1905, and Manono in the Congo followed in 1920. However, with the added production of Maniema in 1942, Congo production took the lead. From the same areas, the export of columbite-tantalite commenced in 1933. At Broken Hill (Kabwe), miners extracted zinc in 1913, and at Kipushi, in the Congo (Zaïre), production started in 1936. In 1913, mica mining commenced in Madagascar, and was followed in 1920 by South Africa and Southern Rhodesia.

As to diamonds, following their 1909 discovery in German South-West Africa, geologists found the gems in the southern Congo in 1913, in Angola in 1917, in Ghana in 1919, in Sierra Leone in 1930, and off the coast of South Africa in 1960. Following the early beginnings of manganese mining at Nsuta in Ghana (1916), and after World War I in South Africa and Morocco, operations started in the Congo in 1956 and in the Gabon in 1962. The second Iron Age started slowly in South Africa in 1934, in Liberia in 1951 and in Sierra Leone in 1953, but with our decade it has gathered impetus in Liberia and Mauritania, and other deposits are awaiting the capital investments permitting their exploitation.

Let us now take a look at the remaining 'late comers'. Pyrite production in South Africa did not start until 1938, and the manufacture of sulphuric acid in Katanga also started late. Conversely, uranium production in Katanga (Shaba) lasted from 1922 until 1961. The gold mines in South Africa initiated uranium recovery in 1951.

Production of oil at Gemsa did not start until 1911, even though people saw oil seep earlier from many areas and already used Egyptian bitumen. Although drills occasionally tested the Sahara as early as 1871 and the Gabon Basin in 1931, the exploration of oil needed another stimulus. In 1948, Conrad Kilian, the Alsatian geologist, deposited a sealed document with the French Academy of Sciences in which he revealed to those who could read it after his death that the Sahara must contain oil. Another geologist, Nicolas Menchikoff, shared this view and the Suez crisis gave impetus to accelerated exploration. Following the discovery of the small Djebel Berga gas field in 1954, the year 1956 saw the finds of Hassi Messaoud, Edjeleh and the gas of Hassi er Rmel. In 1957, Shell found oil in Nigeria. In 1959, Esso located the Zelten fied in Libya and Oasis followed suit with Gialo in 1962.

The structure of the industry

Prior to the industrial revolution, simple, labour-intensive operations, and primitive communications and marketing systems effectively limited output. On the other hand, the building of railroads and river navigation systems, i.e., the opening up of Africa, was generally based on mineral prospects. However, two forces triggered colonial penetration ; these forces were the rapid capital formation in Europe at the turn of the century, and the dynamics of nascent industries, liberating capital and talent which individuals, and later companies, invested in Africa. Colonialism also wiped out traditional cultures and changed the face of the continent. As a result of technocratic management, wages rose rapidly from their previous subsistence levels to considerably surpass the salaries prevailing in other African industries or in many other developing areas ; however, compared to the western world, they remained tiny.

As we have seen, the industrial revolution started from three centres in Africa : the South, the Maghreb and the Copperbelt. It spread, however, fast. The period following the Great Depression was spent in regaining the lost ground, but then rapid expansion started. Meanwhile, emphasis shifted by 1960 to oil, in the Sahara and Nigeria. Most outputs peaked between 1978 and 1980 and many among them are gathering momentum again.

Private capital is retrenched in four countries of the south, in Liberia and Kenya. Yet more and more dominant state companies seek joint ventures and/or management. The largest, Sonatrach, followed by NNPC, Nigeria and NOC in Libya own together 60% of the hydrocarbons. EGPC, Shell, Amoco and Elf represent another 15%, but Mobil, Socal, AGIP, Occidental, Texaco, Deminex, Petrofina etc. have smaller stakes - and management rôles.

Cobalt and most iron ore, copper and phosphate also belong to parastatals, so do 60 to 40% of bauxite, zinc and lead, within joint ventures or among different companies.

Requirements of state companies are variable ; several of these, for instance ZCCM, Gécamines and Yskor publish annual reports, but such groups as Rössing Uranium, NOC of Libya and the state mines of Ghana and Moçambique supply only limited data. Further, chartered accountants have annotated many balance sheets and profit and loss accounts, although only some practices have been questioned. Nonetheless, currency conversion and import restrictions as well as artificial exchange rates create incentives for maintaining payments or earnings abroad and according to observers some countries or individuals still export capital.

Ownership or control of industries is in a transition in many countries, both capitalist and socialist. So is mineral and petroleum legislation and in several states even their practical application is changing. Among others, particularly attractive laws or decrees have been applied or promulgated e.g. in Angola, Congo, Egypt and Moçambique, compensating thereby the risk. Such countries as Botswana, Cameroon, Gabon, the Ivory Coast, Kenya and Zimbabwe have always encouraged productive foreign investment. Moreover, in Morocco ONA, a state agency provides guarantees to and/or joins private investors.

Private capital

Commodities other than those already mentioned are largely the preserve of the mining finance houses. The Oppenheimer groups De Beers and AAC largely co-own each other and control JCI. They manage with brio half of the South African mineral industry, followed by the Afrikaner house Gencor, a government favorite, and Barrand. Their activities and structure are described under heading 82 and in Table 107. The 'houses' and their affiliates declare often substantial profits and pay taxes- unlike many parastatals. They have also created the practice of low cost migratory workers. Yet wages are higher in South Africa than on most of the continent. The income of whites averages nonetheless five times that of Africans who are in the process of constituting a new working class - the first in Africa.

Most 'houses', led by AAC, state that they insist on equal opportunity and some progress has been made within the constraints of racial legislation. They practice this objective, though, in joint ventures beyond the borders, where they finance education and share boards, and to some extent management, with Africans. True, 20% of AAC's, 50% of De Beers and Goldfields interests are abroad. The latter controls a quarter of American Newmont and AAC owns a third of Goldfields through lackluster Minorco, the Bermuda based family holding company.

The following tabulation illustrates the percentage share of commodities owned by the 'houses' and other groups :

AAC-De Beers-JCL Zimbabwe : coal 40, chrome 30, nickel 70, iron ore 12%, Botswana : coal 100, copper 25%, Namibia : diamonds 100%.

RTZ Zimbabwe : gold 15, Namibia : uranium 47, South Africa : copper 40%,

Newmont South Africa and Botswana copper 30%, Lonrho gold : Ghana 30, Zimbabwe 15%, PGM South Africa 10%.

PUK phosphate : Senegal, aluminum : Cameroon, Guinea.

Cogéma uranium : Niger, Gabon along with manganese.

Aluminum Harvey, Alcan : Guinea ; Alusuisse : Sierra Leone ; Kaiser : Ghana.

Iron ore Thyssen, Gränges : Liberia ; U.S. Steel : Gabon, South Africa.

To continue, Union Carbide controls many of Zimbabwe's chrome concessions, but only little South Africa chromite. Similarly, only 3% of RTZ's profit comes from Africa and even less at Blue Circle Cement or successful Cogéma. The business arm of French CEA is the world's leading uranium supplier ; it also holds a third of Imétal shares through 66% owned Cogérap. Coframines, Socomines and other companies associated with BRGM and CCCD have many minority interests. The Société Générale de Belgique affiliates, Sibéka and Diamang, are involved in Zaïre and Angola along with the two governments. Lastly, the Empain group controls Zaïre tin through Abay and Cogémines.

Aid and finance

While EEC countries provided a half of all bilateral aid, the U.S. and Japan extended respectively a quarter and 14%. Both UNDP and the World Bank are involved in indispensable infrastructure and water development. Sadly enough, water represents only 5% of all allocations. On the other hand $9B have been earmarked for energy over ten years, mostly for such 'rich' countries as Algeria and Egypt with $3.2B, with the Ivory Coast and Nigeria receiving $1.6B combined. The World Bank and regional banks contributed 62% of these amounts, against 15% of bilateral aid from EEC countries, 13% from Japan and only 7% from OPEC and 3% from the U.S. To continue, European countries, OPEC and Japan are the major contributors of the new $1.1B World Bank - IDA facility for Africa.

Most oil development is financed by the international companies (and their borrowing power), but Africa's share hardly reaches a quarter of its contribution to production.

In barely explored basins such as the Moçambique channel, the Seychelles high and Zambia, the World Bank and IDA even had to provide funds for the compilation of data and airborne geophysical surveys.

On the other hand BP, and to a lesser degree Shell, Total and AGIP are also involved in mining mostly in energy commodities such as coal. Unlike Arab capital, which moved above all into Tunisia and Mauritania, Comecon funding appears to be less than $100M. The U.N. Revolving Fund and EEC's Sysmin are devoted to mining. But out of the $3.3B of transfers to be committed under Lomé III, Sysmin's share is only 10%.

Although minor projects are expected to finance themselves, even the 'large' are small by energy standards ; the Guinea and Senegal iron ore schemes, rehabilitation of copper mines in Zambia and Zaïre do not reach $500 M each, which compares with $2B to be invested by ENI and its partners in the offshore Burri field in Libya and hydropower schemes ranging up to several billion dollars. Supplier credits, usually guaranteed by exporting countries, carry part of this load.

While instability, restrictions and inefficiencies deter investors, declining interest rates improve risk equations. Even better, more and more countries provide tax incentives, which induced, for instance, the South African finance houses to invest $3-5 billion in deep mines by 1990. Mining Magazine's lists demonstrate, indeed, that they already match North America for underground workings both in the category of plus one million ton year and more than 150,000 ty of ore milled (as shown in table 18).

Table 18a Number of mines in Africa, its subregions and other regions.

Number of mines	u.g.	o.c.
	> 1 Mty	
Africa	70	40
North America	60	120
Other regions	70	140
	> 150 kty	
Africa	160	70
North America	180	170
Other regions	330	180

African subregions	u.g. > 5 Mty	o.c. > 5 Mty	u.g. 1-5 Mty	o.c. 1-5 Mty	all >150 kty
South	23	3	27	6	33
Southern	2	3	3	1	24
Central	3	3	6	1	6
Maghreb	2	5	3	2	31
Others	–	9	2	8	19
Africa	30	23	41	18	103

Infrastructure

Apart from inadequate domestic markets, regional cooperation, capital formation and initiative, the lack of infrastructure hampers most the development of bulky or remote deposits, such as the bauxites of Upper Guinea and Cameroon, the phosphates of the Egyptian desert and the iron ores of Senegal.

The following tabulation (table 18b) gives the percentage share of rail and roads per subregion, compared to the surface area, population and gross domestic product :

Table 18b Per cent share of railways and GDP.

Railways %	GDP %	Africa	Area%	Population %
16	26	Maghreb	17	9
13	5	North-East	18	20
2	2	Sahel	15	11
11	27	West	10	27
12	6	Central	18	8
8	5	Eastern	9	15
7	3	Southern	6	2
1	3	Islands	2	2
30	25	South	4	6

In effect, minerals originally created the incentive for the building of most infrastructure in Africa, such as the railway linking Rhodesia (now Zambia) with the Cape and justified such ventures as the 'iron' railways of Mauritania and Saldanha.

Although Las Palmas remains the leading harbor and airport, it has little impact on African business. On the other hand, South African rails hardly match a third of the total length. They transport nonetheless 60% of all goods on the Continent and serve more than ever the hinterland as far as Zaïre since the civil wars in Angola and Moçambique. Zimbabwean soldiers guard, therefore, the Beira railway and pipeline. In contrast, the Richards Bay coal terminal continues to expand. While Zimbabwe has maintained an adequate infrastructure, apart from southern Shaba and Lower Zaïre, connected by a tenuous power line, much surface transport has broken down in Zaïre and delays reach one year along the interconnecting stretches of navigable waterways, rails (and roads) which link Lubumbashi to the coast.

While the Tanzam railway and Dar es Salam harbor are still unable to cope with expanded traffic, expensive trucking to landlocked Burundi, Rwanda and Uganda doubles the cost of petroleum. The RCA is still served by the Congo (Zaïre) river, but the Transgabonais has been rerouted for political reasons towards the south. Consequently, harbor facilities and the railspur will absorb half of the $500 million development cost of the Gabon iron ores.

Recession has decongested Lagos harbor. As a further measure, Nigeria, and on a small scale Cameroon, have expanded their loading facilities and longer stretches of the Niger can be navigated. Rolling stock is being upgraded in Ghana and Dakar, Abidjan and Cotonou continue to serve the landlocked Sahel.

Compared to their area the Sahel and North-East Africa have a tiny share of the infrastructure even though Sudan railways rank second for their length, although not for their maintenance. While telecommunications will be improved in Caïro, transportation along the Nile is efficient by African standards. Libya has its own fleet of tankers, but had to slow down the buildup along the coast and the construction of railways to the Shatti and Tindouf iron ores remains remote. On the other hand infrastructure continues to expand towards the south of Tunisia and Morocco.

Marketing and consumption

International commodity agreements have met success during sellers' markets and periods of expansion. While other interest groupings influenced trade only marginally, the international coffee agreement could be considered as a model for producers.

Although OPEC's share of oil markets has declined steadily, all African members, including 'wayward' Nigeria, have benefited for years. So did Algeria, indirectly, since gas and petroleum products are not regulated. The share of CSO in the diamond trade has also decreased, but De Beers' ability to select the type mined and to stockpile benefited South Africa as well as other producers through agreements with the Diamond Corporation and Britmond. 'Directly into the bufferstock' may have been the slogan of the ITC, but a ton of tin was worth for years 8 tons of copper or 40 tons of lead although most tin is simply dredged.

Oligopolies follow suit : other countries benefited from the weight of South African gold producers, so did the U.S.S.R. from the promotion of emission standards by the platinum mines led by Rustenburg, who curtailed sales when demand was low. China, Comibol and Murchison occasionally tailor antimony shipments. In striking contrast, a few producers fight for diminishing asbestos markets. Phosphates stand in between ; on the one hand high costs weaken the position of American exporters, on the other hand, African countries jockey for sales. As the director of the FAO said, hungry Africa needs more fertilizer and better distribution. Inphos, therefore, tries to promote the right policies for broadening usage from the current 2% world share on the continent which has 65% of the resources.

Lastly, vertical integration saved uranium production in Niger and Gabon, with the help of Cogéma and promoted the use of coal in South Africa. Some other commodities could similarly benefit. Finally, the free market has depressed prices since the share of commodities decreased within the GDP of OECD countries, e.g. copper by 20% in a decade or more. As real or deflated prices decreased producing states increased output in order to maintain their income. Poverty thus feeds on poverty.

Although tariffs on some processed products are higher than on the raw material, they hardly exceed 6% ad valorem. South Africa overcame the 30% U.S. tariff on high carbon ferrochrome and acquired leadership in this field, but similar success would require more cooperation within UNCTAD.

Apart from building materials, Africa exports 90% of its minerals, even South

Africa ships out 85% and the oil producers view uneasily a 10-20% yearly increment on local consumption of 15%, which will wipe out exports in Egypt and later in Algeria. The mediterranean countries rank, in effect, second as consumers of mineral products, following South Africa which uses 60% of the continent's power with 40% of its capacity. Industries also supply in the Republic 95% of mining supplies, which match, as a matter of fact one third of the value of mineral raw materials.

The rest of Africa imports 98% of these supplies as well as most heavy equipment. In effect, vehicles rank first among import items in most of Africa. The situation is the most critical in sub-Saharan Africa which lags third as a consumer of minerals.

Exploration

Apart from Antarctica, Africa is the least explored continent. Nonetheless, crucial groundwater development has a high success ratio. Modern methods have also been applied to mineral exploration and new deposits will be found especially a) under the desert and b) laterite crusts, c) in carbonatite-alkaline complexes, d) offshore and in the mineralogenetic provinces and belts described in the Mineral Resources of Africa pp. 215-245.

UNDP, BRGM, the British Geological Survey and the mining finance houses funded much needed exploration. To these should be added successful Burama in Tervuren, Belgium. The need for detailed exploration can be deduced from the ratio of known resources to the identified reserve base which lags in Africa at 8 to 1 for graphite, 5 to 1 for coal, 4 to 1 for PGM and tin ; at 3 to 1 for chromite, lead and zinc, and for most other minerals at 1.5-2 to 1.

Considering the inadequate infrastructure and their favorable value to weight ratio, the precious should be ovious targets for exploration. While gold continues to arouse interest, gem diamonds and semi-precious stones should also be emphasized. Likewise, the growth of consumption of aluminum, minor metals and special needs should be taken into account in drafting exploration and development plans. Let us repeat some current growth rates in % per year :
industrial quartz crystal 20%, tungsten 16%, gallium 10%.

Between 7 and 4% : aluminum, bauxite, chromite, industrial diamonds, germanium, hafnium, helium, lithium, niobium (columbium) and most platinum group metals, thorium, zirconium.

Between 3.9 and 2.5% : boron, cesium, cobalt, diatomite, fluorspar, fuller's earth, graphite, lime, sheet mica, molybdenum, rare earth metals, rhenium, selenium, tantalum, vanadium and yttrium.

As to oil, only 0.5% of all wells are drilled in Africa and even in the Sahara (plus Suez) basins, their density is a twentieth of the worldwide average (the Near East excluded). Consequently, as much as 75% more gas might be discovered since this was initially not the target of oil exploration.

Discoveries can also be expected offshore since these now represent only 10% of the African total as opposed to 30% worldwide. The Marzuq, central Saharan, Zaïre, east African and Malagasy basins match 60% of the potential and as yet undiscovered oil (and gas) reserves might reach 47 (50)% of EUR in the whole Sahara, 47 (56)% in the Gulf of Guinea and 58 (66)% along the coast extending from Cameroon to Namibia.

Statistical methods facilitate forecasts on the following lines (in Bb) :

	Algeria	Libya	Nigeria
Cumulative production	7	15	10
Remaining reserves	9	20	17
Undiscovered reserves	3	2	9

Let us add that these predictions have already proved to be conservative as far as Libya is concerned. Meanwhile, even the type of field has been computed for the Maghreb and Egypt :

Size of field	known reserves Bb		number of fields		ultimate reserves Bb
	cumulative	average	known	additional	
supergiants	15	7.5	2	0-1	15-22
giants	36	1.6	22	0-25	35-76
large	4	0.25	1.7	15-45	6-10
small	4	0.03	120	350-800	10-25
Total	60	0.37	161	365-871	66-132

1 THE MAGHREB

Arid North Africa's population is only half of that of Nigeria, but with the oil and gas of the Sahara, its advanced economies match a third of the continent's GDP.

11 MOROCCO

Al Mamlaka al-Maghrebia is the world's principal phosphate exporter and the continent's leading producer of lead, and barite (fig. 9, 10, 19).

> GOALS : A broader water supply. Harnessing of more hydropower and tests on bituminous shales in order to save imported fuel. An integrated fertilizer industry and further expansion of building material plants.

The phosphate industry is the motor of exports, industrial and infrastructure developments. It is also the largest industrial employer, but with 25,000 people it barely breaks even. The payroll and the share of phosphate products in total exports are expected to increase with processing from 50% to 70% or \$1.5 B, i.e. 10% of GDP. While Morocco is a geologist's paradise, the contribution of base metals to export earnings is one tenth of fertilizers and the highgrade iron ore and manganese deposits have been worked out.

Electric capacity is expected to double to 1.2 Gw, water supply is also scheduled to expand, and infrastructure is improving in the Center and South.

Table 20 Moroccan production, reserves and resources.

Production		Rsrcs	Production		Production		Resvs
Hydropower	Twh 3	Gw 1	Fertilizer	kt 300	Phosphate	Mt 5	Bt 45
	kt 700	Mt 50	Lead refined	kt 60	Lead	kt 100	Mt 2
Lignite		Mt 40	Cement	Mt 4	Zinc	kt 10	kt 150
Oil inshale		Bt 40	Petroleum	Mb 25	Silver	t 100	kt 1.8
Uranium^p	t 250	Mt 3	Copper Smelter	kt 20	Copper	kt 29	kt 200
Coal			Aluminum	t 200	Gold	kg 700	
Barite	kt 400	Mt 5	Steel	kt 6			
Fluorite	kt 60	Mt 6			Iron ore	kt 150	Mt 140
Feldspar	kt 1		Manganese ore	kt 60	Antimony	kt 2	kt 80
Salt	kt 70	Bt 1	Fuller's earth	kt 25	Cobalt		kt 40
Mineral water	m^3 70,000		Montmorillonite	kt 7	Nickel		kt 6
Mica	t 500		Bentonite	kt 4	Cadmium		t 600
Magnesite		Mt 1					

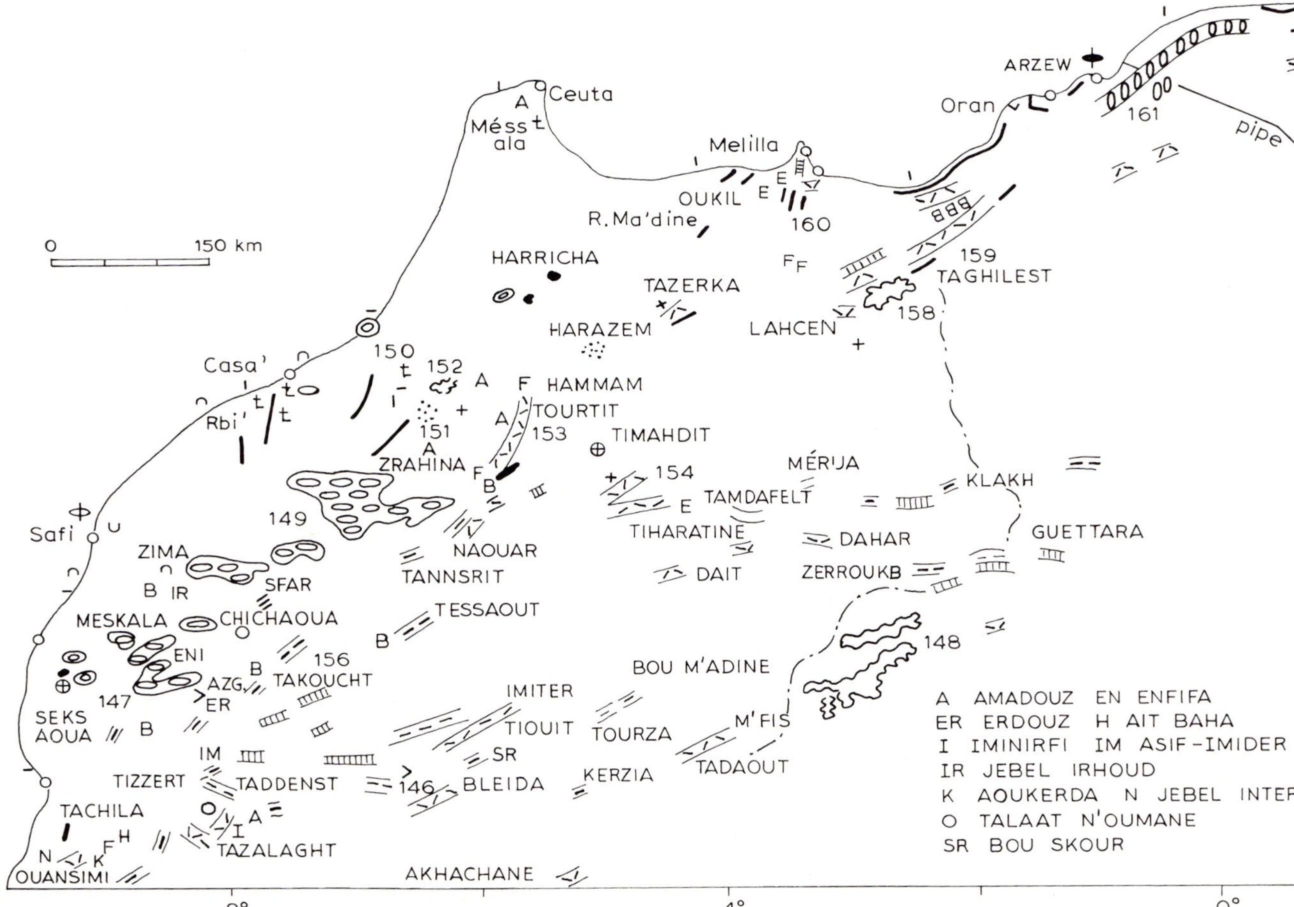

Fig. 9. Morocco and northwestern Algeria. (Legend on page 13.)

The Moroccan state controls the country's largest company O C P, which also owns 65% of the shares Fosfatos de Bou Craa, in the Western Sahara, the remaining 35% belonging to Spanish INI. Through B R P M the government owns or co-owns most mining companies (except for CMT) who employ 15,000 people. It also encourages new mining ventures by Moroccans and foreigners.

Phosphate exports shifted from Comecon to Bangladesh and Sri Lanka (10%), to the EEC (90%) who also take 40% of lead (mainly Italy and France) ; to which trade with Tunisia should be added. Switzerland and the U.K. took most of the expanding silver output (the latter also bought coal) and Nigeria 70% of exported gypsum. The U.S. imported 50% of barites and 20% of manganese, but the country is obliged to import oil, coal, coke, sulfur, and ammonia for the fertilizer industry. The energy import bill now matches fertilizer exports ; the situation should improve though.

Oulad Abdoun phosphates

Table 21 Moroccan phosphate rock reserves (Bt) and % P_2O_5.

	Oulad Abdoun		Ganntour		Meskala		Bou-Kra'		Total P_2O_5		Bt
	Bt	%	Bt	%	Bt	%	Bt	%	Bt	%	P_2O_5
R I E	9.6	31.6	3.4	29.7			0.9	32.5	14.3	30.3	4.3
R I M	4.7	31.6	4.2	29.7			0.7	31	9.6	28.7	2.8
R 2 E	24	31.6	16		10	31		?	50		

To these can be added 22 Bt of resources averaging 24.7% P_2O_5 on the Oulad Abdoun (149)[1] which also provides two thirds of the output. Production of phosphate rock has increased thus : sixties 10 Mt, seventies 16 Mt, eighties 20 Mt, and capacity should increase from 25 to 36 Mt by 1990. Furthermore, OCP marketed concentrates of the following grades (% BPL).

	Calciné	Mt	BPL	Mt	BPL	Mt	BPL	Mt	BPL	Mt
Oulad Abdoun	76-77	.6	75-76	.1	73	1	70-72	11.3	65-66	.1
Youssoufia	74-75	.5	68-69	5	Ben Guérir				60	1.6

The Oulad Abdoun scarp spreads 100 km east to Oued Zem and southwest to Sidi Daoui. At Recette 4 and Meraa el Arech, 6 to 20 m overburden is stripped from selected high grades. These quarries supply a half and 15-20% of the plateau's output. Contractors provided another 10%. Meanwhile opencast operations have been expanded to 20 Mty capacity, and underground should be phased out by 1990.

Among numerous phosphate beds, five average 31.8% P_2O_5 and their thickness adds up to 8.5 m at Sidi Daoui (layers 3A and B are 3.6 m thick and grade 24.7% P_2O_5 only...). Some concentrates are marketable following grinding and screening, but shaly ore is washed and dried at Oued Zem or dried and calcined at

(1) Numbers refer to figure 9.

Khouribga and Beni Idir. Some 100,000 m^3 water are piped each day over 40 km, Sidi Hajjaf is therefore dressed at Jorf Lasfar harbor, a distance of 120 km.

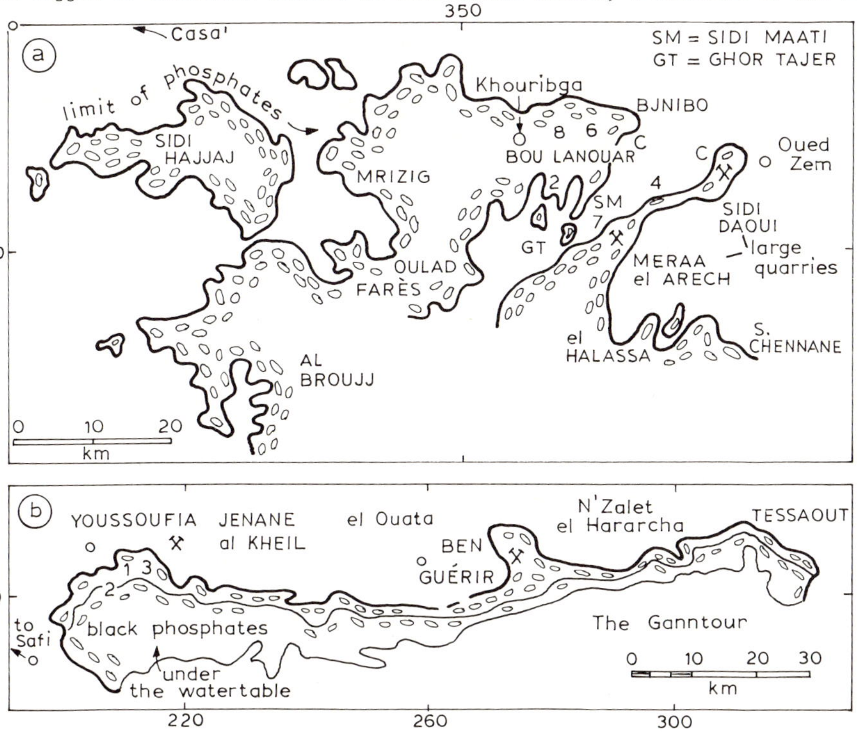

Fig. 10. (a) The Oulad Abdoun: *2-8* workings (recettes); *C* concentrators.
(b) The Ganntour.

<u>Other phosphates and fertilizer</u>

The GANNTOUR extends 120 km from Youssoufia east to Tessaout. Its northern edge dips under the water table and bituminous phosphates of the 2.5 m thick layer are worked underground e.g. at <u>Youssoufia</u>. They calcine easily into highgrade concentrate. Further east at <u>Ben Guérir</u>, 10 beds out of 23 contain 1.3 Bt of strippable reserves, which can be upgraded from 60.5 to 69% BPL. Output is to increase here from 3 to 10 Mty by 1995. The MESKALA spreads further south from Imi n'Tanout north to Chichaoua and production started after six years of development by Soviet experts.

A 100 km long conveyor belt transports crushed ore from BOU-KRA' to the Laiun harbor dressing plant which has a rated capacity of 3.5 Mt of 75% and 80% concentrates. Over 200 km^2 three layers coalesce into a 5.50 m thick bed. However, six times as much overburden is stripped by Fosbucraa. While ore is richer in the Uad Imesleguen, key layers are separated. About half way to Laiun in the uads el Jat and Itgui grades are in excess of 60% BPL, yet siliceous nodules appear in the uad Laabadilla. Lastly, in the northeast, beds are thicker but of lower grade.

FERTILIZER production averaged only 700,000 t in the seventies, but OCP wishes to process a third of its output, i.e. 6 Mt of concentrates. Consequently, considerable capacity is being built up and the Nador facility (640 kt phosphoric acid) is planned for the nineties (Table 22).

Table 22 Moroccan fertilizer capacity in 1000 tons.

Ore from :	Plant	SP acid	P acid	S acid	SP	TSP	MAP	DAP
Light	MC Safi		340	660		330	430	130
Youssoufia	MPI Safi		660	2300				
	MPII Safi	230	500	1700				
Ben Guérir	MPIII Jorf Lasfar		1300	4400				
Sidi Hadjaj	MPIV Jorf Lasfar				200	400		1000
TOTAL		230	2700	9000	200	730	430	1130

Water and soil

Rainfall declines from 1m in the Rif (N) to 300-600 mm on the coast and then to 100 mm south and east of the Atlas. In the early 80's 1 Bm^3y were lost in the sand in evapotranspiration and beyond the borders for 2.5 Bm^3y exploited.

Rocks of the Lias cover 10% of Morocco, but provide 30% of sourcewater with yields of up to 25 m^3s, although some wells provide only 1.5 m^3s. Limestones and dolomites of the Atlas feed the aquifers of the plains and highlands.

Yields are variable, e.g. : Jurassic highland limestone 150 m^3h in artesian wells, the Bahira limestone 150-200 m^3h, with even greater differences in younger sediments. The Gharb, Saïs, Tadla, Ben Guérir, Tiznit, the Tafilalt and Haouz also take their water from these sediments, but they have been overworked. Unfortunately, part of the Rif and Central Morocco have little groundwater and drains and dams were built in order to reinject water.

South of Agadir and in the western Sahara recent and sub-recent dunes provide water, but Laiun is served by artesian wells. Composition is the problem in the Seghiet el Hamra, and in the faulted rocks of Tiris-Tisla where water contains 1-5 g/l of solids.

Sidi Harazem (Fès) a property of the royal family has been the principal bottler of slightly laxative mineral water since 1932, followed by Oulmès and Sidi Ali (non-sparkling) in 1950.

While most of Morocco is covered by chestnut soils similar to those of Europe, strips of brown forest soils and podzolized soils as well as saline solonyets also appear.

Lead, zinc and silver

A few data (in 1000 tons) will show the evolution of lead (and zinc) output : in 1922 : 3, 1942 : 8(1), 1952 : 75(20), 1962 : 110(45), 1972 : 85(20).

The reserve base stood at 3.5 Mt, but out of 500 occurrences only six areas are important. Firstly, Zeida-Bou Mia, Sidi bou Othmane, and the depleting mines of Aouli and Mibladen (154) apparently hold more than half of the lead reserves; SODIM provides a quarter of the production. Secondly, only 10% of the reserves have been attributed to the Oujda region (158) of oued Mekta and Bediane, where CMT's Touissit plant contributed almost half of the lead output. Further, most zinc and silver came from Bou Beker. Sidi Lahcen has a capacity of 80,000 ty of ore, with 3 to 10,000 ty of lead and zinc concentrates. Thirdly, Bou M'adine is said to contain some 20% of ore reserves ; capacity is 60,000 ty of ore, 9,000 ty of concentrate averaging 60% lead and 6,000 ty of concentrate with 55% zinc, and there is also silver and gold. Fourthly, J'bel Aouam (152) and fifthly Draa Sfar (W) where copper also shows, will be served by the Meknès lead concentrator to be built with a 25% participation of Armico. Indeed, SMA produced in the M'rirt (151) 20,000 tons of concentrates (Pb/Zn 4:1). Sixthly the small mines of the Tafilalt can contribute 15,000 t of concentrate to CADEI.

Other occurrences include :

Lead E : Ksar Moghal (E), Bou Dahar (155), Daït, Aït Labbes (Zn), Taouz, Ksar es Souk (SE) ; Demnat Ksiba, Beni Tadjit, Tazzeka Mts (E of High Atlas); Akhachane (SE), Erdouz (SW), jbels Kitrem, Iznit, and Addana (S) with silver.
Zinc : Tiouli, Ouicheddène, Sidi Makhlouf (SW).

After several closures, FPZ expands the Oued el Heiman smelter (N) to 120,000 ty concentrates mostly from CMT who also supply a third of the silver, Zgounder (Ouarzazate S) has a capacity of 56,000 ty and 25 ty of silver. Other silver grades include Tizzert (SW) 10 g/t copper concentrate, Sidi Lahcen (NE) 10 ty capacity, Bou M'adine 300 g/t zinc concentrate ; Erdouz, Assif el Mal 0.2% concentrates, Koudiat el Hamra (SW), Goundafa 35 g/t. SMI at Imiter (see below) provided a third of output out of 4 kt reserves in ore grading 50 g/t. There were also 60 tons at Koudiat el Blida; BRPM also explored for silver at Sidi Flah, Saghro, Siroua and in the jbel Ougnat, for gold in the Jbilet (SW). Further, 30 kg of precious metals plus gold were occasionally exported from Tirouit (S).

Copper, antimony, etc.

With 50,000 tons of concentrates out of 300,000 t capacity Bou Safer at Bleïda (SSE) is the principal copper producer , followed by Somima's Ouansimi (10,000 t) ; 2,000 t are recovered by Zellidja (NE) and Sodecat supplies 5,000 t of concentrates from Assif Asfalou and Tazalaght and 1-2,000 t at Tiouit along with silver. Finally Cadetaf collected concentrates from small workings in the southern Atlas.

Table 23 Moroccan ore reserves (Mt), location and grades of copper, %.

North, NE	Mt	%	Deposit	Mt	%
Alous NE	5	0.8	Talaat n'Oumane	1.5	2
Cemo, Oumjerane	3	2.5	Naour (C)	1.5	1.5
South	Mt	%	Company SSE	Mt	%
Ouansimi	1.5	2	Somifer Bleïda	5	2
Tizzert	1.5	1.5	Somemic Tabaroucht	5	2

South of the trend of Bou M'adine (Zn), Imiter (Ag), Tiouit (Cu, Au) and Bou S'Kour (Cu) (in the southeast) is the alignment of M'Fis, Tadaout, Boukerzia (Cu) and Oum J'rane (Cu) in the eastern Anti-Atlas.

Six trends can be distinguished further southwest : I Argana 60 km ; II Tataout, Tinnfit, Alous, Assif-Asfalou (Cu, Ag), Imider, Oued Arrhen, Tanguerfa 40 km E ; III Iminirfi-Talaaat n'Oumane ; IV Zgounder (Ag), Tirzit-Amadoust ; V Aït Ourhaïne, Tazalaght (Cu, Ag), Aoukerda, Tasserit, Agoujgal ; VI Tailezza, the jbels Ouansimi ; Adrar, -Taourirt and Idm ou Anas.

There are indeed 800,000 tons of resources and plans are afloat for a new unit at Mohammedia harbor. FMC's smelter at Agadir (SW) is to produce 25,000 t blister, 25,000 t cathodes, 6,000 t other metals and 130,000 t acid from the Bleïda concentrates 300 km to the east. We might further mention here that Samaes' aluminium smelter has a 200 ty capacity at Aïn Sebaa.

Antimony output reached 2,000 t in 1962, peaked with 5,000 t and could expand again. Three areas share production a) The Fahama South zone of the Beni Messala (N) has produced 10,000 t of stibnine.

b) The northcentral area I Bled Zaiaa with jbel Timerhdoudine, Mouhajbat, Sidi Embarek and Aberki, II the central Mrirt including Koudiat Tibtahine, jbel Achek, Tirza, the jbel Izougenerz, Asfah, Bou Iskra and Mguedn, III the most important northern Mrirt (151) with Tourtit, Tafgout,Ich ou Mellal and Masser Amane and IV the Oued Beth occurrences Talet Tametot, Ighoud, Igouda and Bou Ichilfane.

The Kef n'Sour veins (c) produce stibnine and lead.

600 t of beryl, up to 80,000 ty of pyrrhotite, the sulfur mineral, and a few tons of tin have been recovered at Tazenakht (S), on the jbel Kethara (W) and el Karit (CW) and other occurrences include:

Arsenic, Co, Ni, Bou Azzer (W), jbel Aouam, Koudiat Cheïba, An Tikkaderine, Oukilal Agourzi, jbel Tisguine, Bou Ijda, Beryl Angarf (SE) Cadmium Zn Touissit, Bou Beker (NE). Ilmenite Oum Rbi, Azemour, Sidi Ishak near Essaouira, Agadir (SW) and Lagwira at the southern tip of the Sahara.

Industrial minerals

Out of a reserve base of 3,5 Mt and total resources of 15 Mt, barite exports increased fivefold since the sixties and seventies even though only accessible deposits are worked. 80 km from Safi harbor, on the jbel Ighoud coarse grained pinkish barite contains 96% $BaSO_4$ and only 2,5% SiO_2. White barite occurs nearby in the extension of the Koudiat el Hamra silver show, and Seksooua is to the south. SMPPC worked underground Isserder, Almezi and Aït Souru, but Tnine has been stripped (SC). Comabar produced white barite grading 98% $BaSO_4$ at Zelma and near Tizi n'Tichka (SC) at an altitude of 2,150 m (see molybdenum), and the jbel Zerrouk is in the east.

Fluorite production reached 4,000 t in 1952, plummeted to 0 in 1957 and increases again. At Achémèche (NC) near el Hammam fluorspar occurs over 10 km and trends 50 km to the jbel Zrahina Pb, Sb, Sn deposit. Proved and probable reserves reached 3 plus 2 Mt grading 62 and 50% CaF_2. At Oued Bout Nokret ore averages 90% CaF_2, Bergamou, Gouiada and Moufrès are nearby and the 2 m thick jbel Tirhemi (NE) vein contains a million tons assaying 85-90% CaF_2.

Salt recovery is back at its 1958 peak of 80,000 tons. S.C.S. works Zima, near Safi (W), waters of the lake have a Baumé grade 6.5-17 and CSM's pans are at Sidi Brahim and Sidi Moussa to the NNE, and SSM produces rock salt at Mohammedia (W). The Taza (NW) beds are 3 and 7 m thick and south of the Tissa salt mountain 2,000 m of salt and anhydrite were drilled. Lastly, some salt is also recovered on the oued Moussa (SW).

Bromine was extracted from salt at Fedala, Frag el Ma exported some graphite, feldspar is worked around Timgharghine along with tourmaline (SW) and other occurences include :

Andalusite: Sidi bou Othmane, Zaïr ; barite Glib en Nam (E), Tazzeka Mts, Sidi Lahcen (NE), Taouz, Tessaout ; diatomite Oulmès, Bou Gafer ; garnet Azegour, Tichka (SW) Sidi bou Othmane, S'bara with wollastonite, Ment, Bou Agra with nepheline syenite, jbel Aouam with opal ; salt Dessenou, Touaba, M'soun, Nsala en Oudaïa (NW), sodium sulfate hamada of Saoura (E), strontium Tirrhist (SC).

Clays and building materials

Serama's Fuller's earth production increased from 15,000 t in 1951 to a peak of 40,000 t in 1962. Smectite layers reach indeed 2 m of thickness at Melg el Ouigdane (NE), Beni Oukil on the ouadi Za, Moul el Bacha, Dar Slimane and Tikkerdine and the bentonite bed of Toouriat Hamet is 12 m thick. Other deposits include the jbel Tidenit and the oued Aguilmane (NE). Montmorillonite is found at Rio Masin and Tamonat Arromani and detergent clays have also been worked on the jbel Rassoul (S) and Tamdafelt (CE).

The principal limestone quarries include the oueds Akrecht and Ygem where black and white/rose marble also occurs. Lime is burnt on the oued Cherrat (NC), at Casa' and Quenitra (NW) and the cement plants mix limestone and shale at Casa', Saîs limestone and clay at Meknès (NC), shells and marls at Agadir (SW). Production has doubled in 10 years and most plants still use oil (table 24).

Table 24 Shareholdings %, and capacity (kt) of Moroccan cement plants.

Shareholding %	Company	Plant	Pc	kt
ODI		Oujda	d	1000
O. Qadiri 22	Cem. Marroquies	Tetuan	s	210
Icex 47, Uniban 17	Cementos	Tanger	dw	240
Lafarge 40	Soc.Cim.artificiels	Meknès	d	500
SNI 34	Cadem project	Beni Mella	w	400
BID,SAI,SFI,SNCE	Asment Temara	Aïn Attig	d	500
CDGIS,SNI 20	Cim.nouvelle	Casablanca	d	1000
Lafarge 50	Lafarge Maroc	Casablanca	w	720
BNDEIO,CNM 10	Asmar Marrak	Mzoudia	d	100
CIMR 13,Cim. Fr. 35	Soc. Cim. d'Agadir		s	720

The gypsum mountain of Sidi Ahmend Tiji rises near Safi (W). Other deposits are inland or at Youssoufia (see phosphates). Mica is worked in the Anti-Atlas, e.g. west of Azegour (see Cobalt), but asbestos shipments from Bou Offroh (SSE) have stopped in 1958. Iron oxid pigments declined from their peak of 1000 t and other deposits include :

Asbestos and magnesite over 160 km between Tissoufra, Nkob, Bou Azzer (Co), Inguijem (Cr), light basalt Outigui, Koudiat Ifrane, granite Oulmès, Tifbat, kaolin Mt Maaza with alum Souk es-Sebt of Sgangane (NE), marble Cherrat (white), Sidi Lachemi, Imi Moqqorn, Bou Acila, Boured (green), perlite Tiedenit, sand 95% SiO_2 Sidi Larbi (W) ; phosphate plateau, Saïs, Mamora, Dabès, Azegour, vermiculite Timgarghine (Be).

Ferrous metals

Iron ore production ranged from 650,000 t in 1951 to 1.3 Mt in 1962, 400,000 t in the seventies and 50,000 t in 1981, but the 480,000 ton Nador (NE) steel project may again provide stable outlets, for the Ouiksane deposits (160) and Atalayane concentrator where Seferif had 40 Mt of reserves and a capacity of half a million tons. Old estimates indicate 60 (50), 240 (150) and 100 (80) million tons of proved and indicated reserves and additional resources for Morocco. In brackets are data for the Agacha deposits, with assays of 54% Fe and 14% TiO_2, probably controlled by the Saharoui Arab Democratic Republic, Polisario, in the southerr Sahara.

Table 25 Moroccan iron ore reserves (Mt) and grades (%).

Deposit	Mt	Fe	SiO_2	P	other %
Oulad Saïd W	20	35	23	0.7	
Aït Amar C	10	45	14	0.6	Mn 0.3
Kettara (153)		51	13	0.03	S 0.5
Bou Ousel	30	43	9	0.3	BaO 15
Imin n'Tourza SE	40	52	7	1	Al_2O_3 7
Tidsi SSE	100	42	7	0.1	CaO 10
Ouarzemine S	10	42	13	0.6	
Aït Ahmane SE,Boulhaut,Keradit (EC)					

Manganese ore production peaked with 600,000 t in the sixties and bottomed out with 80,000 t in the seventies. Sacem (a subsidiary of Imétal and BRPM) at Imini is the last producer but ore reserves in Mt were low :

Ouarzazate	rsrvs	rscs	% Mn	Met.	Deposit	Rsvs	Rscs	% Mn	Metal
Imini (156)	1.7		50	0.8	Bou Arfa (157)	0.4		25	0.1
Tiouine (S)	1.1		37	0.4	" " low grade E		2	12-17	0.2
			20-35	0.5	Boulbab NE		1	12	0.1
80 occurrences		2							

Imini has an underground capacity of 120,000 ty but concentrates of 2 mm 74-88% and 2 mm 74-80% and 80-92% MnO_2 have to reach the Marakech railhead at a distance of 175 km.

The cobalt belt extends over 60 km from Tarouni to Aït Ahmane (SSE). CTT averaged a production of 1500 t in the sixties at Bou Azzer (146), byproduct nickel peaked with 170 t and other nickel shows were Ahmbed and Aït Ahmane. In the sixties 1,700 t of molybdenum concentrates were exported from Azegour (147) (SW) and total tungsten production of Tichka (SW) and Oulmès (W) was a few thousand tons.

Other occurrences include :

chromite : jbel Ingouijem (SE), Ni Beni Bouchara (N), cobalt: Bou Offroh,

Arhbar Oumlil, Tamdrost, Irhtem (SSE) ; Ni : Tamdrost (SSE), Méfis (E), Azegour (SW) ; Ag, Bi, Se : Sidi Embarak ; iron ore : N: Rio Ma'dine, Taghilest, Dra Sfar (W), Tachila (SW), Sebkhat n'Tazgha (S) and in the western Sahara : Dala't Erni (N), Sefariat, el Forfarate, el Khoursiat and Oum Rouessim (S), manganese : Tiharatine ; molybdenite, W, Sn Tarmilet (W), niobium Mo:Sidi bou Othmane (W), tungsten : Isk Imoula, Figri, jbel Afadad, Agadir n'Maïne, Amsiouni (SW), vanadium : jbel Tadaout (EC).

Energy

ONE already uses 8 Mb fuel and 800,000 tons coal and in the nineties total petroleum and electricity demand may exceed 30 Mb and 10 Twh. Plans call for : fuel oil 36%, hydropower 30% (currently only 9%), imported coal 20% and Jerada coal 14%. The Government also hopes that shales will produce by 2000 5 Twh. There are already 30 hydropower plants, the largest on the Oued el Abid and Oum el Rebia, and most of the potential (4.6 Twh), including Lalla Takerkust and Amougguez, will be harnessed by then. Dams also serve irrigation :

Table 26 Principal Moroccan dams and hydroplants and projects[p].

Plants	Mw	Upper Sébou[p]	Mw	Dams	1000 Mm^3	Dams	Mm^3
Bin Ouidane	135	M'Dez	52	Bin Ouidane	1.5	Zaouria (S)	540
Al Masstra	128	el Menzel	148	Arbat (N)	1.4	Oued Grou (W)	500
Afouner	94	Matmabe	240	T'fer (N)	0.8	Ouezza (N)	500
Idriss I	40	Merza	60	Mechra Klila(NE)	7	Lakhdar (SC)	250
O. Makhazine	36			Dechra Oued (C)	0.6	Low Arbat (NW)	250
Im'Fout	31	Dchar Ouest	92	Zemrane (W)	0.6	Oued Massa (S)	200
		M'Jara	240				

Little oil and gas are left at Haricha, Oued Beht, Sidi Rhalem, Sidi Fili(NW). Meskala (SW) has been scaled down to 80 Bcf with only 40% recoverable and estimates for Kechoula range up from 330 Bcf and 20 Mb liquids in the Triassic.

Bituminous limestones average I 6% oil at Dar Caïd Meknafi and Dar Cadi Zelteni (SW) II near Taroudant (C) and III at Tourtit n'Aït Argi. On the Sebkat Tazgha (I) 70,000 bd would require in the nineties an investment of current $45 B. 100 Bt of shale grade 60-90 l/t and contain 35 Bb of oil, but more is to be stripped above thinner seams. Fortunately, the overburden ratio of 6 km^2 at Timahdit (NC) is only 1.1 and a pilot plant is to produce here 50 bd of oil from 50 td of shale. Proved reserves of 3.3 Bt grade 65 l/t and another 3 Bb are near Tangier.

The coal trend extends from the Rabat area (W) to Jerada (160) where production increased from 150,000 t in the fifties to 450,000 tons in the sixties and is again expected to almost treble with further mechanization. The Soviet built powerplant uses local anthracite and Casablanca and Mohammedia also fire coal. Polservice found the principal lignite reserves on the Oued Nja (NE). Others include Khenifra, Skoura, Sefrou, Tamda, Boulemane, Guerif and Rafsaï.

Finally uranium shows near Bleida (S), anomalies strech over 15 km at Imi n' Tanout (SW), ores of the jbel Affaga (C) assay 0.15-0.2% U, with 2,000 tons of reserves, and potential resources may reach 70-180,000 tons, mainly in the north. Phosphates grading 0,013% are nevertheless the principal resource, Ben Guérir is a target and OCP expects to recover 250 ty in the Safi plant.

12 ALGERIA

Along with Indonesia, Algeria ranks first in LNG shipments and fourth for all gas sales. While these are expected to double before 2000, exports of oil and its products should decline to 30 Mb then, with 170 Mb in 1990 when condensates will peak (Fig. 9; 11, 13, 19).

GOALS : To provide water, improve the soil and expand fertilizer production ; to market more gas, LNG, LPG while conserving crude ; to upgrade metals and meet the need for construction materials.

Gas and oil account for 90% of exports, 60% of revenue and a third of GDP since domestic fuel prices are subsidized. While Sonatrach and its affiliates are in control, exploration has met less succes and few foreign partners are left. Gas provides more than half of energy and consumption matches 20% of ouput for oil too. Supply should attain in the nineties their 3.5 Gw and 8 Twh targets. Simultaneously, power should reach 70% of households.

Raw materials of the sparsely populated desert are piped to the coastal industries and gas-line capacity may reach 2 Tcfy before the end of the century, since resources might attain 250 Tcf. Conversely hydrocarbon revenues create options for the exploitation of solids of the Sahara or for alternative imports.

Table 27 Algerian reserves and production.

	Produc.		Resrvs			Produc.			Produc.		Resvs	
Salt	kt	150	Bt	0.9	Gas reinj.	Tcf	1.6	Gas gross	Tcf	3.5	Tcf	110
Soda	t	700			Gas marktd.	Tcf	1.6→3	Helium			Bcf	72
Barite	kt	100	Mt	2	LNG	Tcf	0.7→1.2	Oil	Mb	300	Bb	9
Celestite	kt	3	Mt	6	Condensate	Mb	180	Hydropower	Mw	300	Gw	5
Diatomite	kt	5	Mt	4	Petroleum	Mb	150	Uranium			kt	33
					LPG	Mb	7	Coal	kt	7	Mt	50
Silver	t	3	t	30								
Mercury	t	230	kt	5	Fertilizer[P]	kt	50	Phosphate	Mt	300	Mt	80
Lead mine	kt	5	kt	100	N in ammonia	kt	150	Sulfur	kt	15		
Zinc mine	kt	10	kt	300				Iron ore	Mt	4	Mt	50
Copper	t	150			Steel	kt	600	Kaolin	kt	15	Mt	25
Manganese			kt	500	Zinc smelt.	kt	30	Bentonite	kt	35		
					Cement	Mt	6	Fuller's earth	kt	5		
					Gypsum	kt	250					
					Lime	kt	40					

Although Algeria is the world's fifth producer of mercury, Africa's leading producer of strontium and second supplier of barite, mining represents only 2% of GDP. Nevertheless, Algeria has reached self sufficiency in most steel, fertilizer and construction products. A state company, SNS controls smelting and iron mining and its sister, Sonarem, is responsible for all other mining, and the manufacture of cement. Foreign participation has been invited in these areas too and new plants are built in cooperation with international companies. However, expatriate labor is only a small part of total mineral employment of 30,000. Meanwhile, export patterns (shown in %) are shifting since Trunkline suspended gas purchases.

Commodity	U.S.	France	Italy	Other countries	Commod.	Others	
Gas			90	Tunisia, Yugoslavia[p],	Zinc	U.S.S.R	30
LNG	10-50	30-50		Belgium 10, Spain 10,	Lead	Hungary	100
Oil	50	10	8	Germany 15,	Mercury	E. Germ.	90
Phosphate		25		Finland 15, Hungary 12			

Gas and oil fields

In 1922 Conrad Kilian the Alsatian geologist 'dreamt' of oil in the desert, yet only in 1956 did French Repal strike Hassi er Rmel. By now some 50 fields operate and around 1990 half of the crude will have been removed from the Sahara. Two thirds of the 30 Mboe are gas to which might be added 8 Mboe possible.

I In the remote southwest, gas reserves of 1 Tcf or more are scattered around In Salah, the Tideikelt and the northern Ahnet, including (reserves in Bcf) :

a) Zini 110, Meredoua 50, Gara Azal Matti ;

b) In Salah 450 (120 km^2, 12 m pay), Djebel Brega 400 (200 km^2, 20 m pay), Oued Djaret 140 ;

c) Bahar el Hammar 10, Tirechoumine, Tirebadine ;

d) Mahbes Guenatir 7.5, Bezzane 10, Mouhadrine 2, Thara 1 Bcf.

II Hassi er Rmel (C) originally held 70 Tcf gas and 2 Bb condensate at 2,100 m depth ; 60 m are productive over 1,700 km^2. Permeability and porosity oscillate between 400 and 700 millidarcys and 16 to 20%. 5-600 km long lines pipe gas to Skikda (NE) and to Arzew (NW) along with 150,000 bd of condensate.

III In the Triassic basin (E) 70 km thick reservoir sandstones are capped by 400 m of salt and gravities range from 39 to 40° API.

a) While recent production data remain confidential the Ouargla fields flowed 40-45,000 bd and their original reserves, gravity and depth are shown below.

Field	Mb			Field	Mb	Field	Mb
Haoud Berkaoui EUR	380	40	3300	Guellala	80	Guellala NE	12
Haniet el Beïda	40	39-45	1000-1300	Ngoussa	40	Ouargla	12
Draa el Temra	21	45	1000-1150	Kef el Argoub	17	Takhoukht	12
Bou Khezane	18	41	1000-1300	Garet ech Chouf	17		

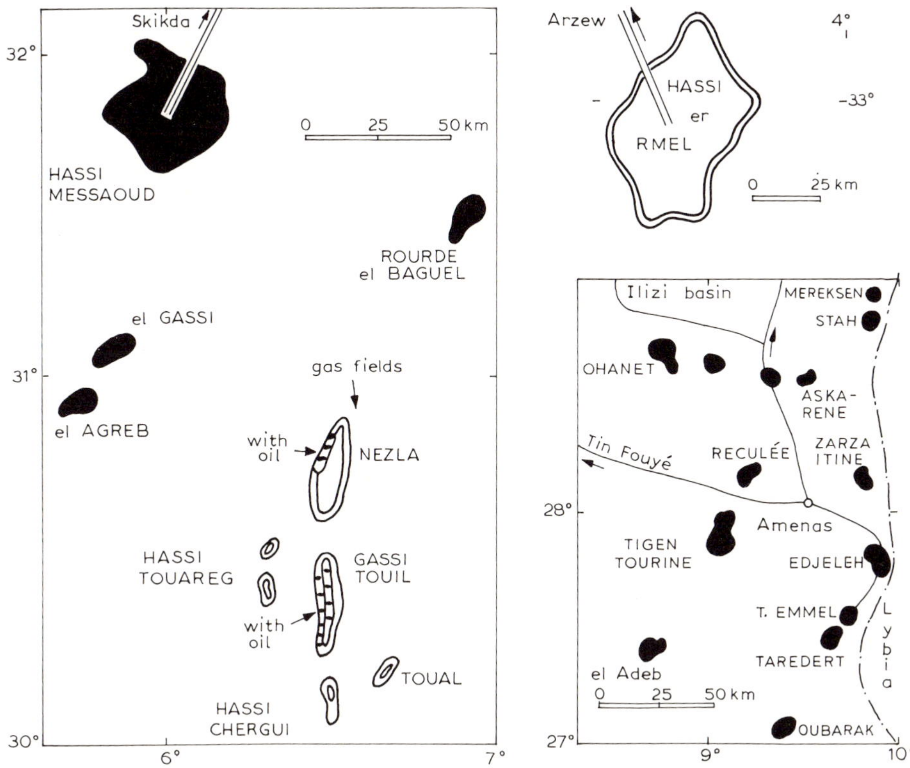

Fig. 11. Sahara oilfields. (Legend on page 13.)

Total has a 41% share in Berkaoui-Ben Kahla; 110 km NW and Moukhag el Kebouch flowed 2,000 bd from 4,000 m depth.

b) There were 9 Tcf gas and 20 Bb in situ at Hassi Messaoud out of which 7 Bb were EUR ; 1,000 km^2 can produce from 120 m and 3,300 m depth 500,000 bd of 46° API oil. More than 7 Bd have been produced from these fields, but Rhourde Chegga only flows 1,000 bd and water invades Bordj Nili.

Field Originally :	Mb	API	depth m	Gas fields
c) Rhourde el Baguel	500	40	3-4000	Hassi Touareg, Sbaa & Ilabou, Rhourde Adra ;
Gassi Touil	350	50	2300	e) Garet el Bafinet, Nezla : condensate
el Aghreb	250		3150	gas cap.
d) Rhourde Nouss 3.5 Tcf	40	40	1900	

IV The Ilizi basin (E) originally held 2 Bb of 35-45° API oil and 12 Tcf of wet gas :

Field	Tcf	Mb	API	Depth m	Field	km^2
Tinfouyé-Tabankort	1.2	700	40	1200-2200	Ohanet	3.5 x 20
Zarzaïtine EUR 1300		1900		1500	Reculée	10 x 15
Edjeleh		1000	37	5-900	Dj.Ouan Taradert	2.5 x 10

Field	Depth in m
Ohanet	2300
Ou. Oubarakat	750
Ikebrane	600
Merksen and Stah 45 API	3150

From the In Amenas collector, these fields and Alrar connect with Messaoud and Skhirra in Tunisia.

Marketing and processing of hydrocarbons

Hardly any gas is flared since 90% are non-associated, but Messaoud is also an important producer. The Oued Noumer plant has a capacity of 310 Mcfd providing 50,000 bd of liquids and Snamprogetti's Rhourde-Nouss plant uses 1.5 Bcfd of wet gas out of which 900 Mcfd are reinjected with the remainder providing 3.5 Mt condensate and 800,000 t LPG. Other important gas suppliers include Alrar and Rhourde Adra and Chouff. The 48" Transmed pipeline which crosses the Strait of Pantelleria to Sicily can deliver 450 Bcf to Minerbio near Bologna. Gaz de France is a traditional customer for 250 Bcfy, but Enagaz and Distrigas of Spain and Belgium have whittled down their costly purchases to 50 Bcfy each. Negotiations with other customers continue.

Algeria also planned to increase its current LNG capacity of 1 Tcf which consists of the 50 Bcf Camel plant at Arzew (NW) and three 325 Bcf facilities, two more at Arzew and one at Skikda (NE). Simultaneously LPG capacity doubled at Skikda, in 1986 Kobe Marubeni and Mitsubishi increase LPG capacity from 8 to 40 Mb at Hassi er Rmel and Arzew attained an LPG export potential second only to Saudi Arabia.

While the production of crude peaked in 1978 with 430 Mb, refined capacity reached 1/2 Mbd at Arzew, Hassi Messaoud, Skikda and Algiers, to be joined by a 90,000 bd plant at Saïda before 1990. Algeria can refine thus half of its oil and it exports twice as much refined petroleum than crude (130 against 70 Mb). To this should be added various downstream facilities planned for this decade :

a 2,800 bd lubrifier plant at Arzew, 40,000 ty linear alkyl benzene plant at Skikda, 70,000-75,000 ty polyethylene plant.

Other resources of the Sahara

Gas contains 0.1% helium ; resources reach 180 Bcf and Algerian production could increase from 160 Mcf in 1987 to 700 Mcf, matching thereby 70% of European needs. Radioactive sediments lie at 2-3,000 m depth and in the southern oilfields, salt and gypsum are abundant. While little byproduct sulfur is recovered, 4 veins with pitchblende and uraninite accompanied by Pb, Zn, Cu and Mo are mined in the hard gneisses of the arid Hoggar (S), 12 Mt of ore grade 0.195% U_3O_8 underground, at Timgaouine an open pit, at Abankor and Daira (plus Tinef). There are twice as many resources, and another 2,000 t at Tahaggart. Similar channel fillings of the Tin Séririne sandstones extend the Niger uranium trend into the Tassili (SSE), - and Mali.

A project also calls for the recovery of 35 t of tungsten from 34 Mt ore from Laouni and other resources of the Hoggar include : asbestos cross fiber Tin Ralles, Imzaden (S), traces of diamonds and gold west of Tamanrasset, platinum and chromium Tibeghim, G'oudène, tin In Tounine, Djilouet.

Salt layers reach 400 m of thickness in the Kénadza-Guir basin (148, Fig. 9). Production peaked in 1946 here at Béchar with 200,000 t coal, but costly Ksi-Ksou was closed in 1963. While Algeria relies on imports, more coal is left at Mézarif (60 km E) and another remnant is the railway which covers a third of the distance to the Djebilet. The upper layer of the Garas of Djebilet (W) is 5-15 m thick and underneath are 6-10 m richer oölitic magnetite ; depending on cut-off resources are :

Table 28 Resources of the Gara Djebilet in billion tons.

Area	km x km		Bt	%Fe		Bt	Fe%	Bt	Fe%
West	10 x 4	West and center	1	57	magnetite	0.8	55.5	1.3	39
Center	20 x 4.5	West and center	2.2	50-57	hematite	1.6	53	0.2	30
East	7 x 1	East	0.5	45	hematite	0.5	48		
km^2	137		3.7	53		2.9	48	1.5	37

Unfortunately the ore contains some sulfur, 0.6% P, and 0.17% Ti. Further on, the Merchui Abdul Aziz has been credited with 2 Bt of medium grade ore. Recent estimates have expanded the 500,000 tons of manganese contained on the Djebel Guettara (NW), penalized though by 1.9% As. This trend stretches to Bou Khdeissar (E). More important are the oölitic iron ores which extend the Shatti belt over a width of 50-80 km and a length of 100 km to In Amenas (E).

Finally nitrated earths have been worked in the Sebkhet Oulad Mahmoud, Kaberstan and Meraguen near Guerara, in the Tidikelt (SC) and at Sebââ.

Soil, water and hydropower

Although rainfall is rare, groundwater is abundant in part of the Sahara. 20-40,000 years old aquifers are partly replenished by infiltration and run-off from the Saharan Atlas : 18.5 m^3sec. into the Late Continental. Some 10 m^3sec. percolate upwards and since 15 m^3sec. are withdrawn by wells (and sources) 5m^3sec. are mined. Further, artesian water is particularly abundant along the rim of the basins and in the center. The Intermediate Continental provides in a broad ellipse 3m^3sec between In Salah, Reggane and Timimoun, 2m^3sec El Goléa, Ghardaïa, on the Nsa and along the Biskra-Ouargla axis, 4m^3sec. at Hassi Messaoud and 1m^3 sec. at el Oued (NE). While it is only fed by 8.5 m^3sec., 0.8 m^3sec. percolate upwards, and 3.6 m^3sec. filter into the sea and towards Mali. Since 15 m^3sec. are withdrawn from this aquifer too, 5m^3sec. are mined.

	Surface 1000 km^2	Resource km^3	Replenishment km^3y	Artesian area km^2
Eastern Erg	330	1500	0.4	180
Western Erg	370	1700	0.9	325
Tanezrouft	240	400	0.02	-

If west of a Biskra-In Salah line these waters contain 'only' 2 g/l of solids, this amount increases to 6 g/l at the meeting point of the frontiers of Tunisia and Libya. Needs are increasing too. By the year 2000 8 m^3sec. will be required at Touat, 11 on the Oued Rhir, and 8 in the Souf, creating thereby deficits of 6 and 3 m^3sec. in the latter.

Smallish hydropower sites are aligned 25 to 60 km from the coast frequently above irrigation schemes. Small Kabylia alone has a capacity of 80 Mw, mostly on five sites.

Table 29 Major Algerian hydropower sites.

Plant/Site	Mw	Plant	Mw	Plant	Mw
Mansouria	100	Cued Fodha	15	Hamiz	3
Darguinah	66	Erraguène	14	Bou Hanipia	5
Iril Emda	24	Ghrib	8	Chabet Saïd	5
Bakhagda	4	Oued Rumel	30	Djendjen	38
Lower Boghni	3	Kediat Amram	30	Bou Sellah	14
Beni Bahdel	7	O.el Agrioun	21	Aïn Kerma	30

Furthermore, the waters of the Chott Chergui (NW) can be utilized on the canals which connect with the Relizane plain. All in all some 300,000 ha are irrigated, and the cultivated land matches the area of the Netherlands, i.e. 20% of Northern Algeria, where chestnut soils resembling those of Europe appear.

Millennia of deforestation and ranching have destroyed though many shallow mountain soils.

Phosphate, fertilizers and brines

To 310 Mt of phosphate rock reserves can be added 55 Mt of resources (Krauss 1983). Subeconomic reserves of 8 Mt and 7 Mt P_2O_5 in rock grading 29.8% and 27% remain underground at Kouif and Mzaïta, which might even reopen. More important are 25 and 100 Mt P_2O_5 in rock averaging 27.5 and 24.3% P_2O_5 at Djebel Onk (172)[2]. Strippable beds are 30 m thick in the Djebel Djérissa, and even more in the Rass Mergueb and Tir. The Bled Djemidma holds most of the lower grade reserves in a 5 km^2 area, where overburden/ore ratios range from 2 and 4 to 1 (see MRA pp 564-7). Finally, in the Kef es Sennoun this ratio is 2 : 1.

Production has increased regularly until 1980 to 2.4 Mt ore upgraded into 1.4 Mt concentrate assaying 63% BPL. Lower grades are for the domestic market and 700,000 t are washed, calcined, and carbonated mostly for export at 75% BPL. Processing requires oil and water which is piped from a distance of 60 km. Although Algeria imports most of its fertilizer, Sonarem and Sonatrach plan further expansions at their Annaba (NE) and Arzew (NW) plants which operate at part of their nameplate capacity. Moreover a new facility is to be built at Tébessa in the nineties.

Table 30 Algerian fertilizer capacities in 1000 tons.

Plant	PAc	SAc	DAP	MAP	TSP	PK	NAc	Am	AmNi	NPK
Annaba	170	530	130	200	245	70		300		150
Arzew							120	270	330	
Tébessa	530	170			280					

The sebkets of Oran, Zahrez Chergui and Zahrez Rahrbi (NW) contained 500, 300 and 200 Mt of salt. Chergui has 91% sodium sulfate and Rharbi magnesium chloride. Epsomite accompanies sodium salts at Nador and Melia, Bou Ziane rejuvenates its salt each summer and in the Saharan Atlas rise the Djelfa and el Outaya salt mountains.

Industrial minerals and building materials

With increased demand for drilling, barite output grew steadily at Ichemoul : 1938 3,000 t, 1950 15,000, 1960 60,000, and 1985 100,000 t. Many of the 16 Mt resources are in a 2 x 5 km area extending from Tala Khalifa via the Zouggara mine to Aïoune el Agha and Aïn Sultane, near Keddara, but also at el Melka, Bou Zega (N) and Mahni (NE). The strontium mineral celestite occurs in a similar

[2] Numbers refer to Fig. 13.

environment. Fluorite, on the other hand, accompanies the base metals.

Africa's largest diatomite deposits grade 74-86% SiO_2 and extend over 50 km and a width of 1-3 km between Macta, Mina and the Chélif River and continue east 30 km towards Mostaganem over a width of 0.5-1.5 km. Production of the el Ksar quarry has, however, peaked with 30,000 tons in the sixties. Similarly, vermiculite extraction rose to 400 tons in 1960.

Production of bentonite started only after World War II at Marnia (NW). Output peaked with 100,000 t in the sixties and other deposits comprise :

Bentonite : Mchounèche, Tighanimine, Sillègue (NE), Tsighaout ; brick clay : Sétif, Annaba (NE), etc. ; gypsum : Dahra in the Chélif (161), Djebel Chegga (N W), Mostaganem, Arzew (NW), Sidi bel Abbès, Haboucha (Mina Mts), Algiers-Babors (C) ; limestone : Filfilla ; onyx : Isser, Tekbalet, Sidi Hamza, Oued Chouly (NW) ; pottery clay : Tell Atlas, Miliana, Babors (C) ; red onyx : Rhumel (NE) ; sand : e.g. Saïd Tlemcen, Sidi bel Abbès, Hadjout ; slate : Chiffa, Djebel Doui, Trara ; sulfur : Seybousse, Djebel bou Kebch, Tébessa.

Cement plants use the dry process except for Zahana and inexpensive gas. Production has increased sevenfold in fifteen years to 5 Mt, and 9 Mt capacity. Another million tons planned at Aïn Touta will hopefully eliminate 3 Mt of imports.

Table 31 Capacity of Algerian Cement plants in 1000 tons.

West : ERCO	kt	Center : ERCC	kt	East : ERCE	kt
Béni Saf	1000	Djelfa	500	Aïn Kebira	900
Zahana	850	el Asnam	900	Hamma Bouziane	700
Saïda	450	Meftan	850	Hadjar Soud	800
Bouïra : Sour el Ghozlane			1000	Batna	500

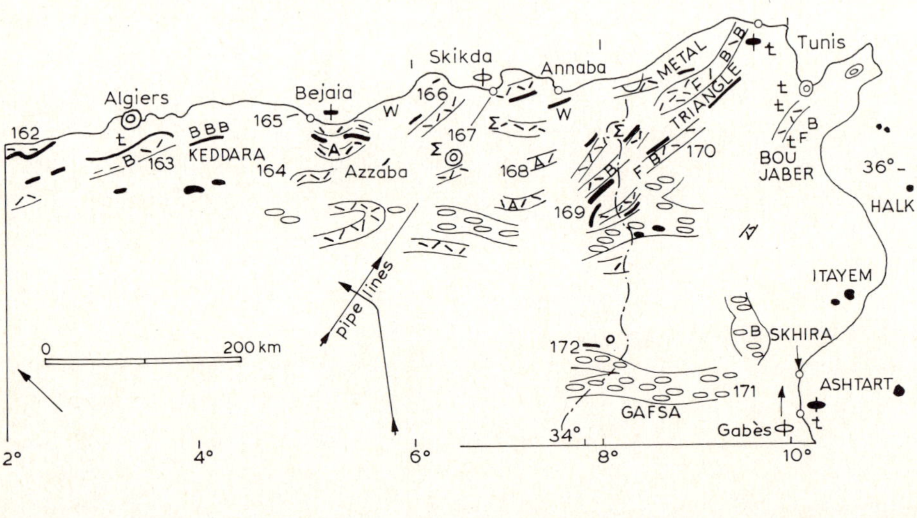

Fig. 12. Northern Algeria and Tunisia. (Legend on page 13.)

Metals

Even before the Phoenicians metal industries flourished and plans call again for increasing iron ore and steel production fourfold to 13 Mt and 2.5 Mt. The open-cast Bou Khadra mine produced half a million tons of siliceous concentrate, grading 58% Fe, shipped via a 25 km tieline to Oued Kebrit station, but only 40 Mt and baritic ores are left. Further south, Khangouet, near the Tébessa phosphate mines provided 200,000 ty, but reserves grading 55% Fe have been exhausted. About a million tons of siliceous (50% Fe) ore remain at Zaccar (162) but little is left at Timezrit (165) and at Benif Saf mine (160). Fortunately, Ouenza (169) boasts 90 Mt of calcium bearing hematite above the water table. 2,500 men extract 2.5 Mty opencast here and the 53% Fe concentrate is in great demand. It is shipped 180 km to Annaba harbor and 0.7-0.8 Mt to the el Hadjar works rated at 1.8 Mt steel. The rod and bar mills are to operate here nearer to capacity. Other projects envisage 2 Mt specialty steels at Bellara (NE Jijel) from Tébessa ores and the manufacture of pipes at Ghardaïa in the Sahara. However, the La Macta project (NW) based on 10 Mt of Djebilet ores located at a distance of 1000 km and the 140,000 ty M'Sila aluminium plant may have to wait. Granted, the 2.3 Mt direct reduction furnaces of el Milia, based on Brazilian ores and the 900,000 ton Moulay Slisse rolling mill might open by 1990.

Mercury output varies considerably. It was halved again after peaks of 1,000 t in 1990 and 840 t in 1981. Out of M'Rasma's (167) resources assaying 0.4-0.8% 230,000 tons graded 1.3% Hg with enrichments of up to 7%. In contrast, 200,000 tons averaged only 0.5% at Ismail. All cinnabar concentrates are reduced nearby at Azzaba (NE) and low grade deposits dot the Aurès e.g. near Bir Beni Salah and Taght. All in all resources may reach thus 30,000 t.

Hammam n'Bails (168) produced 1,000 ty of antimony in the early sixties accompanied by zinc and cadmium was another byproduct of zinc. Pyrite production peaked at el Halia (166) with 45,000 tons in 1938 and zinc and lead, mostly at el Abed (159), in the early sixties. Later Sonarem claimed resources of 1.9 and 0.7 Mt metal with grades of 6 and 2.2%, including seven deposits with 2.4 Mt ZnCu and 0.5 Mt ZnPb minerals in 14 Mt of ores :

Deposit	Mt	Ore	Partners sought	Mt	Ore	Deposit	Mt	Ore
El Abed (NW)	8.2	5Zn,1.2Pb	Kef Oum Teboul	2.4	ZnCu	Ichtmoul	1.3	Pb
Kherzet Youssef	1.6	Zn Pb	Bou Soufa	0.75	3%Cu	Cavallo(166)	1.75	Cu
Ouarssenis Mts (163)	0.4	Zn Pb	Oued el Kebir	0.6	2.5%Cu	0.1Mt 5.5%Pb4.9%Zn (Ag)		
Aïn Barbar, NE Annaba			Sidi Kamber, NE Azzaba			Dj. Gustar (164)		

Targets of 10,000 tons and 4,000 t Pb have been met by el Abed (NW), the principal producer, Kherzet Youssef and Aïn Barbar are smaller mines and Guerrouma assisted by Tsvetmetpromexport of the U.S.S.R. appears to ship little zinc to the 40,000 ty Ghazouet electric smelter which has extra capacity. Lastly, minor occurrences include :

antimony : Kheneg (Aïn Kerma), Taya Mts ; arsenic : Dj. Debar, Aïn Achour (Nador) 30,000 t ore 12-23% Pb, 12-25% As ; mercury : Taght (Aurès) ; silver : Oued Zounder, Mesloula ; tungsten : Béléliéta (Azzaba) ; uraninite and molybdcnite : Aïn Sedma (Kabylia) and Ichcrridène.

13 TUNISIA

Gas, oil, and fertilizers fuel the fast growing economy.

Table 33 Tunisian reserves and production, including projects^p.

Production			Production					Prod.	Reservs	
Cement	Mt	3.5	Petroleum	Mb	15	Oil	Mb	40-50	Bb	2
Lime	kt	500	Gas marketed	Bcf	15-30	Gas	Bcf	30-40	Tcf	4
Clays	kt	350	LPG project	Mb	1.5	Fluorite	kt	40	Mt	10
Potash chloride	kt	30	Fertilizer^p		600	Barite	kt	25	Mt	4
Potash bisulfate	kt	10	" " NPK		150	Phosphate	Mt	1.8	Mt	80
Salt	kt	400	Steel		250	Silver	t	2	t	400
Gypsum	kt	80	Rolled prod.	kt	400	Iron ore	kt	300	Mt	20
			Lead smelter	kt	10	Lead	kt	5	kt	200
Cadmium	kt	1.5	Galvanized prod.	kt	40	Zinc	kt	7	kt	300

GOALS : Intensive exploration off and onshore for hydrocarbons. Conservation of highpriced oil, less flaring and more distribution of gas. Further transformation of low grade phosphates, the extraction of uranium from these and the creation of an integrated fertilizer industry. Utilization of hydropower in the north and the extension of infrastructure towards the south.

The State oil company ETAP is associated with Agip and ELF and exploration is carried out by joint ventures where the foreign partner bears the cost of unsuccessful drilling. Phosphate, other mining, cement and the steel mill are controlled by the state via CPG and Sotémi. Even though the mineral industry contributes a quarter of GDP, it represents only 10% of net exports, since Tunisia exports ingeniously highpriced oil and imports cheaper crude. Further, fertilizers compete with tourism as a source of exchange and exports shares reach in % :

	Italy	Germany	Others		France	Italy	Others	
Oil	20	30	US 25	Phosphate	30		Turkey	15
Lead	20	40		Zinc	20	30	Yugoslavia	15

As in the rest of the Maghreb, the major centers are on the coast but the new oil and gasfields, the new refineries and fertilizer plants are further south in the Gulf of Gabès and the phosphate deposits in its hinterland, i.e. in the geographic center.

Oil, gas, and hydropower

Further south, most of Borma (S) is in Tunisia. Sitep (Agip)'s 70 wells have already pumped 500 Mb of 42° API oil, 70% of the country's deliveries along with 45 Mcfd gas from a depth of 2,600 to 3,000 m. At Ashtart (CE) Sofratep (Elf)'s 12 wells pump 29° API oil from 3150 m with secondary water injection ; 80 Mb a third of total EUR, are left. At Isis, ESE of Lampedusa which brought fame to the 'Leopard', Total develops in 100 m of water 100 Mb with a potential of 40,000 bd. Further north Shell's production control system upgraded Tazerka from 10 to 40,000 bd and smaller fields include :

South : Makrerouza, Bir Ali ben Khalifa gas, Chouech es Saïd 650 bd, 3,500 m depth, Sabria 25 Mb, an Amoco wildcat 1,800 bd, 13 Mcfd 3,900 m. Center offshore : Segma's Didon 3,500 bd, Elf's Miskar 3,600 m, Iugurtha , Elyssa, Hasdrubal, Salammbo. Onshore : Biban, 1,500 bd 2,700 m, CFP's el Itayem, 2,700 bd 44° API and Sidi Behara, Guebiba and Gremdo 25-30 Mb.EUR Northeast: Halk el Menzel, Birsa Oudna, and Petrofina's Laarich, Makruga and Debbech 1,900 bd. Northwest : Douleb Semmama 1,200 bd.

Half of the oil is used locally ; a new $1.5 B refinery is therefore planned at Taberka (N) and gas consumption is to expand sixfold this century, STEG and SNDP are to convert to 24 Mcfd of Borma gas into 800,000 b of LPG and products. Algerian and Borma gas of the Trapsa pipeline is stored at La Skhirra (CE). Kuwaiti and Saudi capital (BTKD and Stsid) contributed to the financing of Paktank. Tunisia even draws 20 Bcf from the Transmed pipeline.

Most hydropower is restricted to the Medjerda basin (NW). Capacities include 7.5 Mw at Trajan's Bridge, 9 Mw at el Lil and 13 Mw at Sidi Salem. Downstream, the Ben Metir dam and Fernana Falls are conveniently located near Tunis.

Phosphates

Out of moderate reserves, but substantial lowgrade resources - 300 Mt and 1.3 Bt of rock grading respectively 28.5 and 27% - recovery of concentrates is scheduled to increase from 4 to 10 Mt by the early nineties. Production of acids and compounds should treble earlier to 2.5 Mt. Four groups can be distinguished :

a) At Sta Ouertane (170) estimates range from 1.5 Bt grading 15% P_2O_5 to 3 Bt assaying 12-15% depending on cutoff point and delineation. Semipso, a joint venture of Kuwaiti PIC (50%), the state (25%), ICM Siape, and other Tunisian groups, plan to mine opencast and ship from here 700,000 ty of concentrates through a new railway and harbor to Cap Séret (N).

b) Kalaa Khashba north of Gafsa with 12 Mt grading 27% P_2O_5 and a capacity of 240,000 ty.

c) Mezzouna, Maknassy, Meheri Zebbeus and the Djebel Mechaib branch between Sfax and Djérissa to the northwest of Gafsa where

d) largest reserves were available at Kaf es Schfair, Redeyef and Mrata in the

Gafsa zone (171) proper (grades 25-26%) and at Mdilla and Sehib (29%). Kef es Schfair is mainly worked opencast, although some stripping is done at Moularès and Mdilla. Most others, however, were underground. These mines also had the largest capacity, 1.5-2 Mt each or 10 Mt aggregate. Since mining started in 1899 and recovery rates of older installation is 69-74% only two layers grading 25.9-27.2% P_2O_5 have been worked to date out of a maximum of nine. Further, two new areas have been opened up for stripping : Kef ed Dour, an extension of Kef es Schfair (Fig. 13) with a capacity of 1.4 Mty and Oum el Khjer; Djellabia-M'zinda (1.5 Mt), the extension of Sehib will follow later.

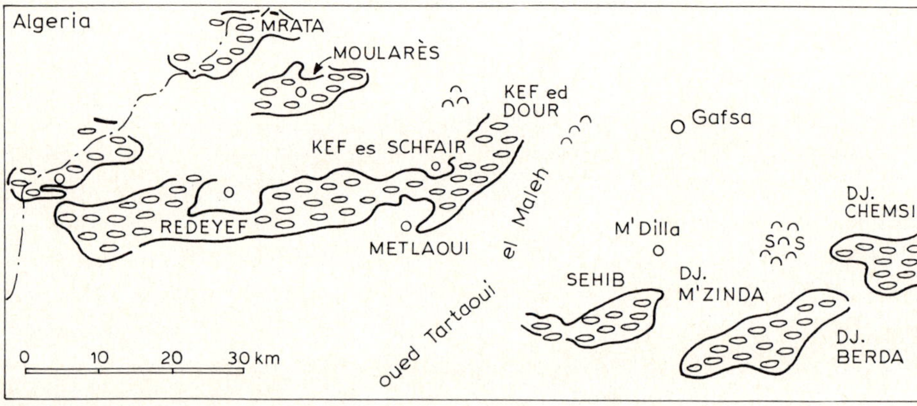

Fig. 13. The phosphates on Gafsa.

Domestic sales will increase to 4 Mt, but washed or dry-separated concentrates grading 30-31% P_2O_5 are exported by rail through Sfax and Gabès from the Metlaoui Moularès and Mdilla plants of CPG.

Fertilizers

The lowgrade 70% of the output is used in the fertilizer plants, as shown in table 34.

Table 34 Capacity of Tunisia fertilizer plants (projects^p, targets^t).

	PAC	SAC	SPAC	MAP	DAP	DCP	TSP	NPK	AmN
SIAPE	160^p	520^p	300^p				570		
ICA	160^p	480^p					400^t		
ICM	440					60	100		
SAEPA	330				330				330^p
EG				100	330^p			425^p	

Siape at Sfax, the biggest manufacturer of triple superphosphate, also plans to produce superphosphoric acid assaying 72% P_2O_5. ICG a joint venture of Siape, ICM, Kuwaiti (BTK) and Saudi capital (Stsid), produces acids at M'Dilla for the

Numbers refer to Fig. 13.

domestic market and will supply TSP for export. Meanwhile, the other plants, located at Ghanouch near Gabès, branch out into nitrogen and potash fertilizers. Lastly, Office National des Mines has set up the Comité de Potasse to investigate with the help of IBRD, Sdics and Potasses d'Alsace the feasibility of a potash facility with an estimated capacity of 140-400,000 ty to be reached this century at Zarzis on the southern coast. Plans call for the production of 60,000 ty of potassium chloride, 20,000 ty of bisulfate, magnesium compounds, bromium products and soda before 1990. Brines under consideration are the Adibate sebkha on the Libyan border and further north, the large Chott el Djerid for a smaller facility. Investment requirements may be in excess of $100 M in current funds.

Water and soil

a) Northern Tunisia. Only a narrow band of the Metal Triangle gets 500 mm of rain. In these mountains of the Tell, there is, however, no continuous aquifer. While up to 120 m thick limestones are found locally, e.g. in the steppes east of Kef, there are no large tracks of arable land here. Further east, 2 km thick sediments of the plain yield 1,000 m^3h of water at the Kasserine fault.- Six sources in the Zaghouan limestone flow 2,000 m^3h, further south, the Continental sediments of the Kairouan plain retain water. So do coastal beds.
b) The lifeline of Central Tunisia are the artesian wells of the Late Continental with yields of 350 m^3h in the oases which dot the exploitable rim of the basin east of the border and then arc to Gafsa and Gabès. In contrast, the Intermediate Continental only locally provides 50 m^3h in the south, it supplies water though to the Fedjaj, between Tozeur and Gabès, and is less overworked than in Algeria. All in all 8 m^3sec will be required by the year 2000 at Djerid and Fedjaj each with hardly any deficit and 13 m^3sec in the Djeffara (SE) including a withdrawal of 1 m^3sec. In effect, the Chott el Djerid and the sea drain these Continental sandstones but a third of the total Tunisian water supply of 700 Mm^3y are thought to enter deep aquifers.

Soils only develop in the oases here, along with weak psamments on the coast. Pale soils cover the center followed by chestnut soils in the Atlas. A third of the land is already cultivated and with fertilizers some of these revert to their ancient productivity.

Building and industrial materials

Salt is found in the sebkha and chotts or arid salt lakes of the Center and South, but marine salt provided most of the 160,000-200,000 ty won in the sixties, which is expected to exceed half a million tons before 1990.

Sulfur occurrences extend along the sebket of the north into Tébessa, Algeria. They accompany anhydrite and gypsum. Thick beds of gypsum and limestone are also found in the Gafsa phosphate deposits, but supplies grow only slowly.

Production of cement has trebled in ten years. Most plants are in the North; all are state-owned or joined ventures (%) as shown in table 35.

Table 35 Capacity of Tunisian Cement Plants including project[p].

Company	Plant	Mt	State	Company	Plant	Mt
Ciments de Bizerte		0.7	100	Soc. des Indus-	Enfida	1
Les Ciments d'Oum el Khebil		0.4	-	tries Cimentières	Tadjerouine	1
Ciments Artificiels	J.Jelloud	0.4	75	Soc. Ciments Tunisiens	J.Oust	1
Soc. des Ciments de Gabès		0.5	93	Soc. Ciments J.O[p]	Bir Mcherga	1
Soc. Tuniso-Algérienne des Ciments Blancs					Jebel Oust	0.2

Practically all remaining minerals come from the "Metal Triangle" (NW). Barite is widespread in this area at Rhar ed Deba, Zeflana, Sidi Amor ben Salem and jebel Agrab where it accompanies lead - and at Staa. Bou Jabeur contained in 1981 2.4 Mt of barite, 1 Mt fluorite and 350,000 t lead and zinc in grades of 30,12 and 4.5%. Although output declined, Fluobar's plans call with Armico's assistance, for 40 kt barite, 50 kt fluorite and 3 kt lead-zinc concentrates at el Djedidi, el Giblian, Zriba where celestobaritine occurs too. Strontium also accompanies lead and zinc at Fedj el Adoum, Bazina, Mezzouna, Bechater, Kebbouch, Dogra Zag and Tir, and it shows in the Gafsa phosphate basin. Fluorite is abundant in the djebilet el Khol, in the djebel Oust veins and at Hammam Djedida, all in the Zaghouan area (N). The 1-2 m thick Staa bodies extended 20 to 60 m and at Zriba, a meter thick and 500 m long mass contained 50% CaF_2 ; total length was 2 km over 1 to 7 m width with 6 Mt of reserves. Hopefully SECA's aluminum fluoride facility will also increase its capacity from 20,000 to 115,000 ty at Gabès with the support of the Arab Mining Co. which takes almost 30% of output.

Finally, Tabett ben Ksour, between the old Oued Maden mercury mine and Fedj Hassène, occasionally provides some arsenic.

Metals

Production of lead, zinc and silver varied thus : fifties 27,000, 5,000 and 3 ty ; seventies 17,000, 10,000 and 1 ty and now 5,000, 7,000 and 2 ty in response to irregular markets affecting small workings, some of which have been opened and abandoned several times since the war. A more elaborate mine, Fedj Hassène (170) currently produces a third of these metals. Bou Jabeur produces another quarter and plans called for the trebling of concentrates to 200,000 ty. Reportedly, Sidi Driss alone could produce 8,000 ty of lead and zinc and other important producers of the past were :

jebel Ressas, Khanguet Kef Tout (Pb/Zn = 1), Sidi Amor ben Salem, Djebel Hallouf, Sidi bou Aouane, Djebel Trozza and El Grefa (lead) ; Garn Alfaya - Koudiat el Hamra (Pb/Zn = 6), Sidi Ahmed (Pb/Zn = 2/5) and the reader is referred for other occurrences to MRA pp 386-7.

Although silver is a by-product of lead, recovery increased. In effect, galenas of Sidi Amor ben Salem contained as much as 0.04% of silver. Jebel Chouichia contributed some copper in the past and the Kef oum Teboul shows of Algeria might extend over the border.

Occasionally some mercury is still exported, but production at Oued Maden peaked in the sixties with 15 ty. The same mine produced along with lead, zinc and copper some antimony and arsenic. Moreover, vanadium accompanied lead at Kef Cheboul.

Although only a reserve base of 30 Mt ore is left, Djérissa (169), the principal iron ore producer, has maintained a production of 1 Mty for thirty years. Grades have declined, however, to 48% along with production and siderite is expensive to crush and roast. Few reserves are left at Tamera, Douaria and Bourchiba, the only deposit of the Nefza group (NW) which is low in arsenic. Meanwhile, the Nefzas still held 10 Mt of resources and Nebour (Kef) 10 Mt grading 40% Fe and 6-10% SiO_2.

Above the phosphates of djebel Ank (W), 200 km from Sfax harbor, 3 m thick beds include 30 Mt of ore grading 53% Fe, but also 0.8% P ! Other deposits include Chouichia, Hemeima ; Djebel Oust, Ras Rajel, Harraba and Thuburnic with manganese. Finally, on the jebels Aziza and Batoum manganese shows alone.

STS produces by direct reduction with natural gas steel and ferroalloys at el Fouladh (N) and Menzel Bourguiba might supply 400,000 ty of rolled products in the future. Meanwhile, Tunisacier has installed a new galvanizing line expanding rated capacity from 40 to 70,000 ty. While the 30,000 ty Mégrine Rhaïd smelter relied for years now on lead concentrates imported mainly from Algeria and Morocco, the 20,000 ty Litophone plant is to use domestic barite and zinc.

14 LIBYA

We associate now the name of Al Jamahariya al Arabia al Libiya al Shaabiya al Ishtirakiya with oil. Yet, after the war a mission decided that the country was virtually barren. Then, prompted by French discoveries in Algeria oil was struck at Atshan in 1958. In 1961 Hunt-BP found a supergiant and Esso loaded 200,000 barrels. By 1965 throughput was the same as today - 1M bd - with \$50 M profits accruing to government. Production peaked in 1970 with 3.7M bd and still averaged 2M bd in 1979 when revenues exceded \$20 B, more than double of current receipts.

> GOALS : Development of water resources and irrigated agriculture, downstream products, petrochemicals, fertilizer, steel and aluminum production based on oil revenues.

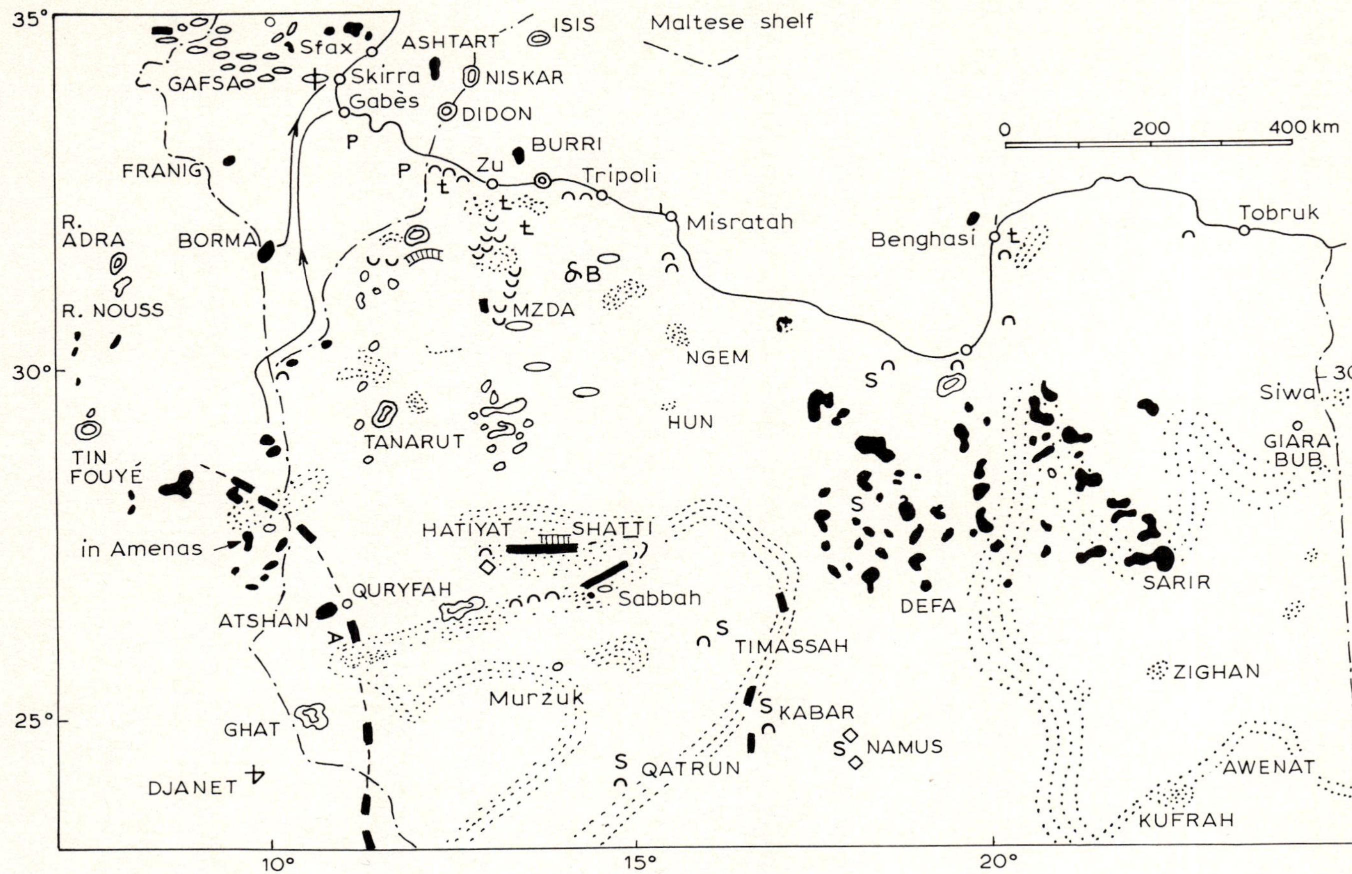

Fig. 14. Libya, southern Tunisia and eastern Algeria. (Legend on page 13.)

Table 36 Libyan production, projects and resources.

Projects			Production			Production			Resrces	
Ethylene	kt	300	Petroleum	Mb	40	Oil	Mb	400	Bb	20
Methanol	kt	300	Gas marketed	Bcf	250	Gas, gross	Bcf	450	Tcf	20
Urea	kt	360	reinjected		130	Gypsum	kt	180	Bt	80
NKP fertilizer	Mt	2	N in ammonia	kt	250	Potash			kt	800
Polyethylene	kt	60	Sulfur	kt	13	Magnesia			Mt	8
Steel	Mt	1.3	Cement	Mt	5	Iron ore			Bt	1.2
Aluminum	kt	120	Lime	kt	250	Steel (plan)	kt	10	Mt^p	1.3
			Salt	kt	10					

Government controls the new projects, mostly in the northwest and much production, including Esso's former fields. NOC also increases its 51% share in the joint ventures and the state levies 17% royalty and 65% tax on the producers. Mineral industries employed 10,000 people, with many foreigners, just as government.

The oil basins

In the west an ENE trend connects Atshan via Bir Tlacsin, Bir el Rhezeil, Emgayet and the Wadi Chebbi gasfield with Burri in 150 m of water off Tripoli. Estimates range here up to 7 Bb and capacity of the single steel platform is scheduled to double from 75,000 bd in 1987. Agip owns 19% of the field after taxes and royalties and its ENI sister companies share in the $2 B investment. Also offshore Al Aqadja holds 600-760 Mb of 36° API crude in bloc MMN, 20 km northwest of Benghazi (N) and exploration continues in the Marzuq basin (SW).

Out of 400,000 km^2, the central Sirte (Surt) basin measures only 200 x 500 km. Yet, it has a potential of up to 35 Bb oil and 30 Tcf gas. With resources increasing eastwards from 3 to 8 and 18 Bb, the Defa-Zelten warp divides the basin into three 'steps' :

I Dahra-Hofra platform with the a) Mabruk-Farad b) Zella-Sabah and c) Raguba-Hedan alignments.

II Zeltan platform with the Hateiba gas field, and the Nasser, Waha and Defa giants.

III Calanscio-Amal uplift with a) Amal, Nafoora, Attifel, b) the Intisar giants and Gialo c) Messla-Sarir.

The oilfields

Out of 900, 700 wells operate in fifty fields with a hundred on artificial lift. Granted, Sarir holds more oil than any other country of Africa, except for Nigeria, as shown by these data : original reserves 7.5 Bb, oil colum 100 m, reservoir area 400 km^2, volume 7 km^3 (5 M acre feet), gas/oil 60-225 cf/b, wax content 19%, only 0.25% S, 37° API. Temperature 111°C, watersaturation 35%, permeability 500 millidarcys - one well/2 km^2.

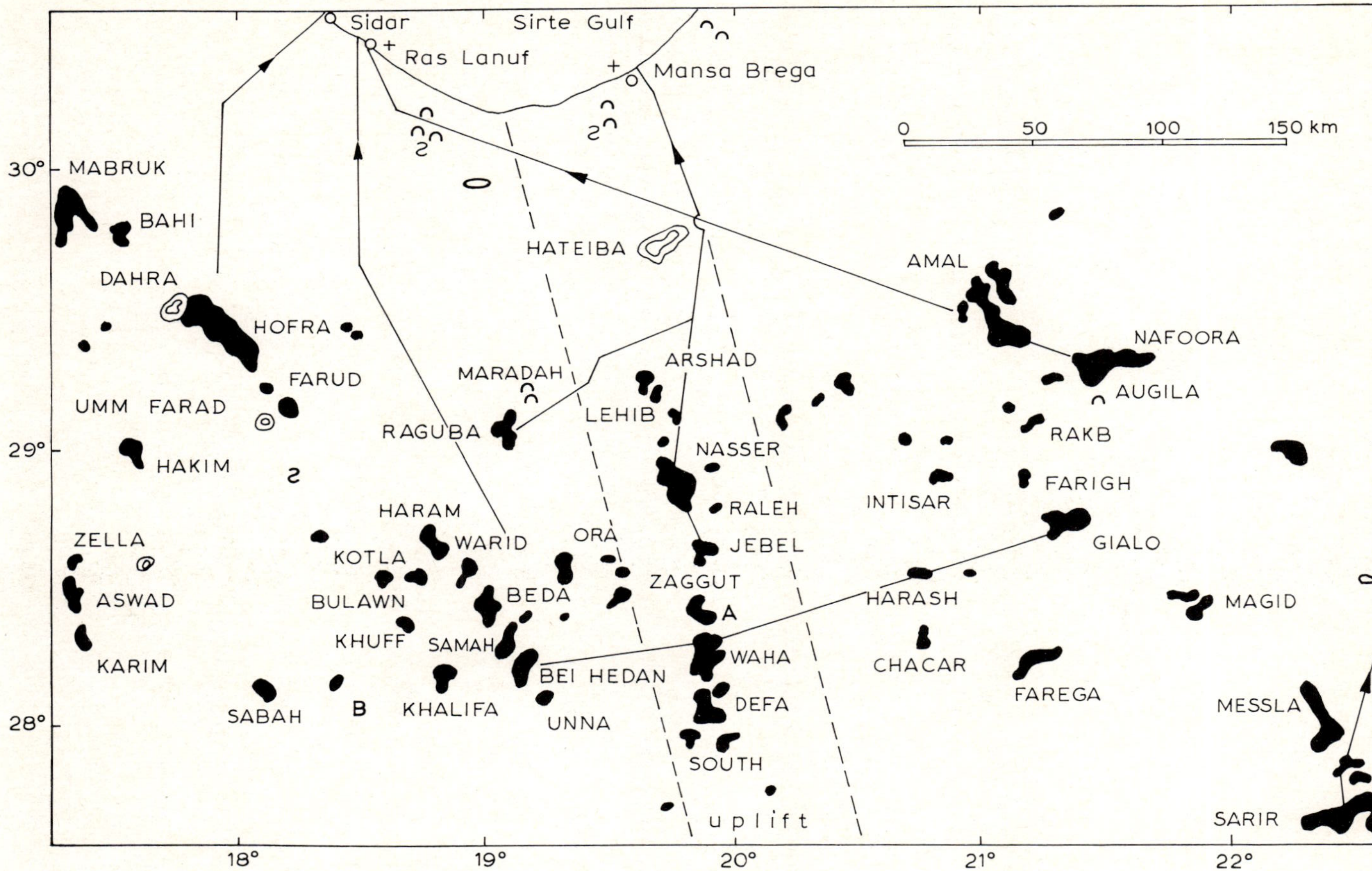

Fig. 15. The Sirte oilfields.

Nearby, Masalah, originally 3 Bb in place, with 1.5 Bb EUR under water drive, API 40°, depth 2,700 m, column 150 m, pay 27 m, area 200 km^2, porosity 16-17%, permeability 450 md. Further, according to NOC 2.8 Bb of additional reserves have been proven at Sarir and Masalah.

Gialo, 125 km^2, originally 4 Bb, up to 9 levels, 850-870 m, 1,900 m depth.

Intissar D originally 1.8 Bb, low-sulfur 39-40° API oil, expected recovery 80% with a hydrocarbon miscible gas drive, 60,000 bd, Ø 45 km, depth 2,600-2,800 m, maximum height 385 m, porosity 22%, permeability 87 md (max. 500 md) and Intisar A 2 Bb with 1.2 Bb EUR, 20,000 bd, 4 x 5 km, depth 2,950 m, pay 300 m, porosity 20-25%.

Amal (NE) originally 4 Bb of 37° API oil, 400 km^2. Nafoora Aguila (E). 1.5 Bb of 35.5° API crude, 40,000 bd from 670 to 3,000 m depth. Bou Attifel flows 160,000 bd, 41° API from 4,300 m. In the central Sirte, Hateiba holds 12 Tcf gas and Arshad and Lehib-Dor Madara nearby flow 42° and 50° API crude from 3,000 and 2,850 m. Nasser contained 2.5 Bb, 120 km^2, depths 1,650-2,350 m, pay 100 m with three highs. Output peaked in 1970 with 200,000 b and a 180 km long 36° line pipes seawater for injection. Waha and Defa had 1.5 Bb and each produced originally 100,000 bd. Waha 36° API crude, 4 x 7 km, 2,000 m depth, but Defa is shallower : 1,600-1,800 m.

The major fields of the western Sirte are :

	API°	m		API°	m		API°	m	1000 bd
Bahi	43	1110	Raguba	43	1600	Zella	48	2000	10
Dahra	39	1000	Kotla	34	1650	Sabah	42	1650	30

Except for Raguba which flowed 50,000 bd, these fields produced 10-30,000 bd, but Fidda peaked with 18,000 bd and Mabruk required secondary recovery.

Gas, pipelines, and refineries

The principal oil producers were the state companies Ageco Sirte (former Esso) and Umm el Jawaly Oil and in association with NOC : Oasis (Conoco, U.S. Steel, and Amerada) 33%, Occidental (including a 25% share of ÖEG) 12%, Agip 13%, Veba (former Mobil) 4%, Wintershall and others (Jakira) 1%. The principal pipelines led to Brega, which had 1.8 Mb of tanks, with a spur to the Misuratah steel works and an LNG plant rated at 9,550 bd. Ras Lanuf had 230,000 bd capacity, a desalination plant, and a 300,000 bd ethylene project. As Sidar's original plan was 300,000 bd and the longest line connected Sarir with Tobruk over 500 km. The Az Zawiya (NW) refinery had a 120,000 bd capacity and also produced 1,200 bd asphalt. Other projects called for refineries at Tobruk (NE) and Misuratah and al Zuwara (NW) since household needs are to treble to 1/2 Mbd by 2000.

Gas is reinjected, sold as LNG from the Marsa al Brega plant, mainly to Italy and perhaps again to Spain, used in powerplants and as feedstock. The petrochemical projects concentrated at Marsa al Brega, Ras Lanuf and Abu Kammash (near Zawiyah) are planned for this decade, to wit 300,000 ty ethylene, 360,000 ty urea ; others envisaged : 1,700 td urea, 1,000 td methanol and 2 Mt NKP fertilizers and at Ras Lanuf 60,000 ty polyethylene. Meanwhile, NOC ships already 350,000 ty of methanol from Marsa, as it does oil : with Libyan tankers.

Water and soil

Weak soils are cultivated in the northwest, near Benghazi and the oases, e.g. 1.5% of the surface. As shown below, 250,000 ha could be irrigated though. I. The only area with more than 500 mm rainfall is the Jabal al Akhdar (N) bordered by another 15,000 km^2 with 200 mm rain between Derna and Benghazi but growing usage of 200 Mm^3y increases here the risk of seawater encroachement. II. The 20,000 km^2 Jeffara (NNW) has the same rainfall helped further by free flowing and pumped irrigation, from 10-25, 25-60 and 400 m depth. Here too, consumption of 600 Mm^3 far exceeds recharge of 135-360 Mm^3y. III. Similarly, in the northwestern quarter, the quality of water deteriorates rapidly and ever deeper wells are drilled. Usage quadrupled to 250 Mm^3y and might reach 350-400 by 2000. Good water is thus scarce between Tripoli and Misratah but two aquifers of the Wadi Shaybah and Ras Hamia hold 13-29,000 years old water, which constitue semipermeable layers in the Nefusa.

IV. In the Marzuq basin (SW) water is 20-40,000 years old, and a puncture of 15-20 Bm^3y would lower the level by only 1 m. In the Marzuq-Traghan-Tmassah and Sebha oases, artesian water rises from Nubian sandstones and in the 200 km long Shatti valley water has percolated upwards into Devonian sandstones. In the eastern Shatti a withdrawal of 100-120 Mm^3 from the lower reservoir would cause a decline of only 5 m and 45 Mm^3 can be withdrawn from the two aquifers of the Wadi Tanezzouf near Gat (SW). New projects at Marzuq, Wadi Ahrir and Sebha call though for a withdrawal of 200 Mm^3.

V. Surprisingly, the withdrawal of 40 Bm^3y from the thick unconformable aquifers of the Kufra-Sarir basin (SE) would only lower the level by a meter. Along longitude 22°E water is at 30-50 m depth and a uniform dewatering of the Sarir would withdraw 2 Tm^3 of water from the upper 150 m of the aquifer... There are but 150-550 ppm solids and the oilfields use three quarters saline water for injection and only 20 Mm^3y fresh water. At Kufra (SE) 120 Mm^3 are withdrawn from subsurface recent and Nubian sediments and the settlement of 5,500 people requires another 100 m^3. Tazerbo (E) calls for 120-150 Mm^3y and in theory similar projects could operate every 50-70 km. Most water will be pumped, however, to the coast at an expense of $3 B.

Table 37 Withdrawals including projects (Mm^3y).

Usage	Mm^3y	Kufra Sarir	Marzuq	NW	Jefara plain	Jabal Akhdar	Total
Domestic	+ industry	60	30	50	200	180	520
Agri-	private	10	560	110	820	160	1660
culture	projects	690	470	300	120	70	1650
Ulterior	plans	540	200	110	-	100	950

Food and chemicals

Tripoli and Benghazi are the principal salt producers. Other coastal mallalah include Marsa, Shiqqah (E) and Tawurgha (N). Inland a tenth of the surface of some others contain more though :

Salt pan	area	km^2	cm	Mt salt	kt KCl	with trona		lake bottoms	Qasrs
Kammash	NNW	50	6			Timassah	S	Maatan	Larocu
Adri	SW	35	40	2	160	al Kabir	SSW	Nashusha	Bayah
Maradah	N	150	50	7.5	600	al Qatrum	S	Fredga SW	Khulayf

Pans reconstitute themselves and the Kammash and Maradah chlorides are already used for synthetics.

Byproduct sulfur is produced from crude and beduins use the sulfur of springs rising from the Sabkah Kābirāh in the oilfields, in the Jabal al Harūj al Aswad and the Wām and Nāmūs volcanoes where it is used to treat camels. Further, phosphatic beds up to 10 m thick have been found in the oilfields at depths of more than 1000 m and west of the Hun graben (N). Bentonite occurs in the same Jabal escarpment and covers 10 km^2 of depressions south of Sebha (SW). Finally at Sardalas (SW) up to 10,000 t of alum is recovered each year.

Building materials

Cement imports are to decline as capacity increases from 5 to 8 Mty, including the floating plant of 400 t/h. All plants use oil and the dry process,

Table 38 Capacity of Libyan cement plants.

Company	Northwest	Mty		Northeast	Mty
National Cement	Al Khums	1.4	Lime Cement	Hawari	1
Souk el Khams General			Sulfur resistant cement	Benghasi	0.27
Zilten	Zilten	1	el Fataian	Derna	1
Az Zawiyah oil Refinery		0.1			
			Sebha[p]	Fezzan	0.5

the cement rock of Baninah (NE) and the 40 m thick Qarqarish limestones which is quarried over 15 kilometer west and southwest of Tripoli (NW). Conveniently, the marly limestones of Ras el Manabia and Ras el Margeb (N) are underlain by clay. To continue, the Benghazi lime kilns have a capacity of 200,000 ty.

Metals

Two locations, az Zawiyah (NW) and Amikal al Khams have been proposed for a 100-130,000 ty aluminum smelter - a joint venture of the Jamahariya and Yugoslav Energoinvest who would provide the alumina. Requirements are 170,000 by residual fuel oil, 250 Mw at startup - and \$1 B. This project might materialize in the nineties and works of the Misratah Iron and Steel Co (N) would cost >\$3 B. Phase I envisions 1.3 Mty of steel, phase II , 3.7 Mty and additional investment of current \$1.3 B. Ultimately, the works would switch to Sidadah limestone and Wadi Shatti (SW) minette, 1,300 km away, where six blocs out of 14 contain 43-49% Fe : blocs 2-3 between Tārūt and Agar (W), 20 km long, 6-8 km wide; east of Dabdab, blocs 9, 10 and 12, 18 + 4 km long, 5-8 km wide, and a smaller bloc 6, near Brach (C). Silica averages 20-40%, P_2O_5 0.8% or less and S trace to 0.4%. Resources can be presented in two ways :

Fe %	Thick m	Indicat. Bt	Thick m	Inferred Bt		Fe%	Thick m	Indicat. Bt	Thick m	Inferred Bt
45-50	3-7	0.7	6-7	1	or	40-45	2-10	1.1	6-11	1.5
48	1.8 Bt, 16% SiO_2, 0.3% S, 3% Al_2O_3					44	2.6Bt, 21% SiO_2, 0.3% S, 0.3% P			

The trend extends west to Serdeles and Ghat and to the jabal ben Ghenema (E 14-16°, N 23-25°).

Veinlets and seams are also found in the iron ore in bloc 8, east of Brach (SW) and near Sheb. Additionally, manganese also shows in the Jabal escarpment (NW) at Wadi Urāri and at Ulad Mahmud (NW).

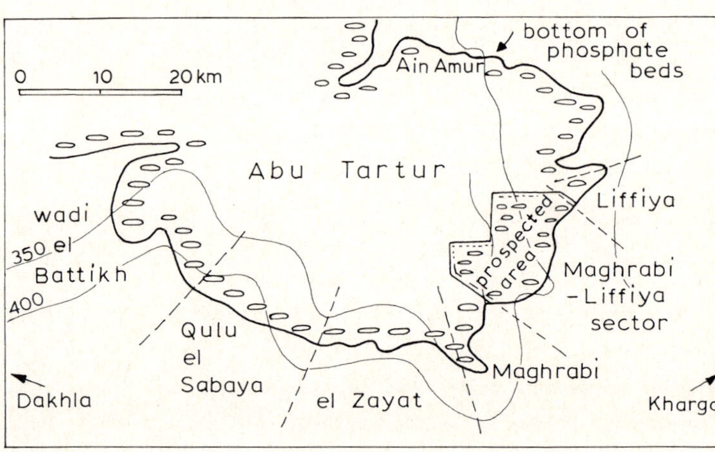

Fig. 16. (a) Abu Tartur. (Legend on page 13.)

2 NORTHEAST AFRICA

With a fifth of the area and of the population the Nile Valley and the Horn of Africa (Fig. 16, 18) supply only 7% of the continent's mineral product - most of which is Suez oil. Further on, smaller deposits follow the Red Sea as far as Ethiopia and another belt connects Harrar with northeastern Somalia. While Egypt continues to industrialize, the key problem of the area remains water, soil (and overpopulation).

21 EGYPT

Al-Jumhuria al-Arabia al-Muttahida is the continent's fifth mineral producer and its second consumer of resources. Its 50 M people are squeezed into arable land the size of Belgium and the population of Caïro alone is bigger than that of most African countries.

> GOALS : To provide pure water, fertile soil and ample energy for rapidly growing needs, that is more fertilizer, more gas, building materials and steel.

Minerals match a fifth of GDP (out of which hydrocarbons represent 60%), but only a third of revenue and three quarters of visible exports. Yet, by 1990 oil exports might cease, since the demand for subsidized energy is increasing at a yearly rate of 15-25%. Fortunately, steel mills already use gas. Should utilization increase, it might cover the gap and also provide for the expansion of the fertilizer industry.

The mineral industry proper only employed a few thousand men, but if processing and energy are included, some 50,000 were on this payroll and Egyptians man the mineral industries in many Arab countries. A quarter of phosphates and oil is exported, the first to Indonesia, Albania, China and France and the second to Italy and Israel. Amoco's (Standard Oil, Indiana) joint venture with the state's EGPC, GUPCO (also Fayoum) produced 60% of the oil, followed by Petrobel's 25% (ENI and EGPC), and SUCO which groups EGPC, Shell, BP and Deminex. Total and Conoco are also involved so is Phillips (WEPCO). Then, production sharing arrangements have been extended to ventures in solids with their provisions for cost recovery, a fair deal and a stable investment climate. Indeed they will, hopefully further import substitution which acquires ever more urgency as the $5 B trade gap increases.

Table 39 Egyptian production, reserves and resources.

Production			Production	Reserves
Petroleum	Mb 150	Oil	Mb 300	Bb 4
LPG	Mb 2	Gas produced	Bcf 150	Tcf 10
Coke	kt 900	Gas marketed	Bcf 120	
Sulfur byprod.	kt 2	Coal		Mt 27
Sulfuric acid	kt 45	Lignite		Mt 2
		Hydropower	Gw 2.5	Gw 4
N in ammonia	kt 700	Uranium		kt 5
Urea capacity	kt 480	Uranium resources		kt 100
N nitrate capacity	kt 330			
Ca nitrate capacity	kt 330	Phosphate	kt 200	Mt 260
N sulfate capacity	kt 75	Phosphate resources		Mt 520
P_2O_5 in fertilizer	kt 65	Copper secondary	kt 2	Mt 30
		Iron ore	Mt 2	Mt 170
Steel	kt 600	Manganese		kt 800
Ferrosilicon	kt 6			
Aluminum	kt 140	Soda ash resources	kt 4	Mt 60
		Soda sulfide	kt 3	
Cement	Mt 4	Barite	kt 3	kt 10
Fireclay	kt 500	Fluorite	t 10	kt 15
Lime	kt 90	Feldspar resources	kt 7	kt 15
Limestone	Mm^3 9	Kaolin	kt 100	Mt 120
Sand	Mm^3 6	Gypsum	kt 800	Mt 120
Gravel	Mm^3 4	Talc	kt 5	kt 700
Sandstone,&	1000 m^3 100			
Marble	1000 m^3 20	Resources		
Quartz	1000 m^3 10			
Granite	1000 m^3 5	Chromite		kt 3
		Coltan		kt 20
Salt	Mt 1	Ilmenite		Mt 1.8
Bentonite	kt 3	Lead		kt 20
Diatomite	t 300	Monazite		kt 200
Asbestos	t 300	Nickel		kt 1.5
Iron pigment	t 150	Tin		kt 5
Mica	t 50	Titanomagnetite		Mt 45
Magnesia[p]	t 100	Zinc		kt 140
		Zircon		kt 360

Suez oil and gas

Estimates range from 3.3 to 5 Bb oil and 7 to 12 Tcf gas and they may even reach 15 Tcf by 1990. Gemsa was discovered in 1909 and production developed thus (in Mb) : 1927 1.3, 1962 30, 1975 100, 1980 200, with 350 planned for 1988. Eight fields out of 50 provide 85% of the output from depths of 2-3,500 m in the south and as little as 1,000 m in the north ; 500 wells are on artificial lift and only 100 flow freely. Oil is waxy and contains 1-2% sulfur ; its specific gravity decreases towards the south and with depth and the sweeter, lighter blends of Gupco were popular.

Table 40 Gulf of Suez oilfields, projects[P] and comments.

	Discovery year	Depth m	1000 bd	API°
Belayim	1953	2100-2600	160[P]	30-22
Morgan	1965	1500-2000	170	27-32
October	1978	3300	120	28
Ramadan	1974	3400	90	32
July	1973	1500-3000	80	31-35
Shoab Ali	1978	1800	30	34
Ras Budran	1978	3400	30	26
GS 315	1981	1900	20	30

Comments on principal oilfields (from north to south)

Company	Field	1000 bd	Remarks
Suco	Ras Budran	30	reinjection gas lift
Petrobel	Abu Rudeis		
	October	100	gas lift
Petrobel	Belayim	180-200	target
	Ras Fanal	10	along west coast
	Ras al Bihar	12	
Gupco	July	70-80	improved recovery
	Ramadan	90	steady production
	el Morgan	200	target
	Shoab Ali		new lease on life
Suco	Zeit Bay	15	target 50
Conoco	Gaysum	25-40[P]	gas 1.25 Tcf
Disc	Ghubba al-Zeĭt	6	
Gupco	GS 195	30	target

Ramadan had 1500 Mb of reserves, el Morgan 1000 and Belayim 600, along with most of the 160 Mcfd of associated GAS gathered by the Ras Shuqair collector. While Asl, Sukr and other small EGPC fields in the north flow heavy oil, exploration extends south of the Gulf, beyond Total's Umm Agawish wells on Tawila which flow 7000 bd.

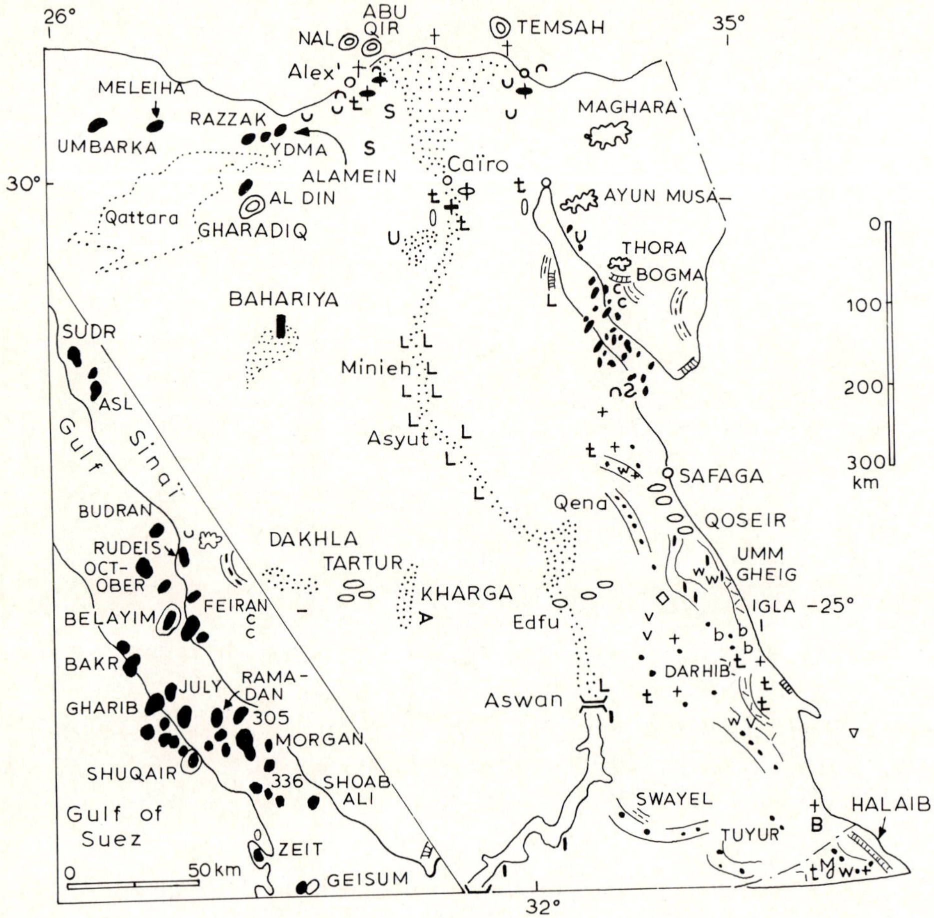

Fig. 16. (b) Egypt with NE Sudan administered. (Legend on page 13.)

Other fields, pipelines and refineries

Gas predominates in the delta on- and offshore, yet 40,000 bd will flow from 200 Mb in the Western Desert and at Abu el Gharadiq at 3,200 m depth estimates range from 0.4 to 3 Tcf plus 15 Mb condensate. Gas finds now extend beyond Qattara to Abu Sinan and the border.

Company	Field	Mcfd	Company		Mcfd	1000bd	Company		1000 bd
EGPC	Abu Qir	100	Mobil	Temsah 2	25	1.4	Wepco	Alamein	3
Petrobel	Abu Madi	50	Faium	Gharadiq	70	10	Wepco	Ydma	2
Texas	Salam	35	Bapetco	Badr al-Din	5	6	EGPC	Razzak	3
			Agiba	Meleiha[P]		20	Wepco	Umbarka	1
							Texas Int.	Khalde	6

Major pipelines include

gas	Abu Madi Talkha	fertilizers, Mehala el Kobra spinneries
gas	Abu el Gharadiq	Helwan and Turah steel and cement, 335 km
	Shubra al Khaïma	Tanta, 230 km
gas	Abu Qir	Damanhour urea, Kafr el-Dawar power plant
oil	Sumed Suez-Pt Saïd	mainly from Gulf 500→750 Mby
oil	Others Suez, Caïro	refineries 50-100 Mby

Refining capacity has increased thus (in Mb) : 1972 50, 1982 120, 1986 200... and should reach 300 Mb before 1990. Refineries are located at Alexandria, Tanta, Ameriya and Musturod near Suez which is to provide 40 Mby of middle distillates, Asyut on the Nile (C, 11,5 Mby) and Abu Rudeis (W, 10 Mby). Others include

Petrobel Belayim 800 bd gasoline from 35 Mcfd gas
Ras Bakr - Ras Shukaïr 1500-3300 bd LPG and 5000 bd naphta from 80-120 Mcfd gas
el Gharadiq 400-500 bd condensate 1500 bd LPG capacity.

A petrochemical industry is also taking shape with a 40,000 ty alkylbenzene plant at Ameriya (N) to be followed by 80,000 ty polyvinylchloride, 100,000 ty vinylchloride and 60,000 t chlorine and caustic soda. Further, a 200,000 ty ethylene and 180,000 ty of polyethylene plants have also been envisaged.

Other energy

Generating capacity has increased from 1 Gw in 1969 to 5 Gw in 1985 and should double by 1990. By the year 2000 at least 350 Mboe will be required and the depletion of proved hydrocarbon resources can be foreseen. Consequently, plans call for two 1 Gw reactors to be built at Daba (N), hopefully in the nineties. Uranium shows in the el Erediya vein and El Mascat might provide 30-50 ty, but the jebel Qatrani is smaller. Further, the Sebaiya, el Qoseir and Abu Turtur phosphates grade 0.008, 0.004 and 0.003% U_3O_8. Although monazites of the Delta contain thorium, a more immediate option are imports from Niger.

Admittedly, the share of hydropower continues to decrease and two thirds of the potential have been already developed. In addition to the Esna (100 Mw) and Asyut schemes, the old Aswan and the underutilized High Dam might normally generate 2.5 and 10 Twh from a nameplate capacity of 345 and 2,100 Gw. Wadi Halfa, in the Sudan, has a 420 Mw potential and a 70 km long tunnel could bring seawater to the Qattara (NW) depression providing 230 to 600 Mw. A million tons of coal are already imported and plans call for the building of 125 to 600 Mw plants at Ein Mossa (Ayun Musa) in the western Sinaï where brown coal seams extend over 15 km at a depth of 450 m. At Maghara , to the north, 65-135 cm thick seams include 50 Mt with 7270 cal/g, 50% volatiles, only 6.5% ash and 3% sulfur. Production was 20,000 ty before the Sinaï war and some 27 Mt could be extracted at 600,000 ty, part of it to be mixed with imported coke at Helwan. Finally, Buda Thora in the southern Sinaï holds 75 Mt grading only 30-40% fixed carbon, 7-27% volatiles, 40-50% ash and 0.4-1% sulfur.

Water, soil and fertilizer

There are thousands of shallow wells and some hundred less than 700 m deep wells. Although solids decrease with depth, they range from 200 to 600 ppm ; temperatures may reach 40° and pipes corrode.

The Blue Nile contributes most of the floodwater - 3000 m^3s - as well as 2 kg/m^3 silt (max 3) which now sediment in Lake Nasser, instead of fertilizing the light brown and sandy permeable soils of the valley and the Delta, the fruit of millennia of cultivation. Salt, therefore, invades the Delta. In effect, only 3% of the land is cultivated, but 5% could be irrigated.

In the "New Valley" of Dakhla and Kharga some 30% of the water comes from the Nubian sandstone. Total yield is in excess of 300,000 m^3/d from depths of 2-800 - maximum 2,500 m, derived partly from the Tibesti-Ennedi. There is little water, however - with up to 3,000 ppm of solids - in the Eastern Desert and on the coasts.

Production at I Qoseir on the Red Sea (E) and II Idfu and Qena (C) ranged from 3 to 700,000 t phosphate rock. Plans call for 1.5 and even 4 Mt in the nineties when III Abu Turtur in the Western Desert should come on stream.

I. About 300 Mt of rock are reportedly left in 15 occurrences such as el Higab and Umm Huweita. At el Hamrawein, Misr's calcination plant will rate 600,000 t grading 33% P_2O_5, but 40 Mt of rock assay only 17%. At Abu Shegeila the Red Sea Phosphate Co. is expanding to 400,000 ty of concentrates.

II. Ten occurrences are said to hold 200 Mt of rock and perhaps 1.5 Bt of resources grading 21.6% P_2O_5. At el Mohammed, el Nasr has an underground capacity of 500,000 ty with 21-25% P_2O_5 and Abu Zabaal expands its 120,000 ty capacity underground at Sebaya and in the Kom-Mir quarries to as much as 3.25 Mt rock. Part of this is shipped to Caïro where 66,000 ty phosphoric acid are produced. TSP is to follow later.

III. Out of 3 Bt of rock averaging reportedly 28.7% P_2O_5 at Abu Turtur, Soviet geologists have delineated in the Maghrabi-Liffya segment 1 Bt grading only 25.2% P_2O_5. Half of this rock could be mined underground profitably and another 30 Mt opencast, and concentrated to 29%. Plans even call for 2 Mty production in the nineties with an investment of $1 B.- Egypt is an important producer of nitrogenous fertilizers and integrated plants have been considered for Asyut (C) and Safaga (E, 300,000 ty too, for Abu Turtur - see fig. 16 b).

Industrial and building materials

With a targeted capacity of 1.5 Mty, salt production has increased from 600,000 tons in the sixties to 800,000 tons from the seventies on, that is slower than population. Marine salt is recovered at Alexandria, Mersa Matruh (NW), Rosetta (N), Idku and Port Said (NE). Considering rapidly growing demand, the Lake Qarun project (NW) calls for the production of 200,000 ty of salt, 20,000 ty magnesia and 7,000 ty of sodium sulfate a crust of which also covers Wadi Natrun (NW). Natron and magnesia also precipitate at Baharia (NW) (see 'iron') and workers also rake it off at Harara.

Cement production increased fivefold in twenty years and there is 50% extra rated capacity. Nevertheless, Egypt imported half of its needs. The Helwan and National companies are furthermore expanding at Menya and Beni Suef (1 Mt).
- The state owns most companies which control together 80% of the capital of Suez cement.

Table 41 Egyptian cement capacity.

Company	Plant	Mt	Company	Plant	Mt	Company	Plant	Mt
Alexandria Portland	Ameriya	1	Suez Cement Co.	Qattamia	1.4	Helwan Portland Cement	Helwan[g]	4.5
	El Mex	0.7		Wadi Hagoul	1		Asyut[g]	2.7
			SECPT	Tourah[g]	1.5	National Cement	el Tabbin	2.5

[g] Gas fired wet process plants.

There are vast reserves of limestone, mainly on the Nile and in the Gulf, where gypsum is also abundant, but the principal quarries are el Hamman (N), at Fayoum (NW), Ismaïlia and Wadi Rina (NE). In twenty years production of gypsum doubled, clay and kaolin have trebled. Largest reserves are in the southern Sinaï (105 Mt), at Kalabsha (16 Mt, SW) and Abu Zeneima (0.5 Mt, NE) and other occurrences include :

I Building materials

Asbestos and vermiculite Baramiya-Sabatit, Hafafit (SE) ; basalt Caïro-Baharya ; clay Qena-Aswan (SC) Asyut, Beni Suef, Giseh (N), Ismaïla (NE), west of Alexandria ; dolomite Gulf of Suez ; granite 30 ; gravel and sand 200 ; marble 30 ; sandstone Aswan and 30 other quarries ; gypsum Ras Malaab (NE), Maryut (N), Manzaba, Ballah ; kaolin Abu Daraba (SW), Mussaba Salama, Abu Natash ; quartz Eastern Desert (92-99% SiO_2) ; sand Maadi, Gulf (95-98% SiO_2), Wadis Koba and Budra (NE).

II Industrial materials

Alum and epsomite Kharga, Dakhla ; barite Siwiqat el Soda -SE), E-C Desert, Red Sea Hills ; bentonite Lake Qatrun (NW) ; diatomites Fayoum (CW) ; feldspar and fluorite Eastern Desert ; graphite Wadis Bent, Sitra and Abu Geraiya ; iron oxide pigment E & W Desert ; magnesite SE, Umm Salatit ; peridot St John's Island (E) ; pyrite E. Desert, Gemsah ; steatite Kahfa, Abu Khug, Nagrat, Nakah ; sulfur Gemsa,

Ranga ; talc Wadis Gulan el-Atshan and Kharit, Derhib.

For further details, the reader is referred to the Mineral Resources of Africa.

Metals

Iron ores, the only significant metallic deposits, include :

BAHARIA : El Gedida 120 Mt, 55% Fe, 1% Mn, 0.9 S, 0.2% P, Jebel Ghorabi 50 Mt, 47%, el Harra 33 Mt 42%, Nasser 27 Mt 44% plus lower grades 110 Mt 28% Fe. OTHERS : Wedan el Faras (NW), Giddi, Umm Bogma (Mn, NE), 30-120 km south of Qoseir (E), 50 Mt, 32-52% Fe and east of Lake Nasser : Aswan, Hawan (31% Fe), Kalabsha, Garf Husein, Kurusko, Abu Simbel, etc.

Production has trebled from 150,000 t in 1956 to 500,000 t in 1962 (400,000 in 1972) and now excedes 2 Mt at el Gedida where 11 m thick beds are stripped with an overburden ration of 2 : 1. Two to four Mt of ore can be transported by rail from Baharia (NW) over 350 km to the Helwan (N) works of ISC, where production has sharply declined from its peak of 700,000 ty pig iron, and 900,000 ty steel, but a new melting and casting shop supplies 160,000 t of billets and projects call for 2 Mty in the nineties. Meanwhile, el Dikheila 's ultimate capacity will be 750,000 t bars and the Alexandria National Steel Co, a cooperative venture of the government, Nippon Kokka, Kobe Steel and Toyo Menka Kaisha has received World Bank and IFC support. It will reduce imported iron ore with Abu Qir gas through the Midrex process.

Egypt also imports alumina from Australia for the Nag Hammadi smelter (E), which uses Aswan power. Furthermore, ferrosilicon capacity is expanding at Idfu on the Nile by 56,000 tons, based on the Umm Higlig quartzites, 120 km to the east and Helwan coke, but ferromanganese output ceased with the Sinaï war. In 1962, 350,000 t of manganese ore were recovered at Umm Bogma (NE). Some ore grading 20-40% Mn is still left here, on the Gebel Musa and Abu Zeneima, further east, but only 100,000 t assaying 45% Mn remain near Halaib (SE corner). Further north, the Atoud, el-Baramya, el-Sokkar, the Umm al-Rouss and Okoud gold mines reopen.

Between 1933 and 1955 ilmenite, rutile and zircon production ranged around 20,000, 1,000 and 3,000 ty in the Delta and in the sixties 100,000 ty of titanomagnetite have been exported from Abu Ghalaga (SE), mostly to Italy. Although resources of zinc e.g. are reportedly large, they remain dormant mostly in the Southeastern Desert :

Beryl Zabara, Sikait, Nugous, Umm Kabu ; chromite Baramiya, Ras Shaït, Abu Dahr, Wadi el Ghadir. Sands of the Wadi Araba and in the jebels Sarabit, Nekhal and Khaden (NE) assay 0.04-0.15% Cu ; copper and nickel Abu el Nimrar, Regeiba (NE), Abu Swayel (100,000 t ore 1.5% Ni, 2.8% Cu); gold Umm Udd, jebel Nugrus, Hanglia, Abu Dabbab ; manganese Bir Nassib, Sharm el Sheikh (NE), jebel Alda (C), Wadi Malik (E) ; molybdenite Gattar, jebel Umm Harba, Wadi Dib, Abu Marwa, the jebels Umm Disi, el Dob, el Shayeb ; tantolocolumbite and tin (Ta/Nb 2.4) Abu Dabbab ; tin Al Nuweib ; tin and tungsten Igla, el Mueilha ; zinc, lead Umm Gheig on the Red Sea coast 1 Mt ore 14% Zn, 2% Pb, Zug el Bahar, Ranga, Ras Banas, zinc, lead and copper Umm Saminki, Atshan, Derhib, Akarem. El Arabi where copper was once worked.

22 SUDAN

A quarter of the people of the Jamhuryat el-Sudan al Dimocratiya suffered from the drought. Indeed, like Egypt, Sudan is a gift of the Nile. Water and oil are its most valuable resources.

Table 42 Sudan production (past$^+$), projects and reserves.

Reserves			Reserves				Production		Reserves	
Wollastonite	kt	400	Gas	Bcf	250	Petr./oil	Mb	6	Mb	300
Magnesite	kt	280	Gold	t	30	Hydropower	Twh	1.2	Gw	14
Asbestos	kt	250	Silver	t	260	Cement	kt	200		
Kaolin	kt	2	Copper	kt	300	Gypsum	kt	15	Mt	30
Marble	kt	380	Zinc	kt	150	Salt	kt	70		
N in urea^p	kt	46	Iron	Mt	50	Manganese	t	400	kt	500
N in ammonia^p	kt	110	Fluorite	Kt	200	Mica$^+$	kt	2		

The principal trading partner is Saudi Arabia, but chromite was exported to Japan, gold to the U.K. and manganese to Yugoslavia. The government owned the principal chromite mine and the fertilizer plant. It was a partner of Chevron, but most small mineral ventures were owned by local or foreign investors. All in all, some 3,000 people work in the industry.

GOALS : Peace in the south, development of energy, and rehabilitation of water resources.

Water and soil

Half of the flow of the White Nile is lost in the southern marshes, 14 Bm^{3}y in the Bahr el Jebel and 6 Bm^{3}y in the Bahr el Zaraf. Storage at Baro, embankments and the long awaited Jonglei canal could save 4 Bm^{3}y, half of what is lost on the Sobat and Machar. Storage on tributaries and diversion to the Bahr el Jebel and White Nile could similarly save half of the 14 Bm^{3}y which disperse from the Bahr el Ghazal. Water seeps as far as 100 km from the rivers into the Umm Ruwada beds which are up to 300 m thick. Although rainwater contributes to recharge, this aquifer is often brackish. On the other hand, Nubian sandstones yield more water from 10 to 170 m depth. While some may come from the Nile, the bulk is probably of Pleistocene age. Water seeps towards the northeast where sediments thicken and the Red Sea coast where wells are shallow. Annual floods replenish the Khor Arbat which serves Post Sudan and in the North sands and clays constitute an aquifer of sorts. Further south, 75 m thick gravels of the Gash delta (E) draw abundant water from the highlands of Erythrea.

In Kordofan (CW) 50 m thick sands overlain by clay yield insufficient groundwater but each hafir can store 5-10,000 m^3 when dredged and filtered. Shallow wells serve most of Darfur (W). The Meidob-Tagera threshold has here, at 1,000 m, the highest static level, directing water

a) northeast where the aquifer dips 400 m over 600 km and

b) southeast where it joins the flow of the Nile Valley. This cultivated area is only 3% of the total land surface and the irrigable area, mainly, the Gezira (C), matches 1.7%. The 2 mm fraction of the Badob soils consists here of lime and alluvial and aeolian sedimentation control the soil pattern. Chestnut soils of the grasslands cover Darfur (W), the coast has psamments and the cracked soils of Blue Nile have nutrients. Finally red brown soils appear in the southeast and the southwest resembles the tropics.

Energy

Estimates range from 200 to 350 Mb of oil reserves and up to 1 Bb resources. Furthermore, the Bashayer well in the Tokar Delta on the south coast might hold 10-30 Mb condensate and 130-400 Bcf of gas with a calorific value of 1300 BTU/cf. Two other wells in the Suakin area have tested gas, but more exploration has to be done in the Red Sea and the interior.

If completed and converted, the Sudan-Ren urea plant could use 20 Mcf of this gas piped to Khartoum. Another possibility is building a new 100,000 ty plant at Suakin.

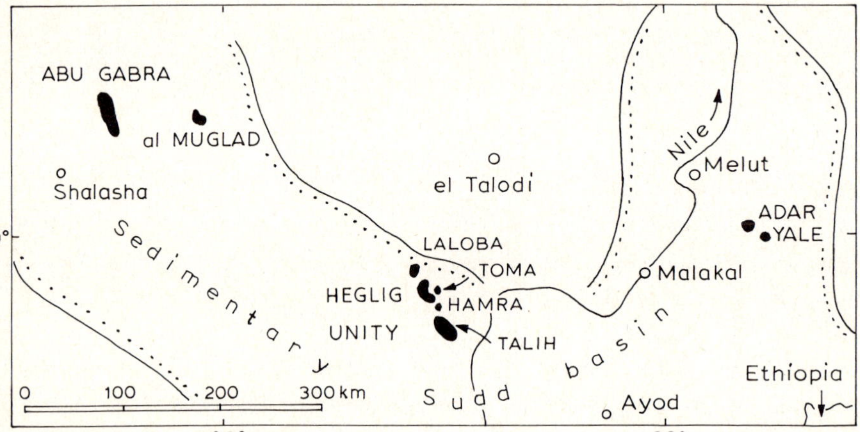

Fig. 17. The Sudan oilfields.

More important are Chevron's discoveries among the Shilluk, where a dozen wells have tested oil - one out of four.

a) The Talih-Unity group of fields, near Bentiu (SC) has a surface area of 2 x 16 km, EUR reserves of 150 Mb and a potential yield of 10,000 bd of low sulfur

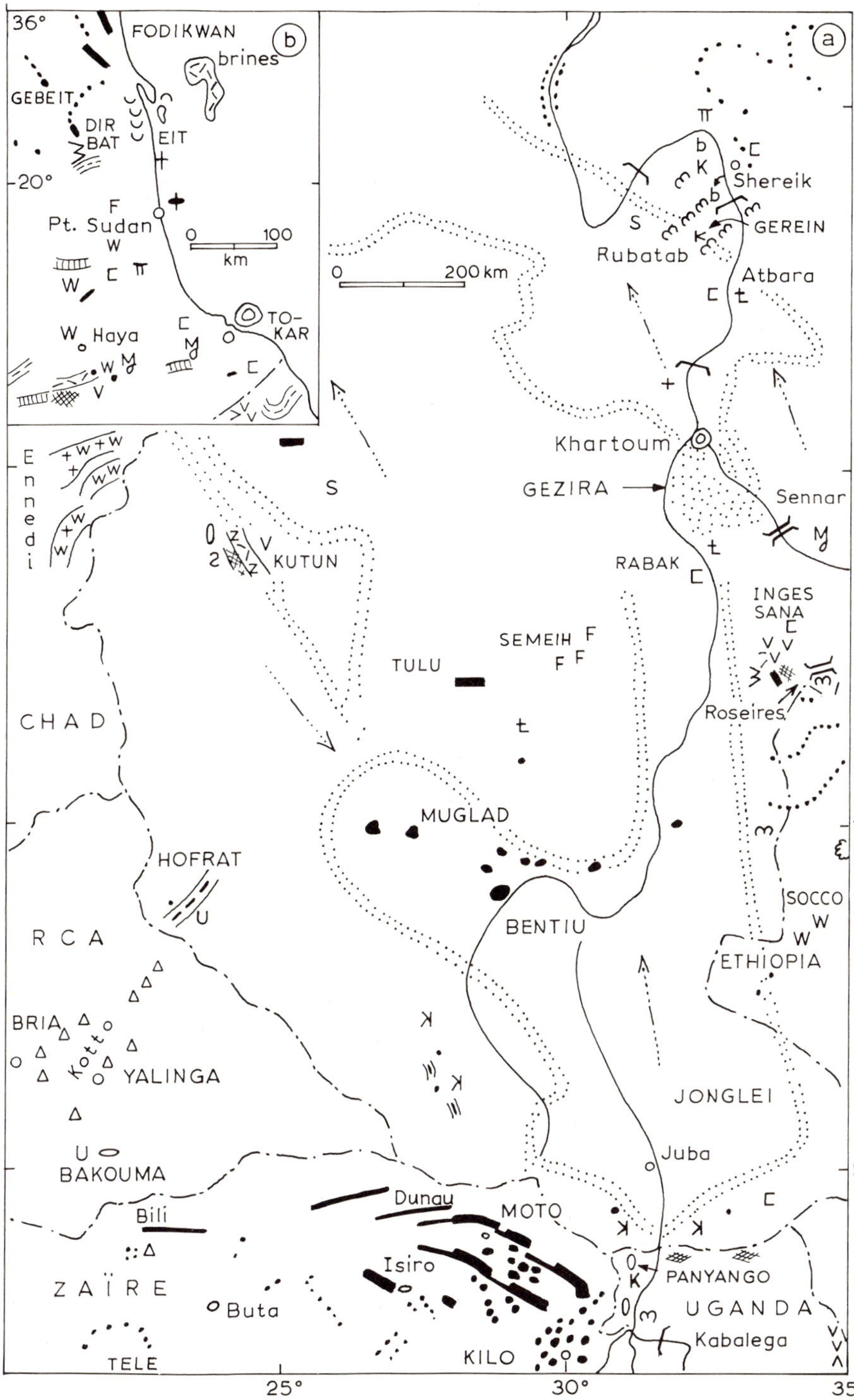

Fig. 18. (a) Sudan and its neighbors. (b) Red Sea. (Legend on page 13.)

waxy oil with a high, 27° pour point. b) Heglig is immediately north of Talih, with Bamboo, Barki, Hamra, Laloba, Kanga and Toma. - Other finds include c) Tayib 70 km northwest d) Abu Gabra and Sharaf near Muglad, 200 km to the WNW, e) Adar Yale 100 km north of Unity and f) Melut in Upper Nile (1500 bd). - Recently, surveys even moved into Blue Nile.

The Port Sudan refinery, a joint venture of Shell and GPC, imports crude and gasoil and exports some fuel oil since demand patterns do not match any more the product mix. Oil and product imports have doubled in five years and by the nineties net imports might attain 15 Mb unless domestic crude reaches the refinery for blending with Arabian light. WNPC, a joint venture of the state, Chevron and IFC wishes to build a 1450 km long 50,000 bd pipeline out of which the state's share is almost half, but 15,000 bd would still be available for export. Even a 40,000 ty bitumen facility has been considered. Similarly, plans call for the building of a 5,000 bd topping plant at Kosti on the Nile, and further discoveries might justify building a new refinery.

Electricity needs increased 2.5 times in ten years and may double to 2 Twh in the nineties. Even though abundant hydropower is available only Roseires and Sennar on the Blue Nile have been harnessed with capacities of 1200 and 110 Gwh, but only half of this is used. Nevertheless, another 10 Mw would generate 70 Gwh more here and a seventh unit another 120 Gwh at Roseires. Finally, another 120 Mw plant operating during the floods would average 270 Gwhy.

Other sites include, from south to north :

Rumela on the Atbara : 20 Mw, Burdana on the Setit, Jebel Aulia 40 km S of Khartoum : 160 Gwh average, Sabaloka 90 km N of Khartoum : 700 Gwh, 120 Mw. In the bend of the Nile : Shereik 240 Mw, Shirri Island 450 Mw, Merowe 750 Mw, and Dal on the border : 600 Mw.

The Blue and White Nile

The private Nile Chromium Corporation has extracted more than 20,000 t ore from the 150 m long and a few meters wide Gabanit lens. In the larger Ingessana Hills, Jebel Jam (Gam) holds 70% of total reserves, grading 48-52%. Operations started here in 1962. To date more than 400,000 t have been produced. Shipments reached 30,000 t in 1962 and 1980 each and rarely dropped under 10,000 t. The Jebel Abu Dam and Chicy deposits are nearby and there are ten lenses at Jebel Kurba 5 km to the southwest. Roads have also been improved to Damazin near the Roseires dam and 20,000 ty ferrochromium plant has been discussed with Mitsubishi.
There is also asbestos :

Orebody	m x m		Mt	%	Orebody	m x m	Mt	%
Kukur	1000	100	1.5	3	Dufur	300 30	2	3
Gasbel	750	100	5	3.25	Bogal	11 seams	0.2	10

Magnesite shows in the same environment near Jam Mine, at Soda and Fadamiya and chromite accompanies on Jebel um Saqata 200 km to the NNE and 40 km to the north at Qala en Nala on the railway line. Talc occurs nearby at Gedareif. There is also gold east of Ingessana along the border and the Mowlem-Africa Construction Co. has prospected the Fazugli district. Finally, low grade manganese shows along the White Nile at Paloich and west of Malakal at Wabait.

Red Sea Hills

SMC quarries 5 m thick gypsum at Khoreit ; the beds extend over 100 km beyond Dunganab bay and cover Mukkawar island. Nearby salt is won from the lagoons, mainly at Ras Roweiya. Sales rose from 60,000 t in 1964 to 80,000 t in 1979 and 80. Partek AB of Finland and SMC have started work on the continent's largest wollastonite deposit. In contrast manganese shipments from the small workings between Ankalidot and Matateib plummeted from a peak of 14,000 t to 700 t in 1976. To be sure, BRGM has located the following reserves and grades :

Ore	Mt	Cukt	Znkt	Ag t	Au t	Cu%	Zn%	Ag g/t	Au g/t	combined	t		g/t
Hassay	13	250	100	230	9	2	0.8	18	7	Kamaweib	8	Au	2-10
Oderuk	2	50	35	30	1	2.3	1.7	14	0.65	& Ganaiet	1	Ag	-18

Other Saudi funded finds include Hadal Awatib also on the Wadi Ariab and silver on the Arbaat. Further north, Greenwich Resources explored Shanakwan and planned to produce 1.7 ty at Gebeit.

Table 43 Red Sea Hills.

Commodity	Locality	km	direction from	reserves Mt	grade
			CENTER		
Gypsum	Khor Eit	70	NNW Pt Sudan	40	45% SO_4
Salt	Dungunat	110	NNW " "		
Wollastonite	Dirbat Well	140	NW " "	0.45	
Ilmenite	Dirbat beach	100	NNW " "		
Gold	Sarakoit	120	NW " "		10-20 g/t
Gold	Shanokwan	80	NW " "		
Gold	Abeir Kateib	100	W Sarakoit	dumps	1.5 g/t
			SOUTH		
Tungsten	Khor Arbeit	110	SW Pt Sudan		50
Manganese	Tomei	100	WSW " "	1	
Manganese	Tehilla	50	SW Tokar	worked out ?	
Marble			W Tokar		
Marble			S Pt Sudan		
Kaolin	Derudet	50	SW Tehilla	1.8	25% Al_2O_3
Magnesite	Asok		N " "	also : vermiculite	
Tungsten	J. Eyab	20	W Imasa		
			NORTH		
Manganese	Halaïb	10-60	WNW Halaib	1	45%
Magnesite	Sol Hamed		W " "	0.2-0.5	37-39%
Iron	Fodikwan	50	S " "	6	60%
	Garb Heit		near Fodikwan	50-100 ?	35-40%
ore	Sufaya	80	SW " "	20	60%
Gold	Oyo	60	WSW Halaib	in 1920 26g/t, 1944 : 7 g/t	
Gold	Gebeit	80	SSW Oyo	7 t	34 g/t

Other areas

In 1964 cement production passed the mark of 100,000 t and has doubled since. Maspio had however a rated capacity of 380,000 t at Atbara and quarried the 50 m thick marbles of the Wadi Kurumut, west of the Nile. The Nile Cement Co. rated 100,000 ty near Sennar and uses limestone mixed with dolomite from Nyefr Rugaig, 60 km south of Rabak. There were, moreover, marble quarries between Shereik and Berber. At Wadi Halfa oölithic iron ores lie at shallow depth. Gold was worked in the far north at Doishat and Dowshat, near the headwaters of Lake Nasser, II at about 30 ancient sites 200 km to the ESE around Umm Nabardi, III 200 km farther south in half a dozen workings of the Wadi Amur and the Jebel Negim might be reopened.

Some 40 pegmatites occur in a 30 km long band between Abu Hamed and Herbe. Other concentrations are at Shereik and along the 150 km long Rubatab belt on both banks of the Nile. Out of 18 lenses 3 contained originally 4,000 tons of mica and in 1976 a ton has been hand trimmed. Moreover, 1,000 to 2,000 ty scrap were recovered between 1979 and 1981.

Kyanite is disseminated on both banks in the Atbara bend and 60 km NW of Atbara 80,000 t of wollastonite bearing reserves and 220,000t of additional resources have been estimated at Jebel Gerein.

In the XIXth century lead and zinc have been worked at Kutum (NW) and old workings cover 400 x 900 m at Hofrat-en-Nahas (W) with 270,000 tons resources in assays of 2.7 to 4%. The mineralized zone extends 80 km to the SSW (see pp 69, 373 MRA).

The 120 m long and 7 m wide Jebel Semeih vein (C) contains only 11-42% fluorite, and the trend reaches Jebel-Dumbeir. Between 1964 and 1970, Sudan reported 70,000 t of iron ore and at Abu Tulu (SC) 35 Mt contain 61% Fe.

Other occurrences include:

Large beryl 30 km west of Shereik and south of Abu Hamed ; lowgrade lateritic iron at Mvolo, Rumbek (S) ; tin and tungsten : Abu Dom (Sabaloka, C) ; graphite : Bongoriv, Yambio-Meridi road (S) ; gold : Thingaïta (far south) ; mica (S) : Khor Buddini (Furg), Longaino, Lufira (Mongella).

23 ETHIOPIA

The north, the east and a quarter of the population suffer from civil war and drought. Socialist Ethiopia includes three mineralized areas : I northwestern and II central Erythrea, the latter extending into northern Tigre and III the 220 km long Salt Valley (NNE). Similar areas extend from IV Harar into Somalia (E), and V in remote Wollega (W) and VI Sidamo (S). The Central Highlands have stone and cement but deforestation creates a 'moonscape'.

> GOALS : Soil conservation, reforestation and management of water resources. Further incentives for oil exploration, evaluation of numerous small occurrences, especially in areas V and VI. Marketing of mining ventures and upgrading of mines.

More than 5,000 people are connected mostly with small workings owned outright or through majorities by the Provisional Military Government.

Table 44 Ethiopian resources and production (including projects[p] and past[+]).

Production			Production			Production			Resources	
Gold	kg	500	Petroleum	Mb	6	Hydropower	Twh	1.1	Gw	9
Platinum	kg	5	Cement	kt	150	Geothermal[p]	Mw	30	vast	
Gypsum	kt	4	Kaolin	kt	10	Gas[p]	Mcfd	220	Tcf	1.3
			Pumice	1000 m^3	5	Urea[p]	kt	200		
			Sand	1000 m^3	600	Lignite	t	500	Mt	10
Nickel	kt	180	Stone	Mt	2	Salt	kt	120	Bt	3
Copper	kt	50	Mica[+]	t	10	Potash[p]	Mt	1.5	Mt	210
Feldspar	Mt	1	Manganese[+]	kt	1.5	Sulfur			1000 m^3	35
Iron	Mt	20								

Soil, water and brines

The pattern of parallel soil belts does not apply to Ehiopia. Controlled by the volcanic Highlands, with their rich red and black soils, an irregular zone of chestnut soils extends along the eastern slopes as far as the Rift bordered by light steppe soils of Ogaden (E).

Deforestation is the century's greatest ecological disaster : each hour 20 tons of soil are washed away, and 600 km^2 of arable land and 2,000 km^3 of forest destroyed per year. Some 50,000 km^2 - an area bigger than Switzerland - is threatened by loss. Aquifers are irregular in these volcanic areas and in mineralized zones I-II and IV-VI underlain by the Basement. To compound calamities the rains failed in the arid north and solids abound in the groundwater of the northern Rift (III). To continue, there is only little - and hard - water under the 170 m thick limestones of Ogaden (E) and gypsum occurs in the same column (see Somalia). The salt ponds of Mitsiwa and Aseb (N) had a capacity of

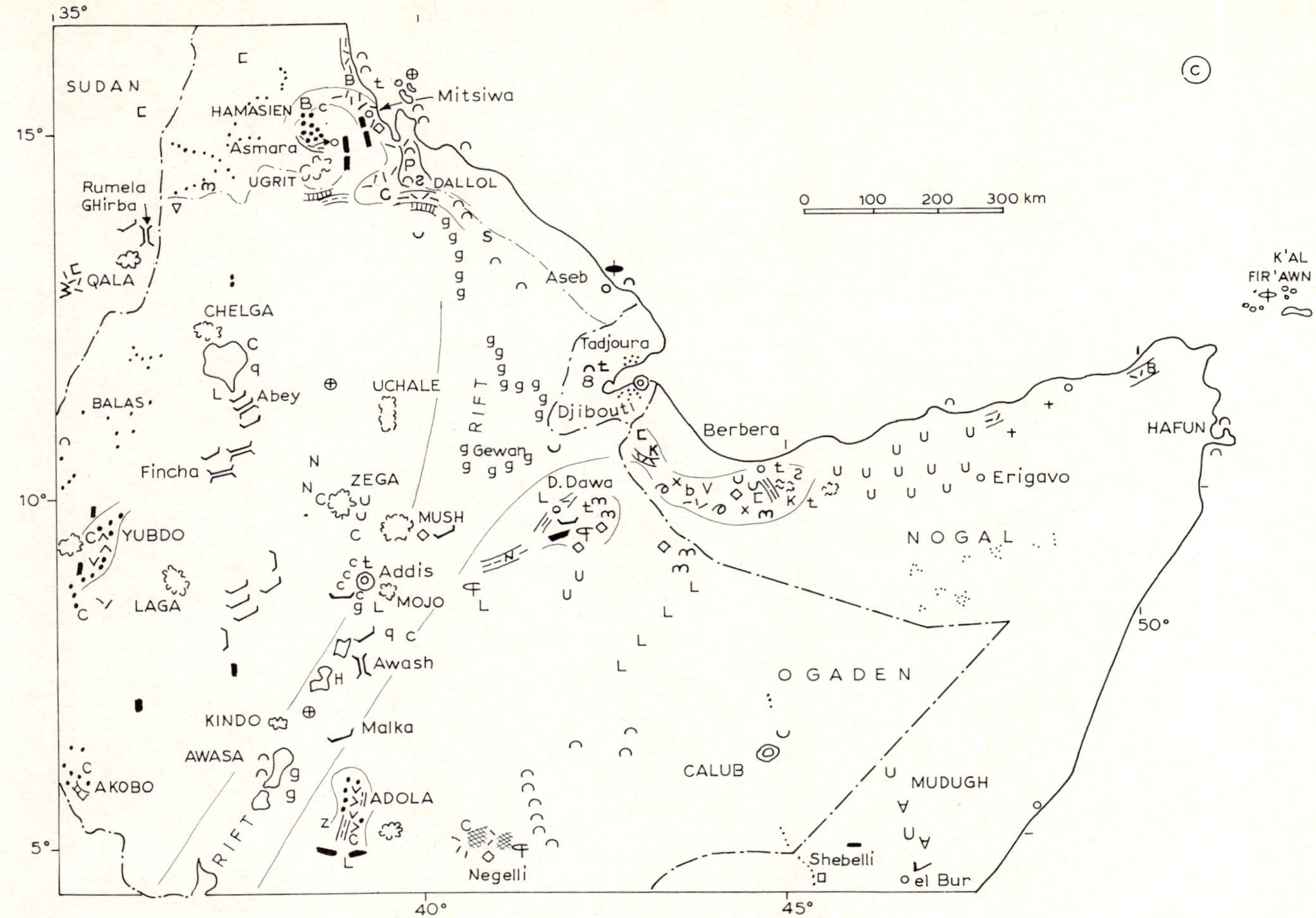

Fig. 18. (c) Ethiopia, Djibouti and Somalia. (Legend on page 13.)

400,000 t, but marine salt recovery rarely exceded 250,000 t. Exports thus declined to 10,000 t. Moreover, Afars recover 10-20,000 t of rock salt in the Salt Valley (NNE) and around the Dahlak Islands several hundred meters of salt were drilled. Finally, in Sidamo (S) the Mega Magado and el Sod lakes contain both salts and soda ash. At Dallol, the center of the Afar Rift (NE), the 1,000 m thick column of evaporites includes gypsum, carnallite and sylvite (40%), the potash minerals; workable grades are 50 m under the surface in a 0.3-8 m thick layer. Compositions are :

		KCl	$MgCl_2$	$MgSO_4$	Br	F_2Cl_6	$CaCl_2$	$CaSO_4$
Brines	g/l	54	480	3	2	5		
Black Mountain	%	81	3.6				0.07	0.03

In 1945 50,000 t of potash were recovered near Mersa Fatu and a 1.5 Mty project has been mooted. Similarly, Dofan has been worked for sulfur and there is also barite, iron ore, manganese and traces of Ni, Co, Cr, Cu.

Energy

Commercial consumption is one of the lowest in the world. Coal could replace some oil imports and with lighter crude, petroleum usage will hopefully stay in the 6 Mb range, i.e. within the capacity of the Aseb refinery. Nonetheless power requirements may excede by 1990 350 Mw and 1.2-1.5 Twh - Out of a potential of 53 Twh most of these are met by a) Tis Abey and the Lake Tana outlet (NW), b) and Fincha (WNW), Awash (SSE) and Abu Samuel (CS), plus c) smaller plants at Debre Markos (CE) - Birhan, Dire Dawa (E) and Jiwa. The diversion of the Amarti will provide another 170 Gwh at Fincha and Soviet sponsored Malka Wakana should be on stream prior to 1990 to be followed, hopefully by Gilgel-Gibe (C). Other sites, including location around the capital, are shown below :

Ethiopian hydropower sites

	direction	Mw	Gwh	Site	direction	Mw	Gwh
Chemoga Yeda	NW	310	1400	Malka Wakana	S	150	560
Alaltu	N	770	1400	Upper Beles	Tana	240	1100
Halele Werabesa	WSW	350	1400	Gibe Daka	WSW	300	730

The search for petroleum continues notably in the Gambala spur (W) and gas exploration will be pursued a) between the Dahlak Islands where oil seeps and the Sudan (NNE) where Mobil C-1 blew out on a salt dome and b) by Soviet specialists around their Shilabo and Hilala oil show and Tenneco's Calub 1 (SE). An 800 or 550 km long pipeline would be required to use this gas at Adis' or Dire Dawa.

Steam and gas escaped from more than 500 hot springs along a 1300 km long arc which connects Mitsiwa (N) with Lake Turkana in Kenya. The EEC and UNDP funded surveys of Corbetti near Awasha, Shalla and Abaya and at the Aluto caldera near

Langano. However, 320° steam was only reached at 2,500 m. In the nineties, the boiling springs of Tendaho (NE) might also be tapped for irrigation ; they extend in fact towards Gewana (ENE). Although the salt saturated springs of Dallol (N) are only 110°C hot, further south fumaroles escape at 220°. The hot water aquifers of Addis which might be tapped for space heating are more accessible though.

Since fuelwood becomes ever scarcer lignites acquire importance when they crop out in five regions :

I Adi Ugri (N) 6-10,000 t, 1100 cal/g. II Chelga (NNW) 70-150,000 m^3, 3000 - 4500 cal/g. III The 200 km long Rift between Weldiya and Debre Birhan (CNE) including Wahi Tita 4000-5600 cal/g, Mush 300,000 t, 3800-4000 cal/g, Guonto, Tiannu Zega Wadeb. IV Soddu (S). V The Alaltu 2700-5400 cal/g Didessa (3300-4400 cal/g) belt including Jars, Nejo 3300 cal/g, Kalati 5200 cal/g and Badessa.

Industrial minerals

Although limestone, travertine and marble are widespread, building stones are mainly quarried near the capital, where most cement is also manufactured. Plants still use oil, and mostly the dry process. While production declined during the last years, rated capacity increased to : Mugher (C) 300-,Addis Ababa 70-, Dire Dawa (E) 38- and Gurgussum (N) near Mitsiwa 60 kt...

Kaolin has been exported from Teraumi, Adi Hausha, Zolot, Bet Meka, etc. on the Asmara plateau (N), but the Addis Ababa area and Haghere Hiwot (CS) are also major sources of silica. Brickmakers also recovered sand at Mojo and in the Chanchan (C).

Barite resources of the Wadis Henet, Meter and Gharsa at Mersa Gulbub (NNE) are extensive ; so are grey asbestos seams at Fawley and Marwa (N). There is feldspar at Bomber (E) and 10 t of cut plus 40 t of raw mica were exported between 1939 and 41 from Shebelli and Tuluhora (N).

Other resources include :

Asbestos : Dubcha, Shamege, Gara, Jabbe (N, with vermiculite), Jatteito, Dubot ; bentonite : Chelga (N), Harrar (E) ; beryl : Chembi (S) ; building materials : diorite : Erythrea (N), Ogaden (E) ; granite : Bizen (N), Jijiga (E) ; 'marble': Agametta (N), Dabus (WNW) ; diatomite : Shebelli, Lake Tabo ; mica and feldspar in Erythrea : C Shilicki, W Amiem, Shiankos, Keru ; Marechi, Aghemeriam (W), Babile (E) ; garnet : Jijiga (E-, Ghedem (N) ; graphite : Adola, Harrar (E) ; gypsum : Dawale (E), Aïsha (E), Belet Won, Mustahil, Gobredarre (E), Mitsiwa (NE), Lake Turkana (S) ; limestone : Labbu (5 Mt), Zegu Wodel, Wolliso , Mojo-Nazareth (EC), Antalo (N), Achim Mt (E), Neghele ; magnesite : Kunni (E) ; peridot : Kod Ali island ; pumice : Ada, Mojo-Awash (E), Bishoftu, Langano, Lake Awasa ; talc (S) : 35 km NW and 50 km SW of Neghele, Ula Ulo, Aghemariam.

Metals

For centuries Ethiopians worked gold. Almost 2 tons were reported in 1959, but recorded production averaged 700 kg prior to its decline in the eighties. Out of this amount the Hamasien gold field (N) and smaller occurrences in western

Erythrea may have contributed 150 kgy. Area V trends south to Beni Shangul and Tullu Kapi (W), where workers recovered a large mass, and points to Akobo (SW). The Adola gold field around Kebre Meriam (S) is, however, more important. Primitive workings of Neghele, Dilla, Yavello, etc. have been supplanted by draglines for a while and new washing plants will be installed one day.

Some gold has also been washed at Yubdo (W) and 2 t platinum have been recovered here and in the Birbir between 1926 and 1965 ; the laterites graded 0.2 g/m^3 and the gravels 0.02-0.06 g/m^3. There was also free iridosmine and the composition of PGM was unusual : Pt/Ir/Rh/Os/Pd = 920 : 50 : 15 : 4 : 1.5. Although nickel, chrome and cobalt occur here too, as well as at the confluence of the Shamege and Ansebe (NNW), the principal nickel deposits are in the south at Tulla, 6,600 t, grade 1.12% to a depth of 6.6 m, Ula-Ulo 4,000 t, 1.33%,Kilta 2,000 t, 1.5% and Dubicha Mika 600 t Ni. Marma covers 0.5 km ; Dubicha Gudda, Monissa and Lolutu assay 0.5-1.2%.
All in all, 10.5 Mt ore average 1.37% Ni and 6.5 Mt 0.5%.

Cobalt, a normal accompanying element assays 0.01% at Ula-Ulo. Chromite occurs here too and along the Barka and Setit (NNW) . Other occurrences include: Ni-Co-Cr : Chabessa (S), Kenticha (S), Budusse (S), (N) : Shawage, Harab, Supit; Cu-Pb-Co : Soka (E) ; cobalt : Kumi (E) ; Au-Cu : Ugaro (N), Adi Rassi, Adi Heza (N) ; Bomo Asosa, Tulu Kapi, Ondonot (W), Akobo (SW), Aghemariam (S) ; copper : Galetti, Katta (E) ; lead : Affratu (E), Gara Ua (E) ; molybdenum : Bissidimo (E), Chiltu (E), Ujima (S) ; tungsten : Sacco bridge (W).

A Japanese group shipped 2,000 t from Debarwa, near Asmara (N) and reported 2.5 Mt ore assaying 3-12% copper, plus lead and zinc. The principal deposit, Adi Rassi (Nefas) contains e.g. 200,000 t metal, grade 1% plus 0.5 g/t gold. Copper also shows in Dankalia, west of the Salt Valley (NNE).

A company exported 1,500 t from Maglalla (NNE) from 1959 on and manganese also accompanies 40-56% iron in 20 Mt of ore at Hamasien (N). Other occurrences extend north to Ghedem and Falket, south into Tigre and include the Kunni-Galetti (E), Tulu Bollale - Gordona (E) and Ghimira-Melka Sedi (SW) trends. Several mines have been worked and a 10,000 ty foundry has even operated at Addis.

24 DJIBOUTI

Bunkering and the foreign legion brought such prosperity that the GNP was higher than that of Ethiopia. The Djibouti franc ranks moreover stronger than the rand and inflation is 0.1% per year... The famous metal sculpture of a palm tree meant to show how arid the land was. Nonetheless, drains collect abundant-but hot and brackish - groundwater and on the north shore the Tadjoura plant fills 4 M 1 1/2 liter bottles of mineral water. There is sweet water in its hinterland at Arsa Gayla and at Lahassa, Illisola and Soubali in the northeast.

The geothermal potential is also excellent since reservoirs contain steam heated to 250° at a depth of 2,000 m. By 1987 a 20 Mw plant will provide two thirds of energy needs in the Hanlé Gaggade plain. Other reservoirs surround Manda Inkir and lakes Abbé and Assal (C) where brines provide salt.

ASED finances the coalfired 60,000 ty Ali Sabieh (SW) cement kiln and exploration of the hinterland continues.

25 SOMALIA

The Jumhuriyat al-Somal el-Dimocratiya is a member of the Arab league and Saudi Arabia is its principal market.

> GOALS : To provide water for the cities and nomads ; to continue exploration for oil and exportable commodities. Gas and hydropower development.

After Britons, Italians, and Holmes's study, UNDP and Brazilian teams assessed the mineral potential controlled by the state.

Table 45 Somali production, potential and resources.

Resources			Resources			Potential			Production		
Gas	Bcf	220	Uranium	kt	10	Hydropower	Mw	200	Hydropower	Mw	5
			Vanadium	kt	4.5	Petroleum	Mb	3.5	Power	Mw	50
			Iron	Mt	60	Cement, plan	kt	200	Salt	kt	30
			Gypsum	Mt	20				Meerschaum	t	10
Past production											
Guano	t	1									
Beryl	t	16									

Water, soil, food

As usual, groundwater of medium quality is found in the shallow gravels of the north coast. Aquifers are not continuous in the northern mountains either, yet wells dot the Dire Dawa - Upper Sheikh axis. At Kaisheloh and Daba Dalol, water is at 2 m depth, and there are thermal springs at Taleh (NE). In the rest of the Shol highlands and the gypsum beds (q.v.) the level is probably at depth, but the Hawd formation and Nubian sandstone probably bear water. In the center, artesian water rises from the Banta-Gialalassi sands as far as the western fault. There are, however, few people and (some radioactive) water in the Mudugh plane (C). In the south, among the Bur outcrops (see 'iron') some groundwater is found in the hollows and in the southwest, Nubian sandstones again provide better aquifers. The Shebelli (CS) is the country's 'lifeline' and the Bardera Dam on the Middle Juba supplies water for irrigation, which is of particular interest since only 1.5% of the land area is cultivated.

Somalia has exported 100-1,000 t of high phosphorus guano recovered mainly on Mait Island near Cape Humbeis and workers recover salt near Kismayo (SW) and Zeila (N). Rock salt shows at Heis and probably also at Daga Der and Darrabo

Hills, 40 km from Berbera (N). In 1935-1936, an Italian company evaporated 300,000 tons of salt at Hordio on the Bay of Hafun (E) and, before the British dismantled the plant, experts rated its capacity at 480,000 tons per year. Indeed, the evaporating basins covered 2,200 ha. Unfortunately funds for rehabilitation were not forthcoming and finally, production declined to 2,000 t in the sixties ; it now ranges from 5 to 15,000 t.

Energy

IDA provided credit for the development of the Afgooye gas field near Mogadishu. The sedimentary column is 7,000 m thick and shows are at 4,000 m depth. In the last 30 years 80 wells were drilled and active exploration continues offshore (NE), near Ras Hafun (E) and onshore (SW). Most promising are the Mesozoic shelves and reefs, the Somali embayment and the Nogal uplift (N).

When funds are available, the Iraqsoma Oil Refinery imports oil for 10,000 bd of products, mostly for local consumption. Oilfired Gezira is the major source of electricity with installed capacity of 100 Mw ; the Bardera (SW) dam will contribute 5 Mw to this total. Bituminous shales are intercalated in gypsum. A seam has also been intersected at Las Anod at a depth of 12 m and geologists have found some coal and lignite in the Hedhed Tug near Onkhot, Daban and Biyo Gora. More important, at Mudugh(C), carnotite fills cracks in soils and below the gypsum near the water table at 6 m of depth. These soils cover Miocene Merca marls and sands overlying basalt, but the best uranium hosts are clayey limestones. Reserves attain here 4,700 t in schists ; Soarmico plans a $300 M extraction plant. Moreover north of Saligh surface radioactive anomalies follow a narrow 240 km long belt in a shallow drainage, perhaps depending on tectonic alignments. Finally, vanadium is an accompanying element north of el Bur and at Bur Acaba.

Other resources

Near Berbera (N), CCCF rehabilitates the Suryo Malable plant of Somalia Cement Co. which has rated 200,000 ty of clinker. Even a second plant has been discussed for Bardera (SW). Limestones are also abundant at Ras Antara (N) and el Wegit near Mogadishu. Nearby, at Brava, Shebeli and Bur, they were also used as building stones, but the quartz breccia of the Webi Shebeli is also attractive.

Anhydrite and gypsum accompanied by some limestone and shale constitue a 600 m thick column which extends in a triangle 250 km east and south of Berbera to Erigavo and Las Anod and at Suria Malableh, 15 km from the harbor, a 130 m thick column includes 2 Mt grading over 95% gypsum and less than 2% carbonate, 6 Mt with 90% and 3% carbonate and 8 Mt assaying 85% and 4% carbonate. Finally, gypsum is also found near Gelits (S).

Meerschaum or sepiolite, a more unusual product is exported to Tanzania ; it covers vast areas of el Bur (C) and assays 16% MgO, 12% CaO, 1% Al_2O_3 and 42% SiO_2. In the south, currents deposit black sands between Gobuin and the old mouth of the Juba River a sample of which contains 15% TiO_2 and traces of rare earths and thorium. At Manja Yihin (E), veins average 1% tin. The mineralization extends west to Dagan Kuled-Asileh and downstream south of Bender Ziada, tin shows in gravels. In 1956, workers collected 16 tons of beryl, some columbite-tantalite and monazite in the Henweina Valley, 5 km south of Berbera and other resources of the same district include :

Asbestos, vermiculite, and mica : Dala'as near Lafarug ; barite, copper, fluorspar, lead : Bihendula pass ; copper : east of Hudiso, west of Lalis ; corundum: NW Hargeisa ; graphite : Ala Ule Tug (Baroma) ; lead and barite : Unkah, Kuh, Arargob ; manganese : Adad-Salawel, Bihen Gaha pass ; molybdenite : Dobo (NW) ; sulfur : W of Berbera ; talc : Mora Mtn.

Iron ore shows at Galan Galo (E), at Tomassoh west of Hudiso and near Sheikh (NW). In the Bur range (C), samples average 35% iron, with maxima of 43%, but there are many outcrops in a 30 x 20 km northwest trending area, notably :

a) 25 km northeast of Dinsor on the road to Baidoa, b) 50 km from Dinsoa, c) 50 km east from Bur Galan and d) 25 km south of Bur Gheluai and 50 km east of Jach Brava. Most important is group b) which comprises half a dozen 5 to 15 km long outcrops.

3 THE SAHEL

Few nomads roam in the Sahara ; yet 30 million live between 15° and 9° in the arid Sahel (Fig. 19) which has advanced south by 100 km in 10 years. The dry spell of 1968-73 killed 100,000 or more, but the last drought has already made as many victims. Seventy percent of the population were affected in Mauritania, 40% in Niger, 33% in Chad, 20% in Mali, and 10% in Burkina ; so is much of Senegal. Infant mortality rates are among the highest and there is disease and malnutrition. Although some policies compounded problems, a better use of water resources would help since requirements are expected to grow from the current 30 Bm^3 to more than 50 Bm^3 in the early XXIst century.

31 THE GAMBIA

Mr Hailey, the American novelist traced his ancestry to the banks of the Gambia, a part of the Senegambia confederation and of OMVG which regulates river flow.

GOALS : To provide more irrigation and pure water for the farmers and to encourage further efforts of oil exploration.

Around Batakunku and Sanyang (SW) 5 Mt of beachsand average 10% heavy minerals over a width of 200 m. This concentrate assays 74% ilmenite (containing 3% Cr_2O_3), 4% high iron rutile, 8-4% zircon and 7.5% staurolite. Between 1957 and 59, 60,000 tons were exported and ilmenite/rutile ratios ranged from 15:1 to 50:1. The usual stone and clay are recovered near Banjul and half way upstream. The medium grade Kundum kaolin deposit is located further east where the highest point (50 m above the sea level) also is. It receives 1-1.2 m of rain but the west is dry enough to grow groundnuts. Soils are less acidic here ; they contain more organic matter and can be upgraded with lime and fertilizer.

Sand and clayey sandstones of the Late Continental are the main aquifer fed by the river at depths of 20-200 m (in the east). The renewable supply is 1Bm^3y and the lignitic layers hold 2.7-5.4 Bm^3 water. Specific yields of 5 m^3/h/m of good to medium quality have been recorded inland, but near the coast and delta are 0.2-1 g/l of solids, mainly lime carbonate. Moreover Banjul's water contains 5 mg/l fluor.

OMVG projects of the nineties call for anti-salt barriers in southeastern Senegal which would open up 80,000 ha of land in the Gambia and 60,000 ha in Senegal, including the a) Niokolo-Koba dam at the Simenti confluence, providing 24,000 ha and 5 Mw, b) Kekréti and the Balinghor anti-salt barriers : 50,000 ha

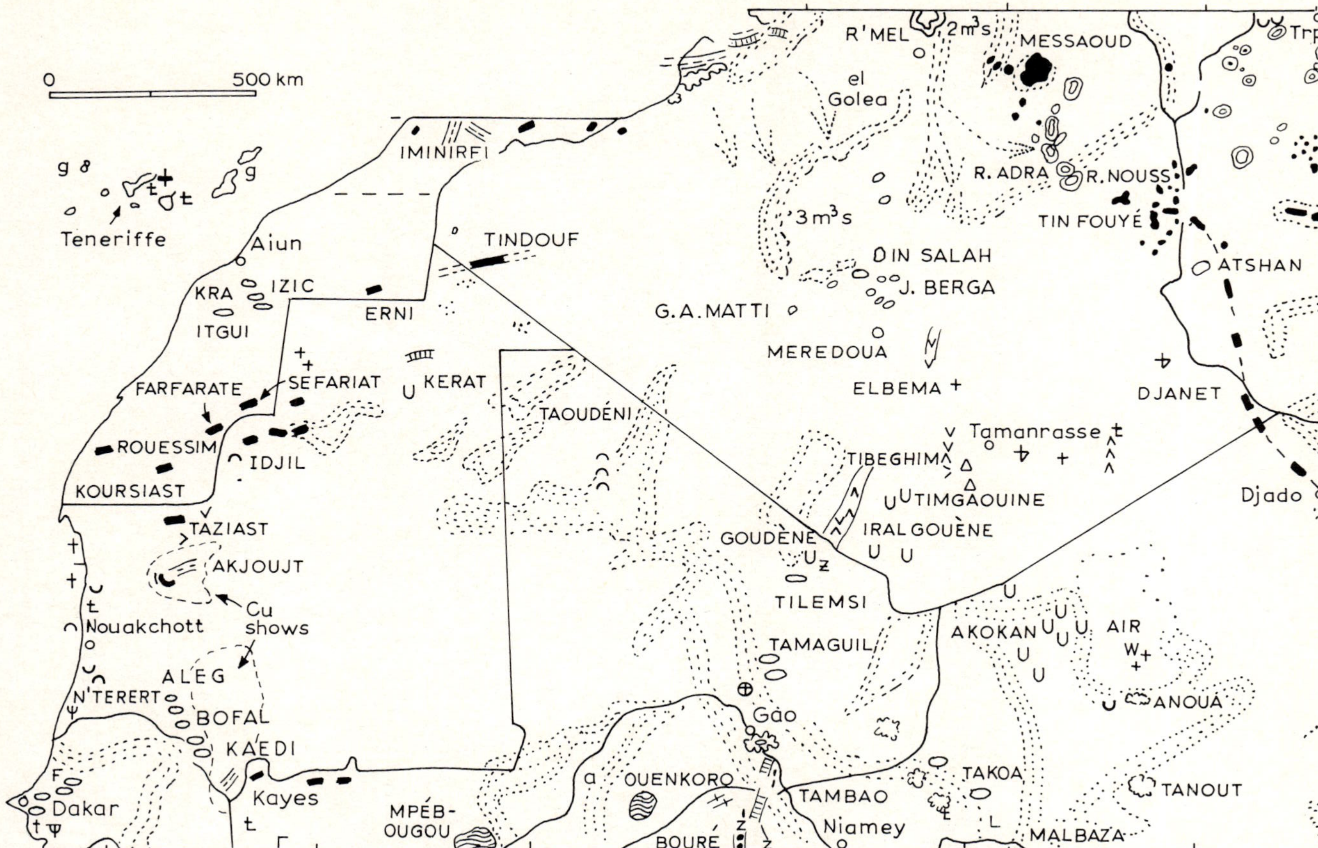

Fig. 19. The Sahara and Sahel. (Legend on page 13.)

and 60-75 Mw power and c) Sambangalou : 30,000 ha and 120 Mw with the Konya and Koukou Foulbé barriers.

To the Maestrichtian lignite clays and the mangrove peats south of Banjul we might add sub-economic oil off Barikama.Unless more is found, therefore, petroleum imports may rise to 600,000 by. Electricity consumption should follow a similar pattern since it increased already threefold to 40 Gwh in the last decade.

32 SENEGAL

One of the most fortunate of the Sahel countries, Sénégal (Fig. 20) has high traditions of learning and three centuries of ties with France : Dakar is the cultural capital of French speaking Africa.

> GOALS : To upgrade phosphates and to build up the fertilizer and cement industries along the coast. To provide better and more water from wells and irrigation and to continue the search for oil, using peat meanwhile ; on a longer term : to extend the rail and work the iron ores of the southeast.

Phosphates represent almost a quarter of GDP but only 15-20% of exports. The state promotes mineral projects as a 50% shareholder of SSPT and CSPT and a guarantor. French agencies and companies such as BRGM, CCCE and Coframines also held shares of CSPT. BRGM and Cogéma were involved in exploration and German Krupp promoted the iron ore project.

Table 46 Senegal reserves, capacity and production.

Production			Reserves			Production P_2O_5	kt
Attapulgite	kt	100	Phosphate	Mt	50	Calcium phosphate	400
Gold P	kg	300	Gold	t	18	Aluminum phosphate	400
Cement	kt	400	Peat	Mt	10	Fosfal etc.	150
Basalt	1000 m^3	170	Gas	Bcf	1.3	Capacity	
Salt	kt	170	Heavy oil	Mb	400		
Petroleum P	Mb	6	Oil	Mb	15	Phosphoric acid	260
Reserves			Hydropower	Gw	4	Sulfuric acid	640
			Gold	t	18	DAP	250
Marble	Mt	1.4	Iron	Mt	400	TSP fertilizer	45
Copper	kt	34	Ilmenite	Mt	5	NPK	90

Most exports reach the EEC, above all France. The U.K., Greece and China also import phosphates and India may become a major outlet for iron ore and fertilizers. More than 100,000 tons of phosphate were used in domestic agriculture and the processing sector was expanding rapidly. Meanwhile a small steel facility has been on the drawing boards for decades. Total employment in mineral based industries may therefore increase from 3,000 to more than 5,000 in the nineties.

Water and soil

There is less than 300 mm of rain in the north, but as much as 1.3 m in the south. People, therefore, move to Casamance (S) and Dakar which receives 600 mm of rain. On this peninsula 50-80 m thick (sub) recent sands also yield 1000-1200 m^3h of water with 440 ppm of salts. Following the Paleocene limestones of Sebitokane, the deep Cretaceous contains more solids than Maestrichtian sands and sandstones which feed many wells at N'Diass. This aquifer continues at a depth 20-50 m north of the Gambia, but dips to 200-500 m along a Longa-Gambia line ; it yields under partial pressure 15-120 m^3h and 150-200 m^3h under normal flow. Unfortunately, solids increase from 550 ppm in the interior to 2500 ppm in the northwest.

The Late Continental is 10-150 m thick at Tambacounda ; it provides the best yields and qualities at Ferlo, the Eocene at Diourbel, and at Nioro (S). Table 47 sums up these resources.

Table 47 Renewable and exploitable water resources in Senegal.

	1000 km^2	thick mm	volume Mm^3y	exploitable Bm^3
Alluvium	4	13	50	0.2 - 0.4
Dunesands	5	41	210	2.5 - 6
Late Continental	105	72	7500	44 - 95
Lutetian limestone	2	16	30	0.9 - 1.9
Maestrichtian	(150)			38 - 80

Most OMVS dams are in Mali (or Guinea, q.v.) but Gourbassi on the Falémé straddles the border. All these would regulate the flow of the Senegal in the nineties, but Manantali is scheduled to be finished in 1988 (see Mali).

We may add the irrigation projects of Kayangé at Niandouba and Anambé in Casamance, since these grass land soils are relatively fertile if some lime and fertilizer are added. Except for the subdesertic north, Senegal was therefore described as a giant groundnut patch.

Phosphates

Plans call for the stripping of 1.5 Mty out of 40-50 Mt rock located under 10 m cover at Kolda and Semmer near Matam (NE) for use by the OMVS countries. Earlier in the sixties, output grew from half a million tons to almost 2 Mt by 1973 out of rock reserves (in Mt) :

Taïba	profitable	certain+probable	possible	P_2O_5%	Overburden/rock m
Keur Mor Falls & N'Domour Diop	20	70	30	31	20:8
Tobène	27	70			35:6
Thiès	Mt	type		P_2O_5%	Others
Lam Lam	20	calcium		28-	Pout 20 km SW
& Pallo	80	aluminum		29	Yen, coast S

The thickness of phosphate beds ranges from 3 to 30 m on the 320 km^2 Thiès plateau and averages 17 m at Pallo. Out of each 1.7 Mt of rock SSPT (50% PUK) extracts through grinding and screening 1.2 Mt aluminum phosphate with a 94% recovery, some of which is used in West Africa, but most is exported to France. In addition to the Lam Lam calcinating kilns, PUK produces in four French harbors phosphal assaying 33% P_2O_5. Part of this is further processed into Polyfos and Baylifos fodders and potassic superphosphal. Finally, SSPT can also supply 50-80,000 t tricalcic phosphate.

On the 240 km^2 Taïba plateau 65% BPL rock of the 30 Mt Keur-Mor Fall Zone is ground, washed, floated and dried to produce a 80-83 BPL (37-38% P_2O_5) concentrate. Recovery grades seem to be low at 30-35% and a 67% BPL product has been added. CSPT (29% Coframines, Cofimer, and CCCE,11% International Minerals and Chemicals Corp.) has an export capacity of 2Mt. The M'bao plant of SIES is rated at 20,000 t phosphoric acid and IDA has backed the recovery of 650,000 ty phosphate from the tailings by ICS, whose shareholders include the state 23%, Indian groups (the major customers) 19%, IDB 9%, SCPA 9%, CSPT 7% and the Cameroon, Ivory Coast and Nigerian governments. ICS/Senchim's Darou-Khoudoss plant near Taïba therefore started production of fertilizers.

Energy

Exploration continues at Longo and Kafountine in Casamance (S) next to the off-shore Dôme Flore heavy oil and smaller light oil fields. Although recovery ceased at Diam Niadie (W), SAP has already proved some gas, under which may lie oil. Meanwhile, SAR's imports of 6 Mby are bound to rise rapidly. Furthermore, power consumption has almost doubled to 650 Gwh in a decade and might excede a Twh in 1990. Moreover, electricity from Sonelec's Tambacounda plant is expensive. Lignite has been therefore surveyed in the Senegal valley, but the coastal screens of peat have attracted more attention and CTS' first plant is to be installed at Biogo with a 30-50 Mw capacity. Reserves in the Niayes as far as the Senegal delta (NW) are cut up into segments : 30 Mm^3 at Mboro and 15 Mm^3 at Kayar or 10 M dry tons combined (3.7 M toe). Between Kayar and Lampoul 23 Mm^3 could feed for 25 years two 30 Mw plants planned around Diojo. Rainwater could be used between July and October, but otherwise seawater washings, filtration and solid/liquid separation could clear out sand and salt to the 1% limit in order to reach 4000 cal/g.

Other reserves are available at Casamome, the Sine-Saloum River and in the southern delta. Marshy terrain is, of course, the problem in the mangrove swamps.

The distant dams of the OMVG would generate 1 Twh or more (see Gambia). Twelve sites of the OMVS, mostly in Mali (q.v.), could supply 6.5 Twh and Senegal's 40% share of Manantali would average 400 Gwh. Finally, Cogéma has recorded radioactive anomalies in the east a) at Nounomfouka near Gouemidi, 200 ppm U and 15 ppm Th, b) in the Daléma veins, c) at Kondoko and Fodakouda (SE) and d) at Lombel and Walidiala. More important, wavellite in Thiès phosphates assays 0.033% U_3O_8.

Other resources

Basalt is quarried at Kombolé, east of Thiès and limestone is abundant in many other coastal areas. The Rufisque 'dry' cement plant of SOAC uses thus excellent marly rock of Bargny (W) ; capacity has doubled to 750,000 ty and high quality marble was located at Ibel and Bondafossi (SE). Clay is worked at Taki and kaolinite at Pout (WSW) along with Fuller's earth. These deposits extend to Nianing and M'Bodiene ; attapulgite also shows inland at Kaolack (W). Prodimat plans to double capacity to 200,000 ty and SSPT strips 3-15,000 ty as a byproduct at Thiès. In contrast, the recovery of marine salt has not expanded as fast, mainly at SNSS's Siné Saloum basins, near Kaolack (W).

At Djiffère (W), ilmenite, rather an oxidized rutile, contained 57% TiO_2 and zircon ratios at the closed Rufisque plant (W) have ranged from 1:3 to 1:6. Reserves remain though (in 1000 tons) :

Area		grade	blacksand	ilmenite	Coastal stretch
Kayar Lampoul	NW	25%	6000	4000	Dakar-Longo
Mbour-Joal	W		650	400	Nianing-Sangomar 100 km
Niafurang-Diogue	S	15%	600	400	Casamance delta

The Bountoum-Kolin gold deposit (SE) is not far from the 100 km long Karakaéné, Kérékounda, Gébouria trend. Grades average 2.5 g/m^3 and 4.5 g/t and reserves reach 12 t at Sabodala. SMS, a joint venture of the state (41-51%) and of Serem (a BRGM affiliate) mines here the oxidized (soft) zone and the hard rock is to follow in 1988/89 with 800 kgy, but luckily there is little overburden. In the same area, preliminary estimates of 90 t gold in the three Moura veins and 23 t in the Sambarougou placers require further exploration (ECA, 1976).

Hopes of the BRGM-Krupp-state consortium rest on contracts for the sale of 6 Mty of iron ore and the building of a 320 km link to the Mali railway. There is next to no overburden and lumps constitute a greater part of the highgrade hematites. Larger sizes grade 64% Fe and 2% Al_2O_3 and sinter 62.5% Fe and 3% Al_2O_3 with the following reserves (in million tons) over 30 km from south to north :

Hill	Kouedekourou E	C	W	Goto & Farangalia	Safa & Kabélea indic[d]:	Karoudiako Nord	Karakaène N'Di	M'Bah
Proved	120	100	26	30		42	21	32
Probable	41	24	7		10	10	7	8
Magnetite				260				

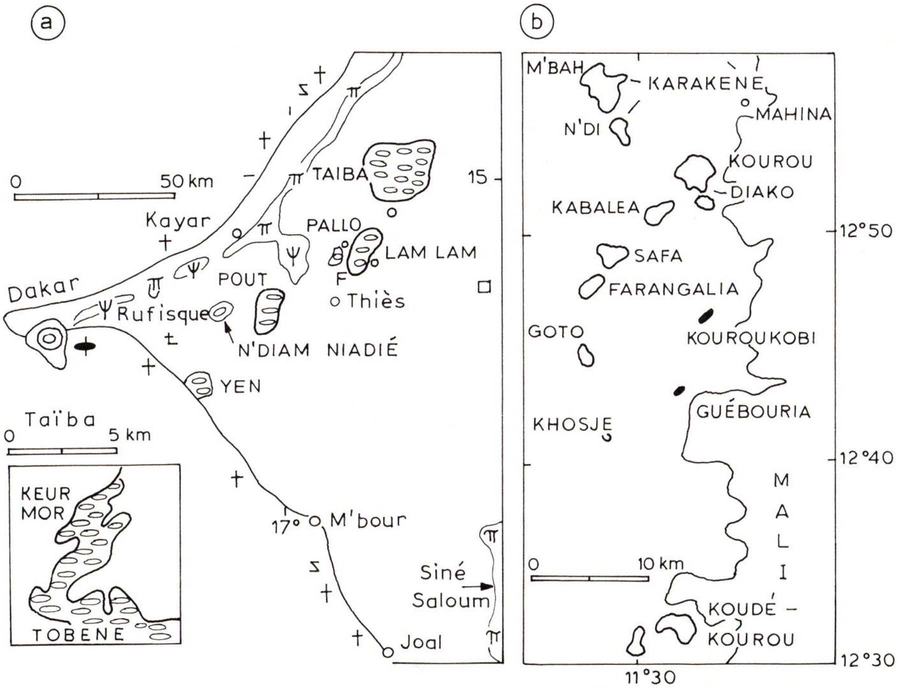

Fig. 20. (a) Western Senegal. (b) The Falémé iron ores. (Legend on page 13.)

Further north, at Gabou (E) on the tip of the Akjoujt copper oval, reserves reach 34,000 Cu, with another 300,000 t of resources. Grades average 1.8 to 2% over 2 and 3 m of thickness and assays in a larger orebody range from 0.8 to 11%. Lastly, other occurrences include in the east:

Copper : Tamborokoto ; gold : copper, lead, molybdenum : Tenkoto ; placer gold : Gambia, Falémé : Kossanto ; graphite : Patassi ; tin, lithium : Saraya (U), tin columbite-tantalite : Diambaloyé, Samékonto, Bafoundou, Falémé with W. Tungsten : Soréto-Saboussiré.

3 3 M A U R I T A N I A

Only on the southern edge of the République Islamique de Mauritanie is there extensive farming and the drought persisting nomads moved to Nouakchott. Luckily though, the ocean teams with fish. Targets, therefore, can be defined thus :

GOALS : Upgrading of iron ores with Arab finance. Meeting of a heavy debt service with agressive marketing. Intensified search for water and oil, working of phosphates, gold and copper and the build-up of the cement, refining and steel sectors.

Table 48 Mauritania production and reserves.

Planned production		resrcs	Commodity		Production		Resvs
Phosphate		kt 20	Cement production	kt 60	Gypsum	kt 4	Bt 4
Gold	t 1.5	t 150	Petroleum capacity	Mb 7	Iron ore	Mt 5	Mt 500
Silver	t 1	t 100			Steel, crude	kt 7	
Copper	kt 25	kt 450	Uranium resources	kt 5	Steel shapes	kt 10	
			Salt resources	Mt 12			

Mineral industries contribute a fifth of GDP and employ 6,000 people to whom should be added transport workers. The state controls all mines, along with Arab capital, but OMRG explores with French, UNDP, IDA and Japanese assistance. France has imported all the copper concentrates and the EEC 80% of the iron ore of which France traditionally took a third to a half, followed by Italy, Belgium (and occasionally the U.K.). Finally, Senegal is developing into a market for steel rods and plaster.

Iron ore

F'dérik-Ségazou, Rouessa and Tazadit hills of the Kedia d'Idjil (NW) rise over 10 to 17 km in the desert. Their reserves of 40, 15 and 90 Mt grading 64% will be depleted by 1992 at Azouazil. The Guelb project was, therefore, inaugurated nearby in 1984 (Fig. 21).

Length, km		Guelbs or hills	Ore Bt
5-25	W	Toumba-Atomaï, Tintekrat el Beïda, bou Derga, el Ajoujt	1
30	ENE	Achouï, Mhirizeb, el Rhein 240 Mt magnetite, 100 Mt oxide, 37-40% Fe (400 Mt waste) concentrable to 250 Mt, Oum Arwagem	0.5
	N	Guelb Lehdid (Iron hill), el Agareb, Kediat Leghnem, Tezerkef	vast
Total		125 km^2 Tiris triangle, centered on el Rhein quartzite + ore	6

At el Rhein 80 ton trucks feed 30 cm crushers, two aerofall grinders (800-1800 th, which use most of the 56 Mw capacity) screens and magnetic separators which produce 65.7% Fe, 1,600 μ GMAS and 65% Fe 300 μ GFM. Water is piped over 60 km, Zouriat town has been expanded and deepening of Nouadhibou freeport facilitates the loading of 2.6 Mty of oxidized sinterfeed, 2.2 Mty of fine magnetite and 1.2 Mty of magnetic sinterfeed produced from 14 Mty of low grade ore. At Oum Arwagem, 66 Mty will be removed from 1990 on in order to produce 33 Mt ore and 15 Mt concentrates.

Following nationalization, SNIM holds 71% of the shares of Cominor ; Arab, Kuwaiti, Iraqi and Moroccan Armico, Kftcic, FED and BRPM another 27% and almost half of the $340 M financing came from Arab, Saudi, Kuwaiti and Abu Dhabi Afesd, SFD, Kfed, Adfaed with the World Bank, French funds and banks, the EEC providing the rest. Further, Safa, a joint venture of SNIM and Kfaed will operate the 2 Mty Nouadhibou pelletizing plant.

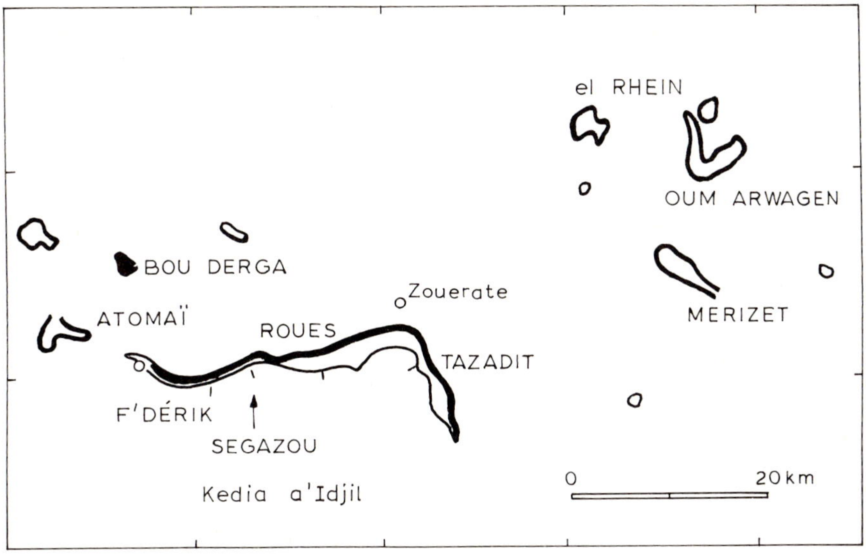

Fig. 21. The iron mountains of Mauritania.

Later on, Tasiast (W), 50 km from the rails, might provide 30 Mty, markets permitting, according to Socomine (BRGM) out of 6 Bt of low grade fine, magnetic quartzite. Lebthenia alone could supply 5 Mty pelletfeed. Lastly, the Legleitat el-Kader includes 15 Mt iron ore assaying 52-55% Fe and 1% P, and there are another 7 Mt grading 68% Fe in the Akjoujt copper deposit.

Water, soil and fertilizer

Three hydrological areas can be distinguished : the coast, the Senegal river basin, the underpopulated granites and metamorphics of the Reguibat and the Taoudéni basin, partly in Mali in the north. The coastal basin has up to 3 aquifers: a) the phreatic and subphreatic water tables in clays and sand with little salt, b) intermediary Eocene sandstones and limestones and c) the lower Maestrichtian sandstone, locally a prime source of water, yet Nouakchott is fed by pipe from Idini, a distance of 62 km.

The Late Continental has the best yields (10 to 60 m^3h), of probably fossile water with some supply from the Senegal river, but in the Trarza near Nouakchott, Eocene limestones only yield 0.1-1 m^3h. As usual, the Intermediate Continental also provides a good aquifer and in the south well known watering points include Moudjeria, Ntakat, Dris-Lakdar and Amaj (SE). The country has 5-10,000 wells out of which 1,000 to 1,500 are concrete faced, with two thirds in good condition. Greatest demands are made on the coast, but the Trarza (SW) is well equipped with wells. So is Bassikounou (SE corner). The greatest number of wells are required along the Senegal (SW) : 1,000, Goudimaka (SE) 400 immediately to the north in the Aftout and the Hodh (SE) : 530. From the Basement of the Reguibat (N) water rises, e.g. at Chegga and el Mzereb. Nonetheless, most of Mauritania is covered by sand, except for some weak soils in the east and along the Senegal where phosphates, discovered by R. Chudeau in 1910 extend 330 km from the sea :

I Over 60 km^2 Tamoust, Beïra 1.50-2 m thick beds at 45-50 m depth

II 30 km SES 70 Mt rock at Bofal west of the oued Guellour and
24 Mt rock at Loubboïra east of the wadi

III To the southeast at Niabina, Foundu, Oulom and at Makhana and Ferrala.

A syndicate of BRGM, SNIM, SSPT and Rumanian Geomin hopes to work II and produced pilot concentrates. Characteristics are :

	o.b./pay	thick,m	P_2O_5%	CaO	SiO_2	Fe + Alox%	U ppm	As,Cd
Loubboïra	4.6	2	19	27	37	6.9		
Bofal	7.4	1.73	20	30	36	5.6	90	20
Bofal concentrate			36	51.5	3.3	1.3		12

Other products

Power supply increased only 20% in ten years to reach 110 Gwh and the 23,000 bd Nouadhibou refinery uses light Algerian crude, pending oil exploration results on the coast, in the Senegal valley and the Taoudéni (N). In the Kerat (N), the 0.4-0.5% uranium shows attracted Iraqi attention. Manganese occurs nearby and peat bogs are at Rosso (S).

In 1981 SNIM's started to produce cement at n'Drahamcha north of Nouakchott and the plant has a 60,000 ty capacity. Between 1973 and 1981 10-15,000 ty of gypsum have been exported from the same sebkha to the Senegalese Rufisque cement plant. Accordingly, Sauria, the Kuwaiti joint venture, planned a new plaster facility with a target of 120,000 ty after the completion of the deepwater harbor. Conveniently, limestone underlies beds grading 85-92% gypsum and there is only 40% overburden. Nearby, in the Trarza (W) sebket surround the Aftout between Toumbas and Nouakchott. Compact salt is being recovered at Twidermi and from eight layers at N'Terert (SW) from January to July. Even so, reserves of 150,000 t are smaller than in the Idjil sebkha (WNW, 11 Mt) where salt and clay alternate though.

Ilmenite and zircon are disseminated along the central coast at Gendert, Lemsid, el Mahara and on Arguin Island. Further north, south of Cape Timiris, 6 Mt of sand contain 2.5-5% heavy minerals.

Hematites of Idjil assay 2 g/t gold and Samin (37% SNIM, rest Armico) reopened the Guelb Moghrein near Akjoujt (W) following losses and the exhaustion of oxides. Since the remaining sulfides require the delicate Torco process, the reserve base has been scaled down to 40,000 t copper and UNIDO financed a concentrator for the dumps :

	Resources	Cu kt %	Au t g/t
Peak production 1973 1.6 t gold, 1 t silver, 22 kt copper Target 19 Mt, 1.2 g/t gold, 1.9% copper, 100 kt 25% copper concentrate and 2.5 Mt tailings assaying 4-5 g/t gold.	Sulfide Mixed	450 1.8 180 2.9	150 0.9 30 3

On the Guelb el-Foulet (W), between the guelbs el-Azib and el-Heirich thin lenses average 20-35% Cr_2O_3 to a depth of 20 m. Other indices include : Beryl-lithium (NW) ; Co, Ni, Bi, W : Guelb Moghrein ; garnet : Cape Timiris ; chrome : Amsaga (NW) ; copper : Inchiri-Idjibiten, Tamarkort, Brakna, Goudimaka (Mauritanides, S), Tasiast (NW, Fe), Ansaga, Taoudéni (Reguibat axis, NNW) ; tungsten Legleitat el bird, Tabrinkort with Cu, 40 km E of Akjoujt ; vanadium : Akjoujt ; yttrium (xenotime) Bou Naza.

34 MALI

The government pursues a positive neutralist policy accepting aid from France and the U.S.S.R. whose geologists discovered a gold deposit, but the bauxite potential of landlocked Mali is compromised by inadequate infrastructure.

GOALS : Rehabilitation of Niger River irrigation and water supply in the south. Further harnessing of hydropower. Expansion of gold mining and detailed exploration for fuel and light minerals.

Table 49 Mali resources, production and projects[p].

Resources			Resources			Production			Reserves	
Lignite	Mt	55	Bituminous shale	Mt	800	Hydropower	Gwh	120	Gw	3.5
Lithium	kt	10	Bauxite	Bt	1.5	Gold	kg	400 → 1800[p]	t	50
Copper/Zinc	kt	15	Iron	Mt	380	Silver	kg	100 → 500[p]	t	4
			Manganese	Mt	3	Phosphate	kt	10	Mt	10
Marble production	m^2	500				Salt	t	5	Mt	50
Clinker produc.	kt	20								

Mineral products go to neighboring countries, notably Senegal and the Ivory Coast, and perhaps a million barrel of fuels are imported, requiring a heavy debt service. Quarry materials are worked by individuals and small groups. As usual, the state controls power supply, the cement plant, and the gold mine, the latter with Soviet assistance. Mineral exploration is carried out by French, and also UN and Japanese groups, and the international oil companies prospect for oil.

Water, soil and fertilizer

Less than one percent of the country is cultivated : the shallow clayey spots of the west are unsuitable for farming which is limited to the dry grassland soils of the peduncle (W), the Niger silts and the Gourma (SE), which require lime and fertilizer. At Tamaguil (E) 20 Mt of rock assay 27% P_2O_5 and in the Tilemsi valley (ENE) 5-10 Mt grade 26-34%. In 1962, 1000 t of phosphate were produced, but with German aid for crushing plant at Bourem Sonarem increased capacity to 120,000 t, with an eventual target of 240,000 to be used in farming the bend of the Niger.

The Sebkha of Tegazza is located in the far north and nomads recover between June and October 5,000 t 99% pure salt from the Taoudéni (N), an increase of 60% since 1962 - capacity is 20,000 t. There is also low quality water in this 'empty' triangle (N) at 10 m depth, but more rises at Tessalit (NE). There are wells at Tabankort and in Tallah, but the Kidal area NNE of Gao is particularly difficult. In contrast, there are 3 aquifers in the south : I The 1.5 km thick Intermediate Continental (W) with relatively little water particularly in the Gourma(SSE) at 2-400 m depth, II The 400m thick Upper Cretaceous and III The 1.5-2 km thick Late Continental out of which 100 m are porous.

The mean inflow of the Niger at Koulikoro with the Bani at Douna equals 67 Bm^3 y, but the outflow is only 36 Bm^3y. The Mali straits (SE) are thus fed by the internal delta and then drained by the lower Niger, although floods swell sometimes upstream to Mopti. To be sure, the Late Continental, mostly covered by 100 m of dunes also yields here several m^3/d. Moreover, waters of the Niger penetrate as far as 180 km from the river at Faguibine Lake west of Timbouctou, 80 km at Timbouctou, 30-50 km between Gourma and Tossaye and 15 km at Gao. The usual problems arise in the Basement covered peduncle and in the Paleozoic limestones. Luckily requirements in the longer southern part of this area are only 15 l/d/man, i.e. half of what nomads need. Yet, on the average a well is required per 200 inhabitants, i.e. 30,000 units by 1990.

Energy

Petroleum imports have increased strongly and may reach a million barrels in the nineties. Similarly, electricity generation increased 50% in ten years to 120 Gwh, mainly in the Bamako area where EdM serves 50,000 customers. With the Dialakoro dam on the Sanzarani, half of the power supply is already hydraulic and Sélingué has an ultimate potential of 45 Mw, with a 2 Bm^3 retention. This would also double Sotaba's potential to 10 Mw with a yearly total of 400 Gwh. Other sites include on the Niger Tossay (2 Bm^3) and Labezanga. Moreover, a power station can be added to the first OMVS site : Manantali. The dam is scheduled to operate in 1988 and regulate the flow of the Senegal at Bakel to 300m^{3}sec and irrigate 450,000 ha. These multipurpose projects would permit 2/3 crops in areas of inadequate food supply (Table 50)and also provide power :

Table 50 OMVS Hydropower projects in Mali.

Site	River	Capacity Mw aver.	firm	Bafing sites (SW)	Capacity Mw aver.	firm
Moussala	Falémé	30	20	Bindougou	50	33
Gourbassi		20	13	Manantali	240	200
				Galougou	280	190
Bandofora	Bakoye	20	20	Petit Gouina	100	70
Badoumbe		70	50	Félou	75	50
Marela	Baoule	24	16			

Meanwhile, oil exploration continues at Gao, Maéna and Arouane, with drilling at Yarba in the Taoudéni (N). At Agamar near Bourem (E), half of the 50-70 m thick bituminous shales heat 1,000 cal/g. There is also lignite at some depth at Tahabant (EC) and Menaka (E).

In the extension of the Algerian trend, Japanese PNC met success in the uranium exploration of the Adrar of the Iforas. Barakoyé, the Taoudéni (N), Tessalit (NE), Tamat (C), Guéret, Hobori-Douentza and Kéniéba (S) were other targets.

Gold and building materials

A few diamonds were found in the Doundé at Sansanto (SW) and at Guindissou not far from the Sekonomala, Toromaya and Kobato pipes and the search for diamonds is on at Liptako and in the Gourma (SSE). In the early nineties 20,000 people washed and dug gold on the Falémé (W) and around Medinandi. In 1963, aeromagnetic surveys located Kalana (SW, Fig.22) in the extension of the Seguiri trend. of Guinea under the laterite cover and by 1977 Soviet geologists contoured four orebodies with 30 t gold in assays of 28-34 g/t to 50 m depth plus 7 t in a dozen bodies of various grades. All consist of quartz veinlets and lenses within complex shapes. Orebody 1 is 0.5-5 m wide, Nrs 2 and 3 lie 10 and 50 m higher and Nr 4 assays 38 g/t ; it might contain 4.5 t gold. The gold/silver ratio is 7.7, arsenic 0.7% ; some copper, zinc, and bismuth accompany. There are some 30 other occurrences within 20 km including Kodieran (Fig. 22), Kalanako, Kalako, Kolenda, Balé and Salamanina, but the quartz veins of Siekorola assay only 0.2 - 0.7 g/t.

Following 20 years of development, mining of 65,000 ty of ore should supply 1.8 ty of gold and 470 kg silver. The USSR loaned Sogemak $10 M at an interest of 10% and a 120 km long power line has been drawn from Sélingué to Kalana, which is now surrounded by claims of Sfmrm. This joint venture of BRGM and Sonarem has located 12-30 t of gold at Loulo.

The Soviet built Diamon plant (W) uses local limestone. If credits are forth-

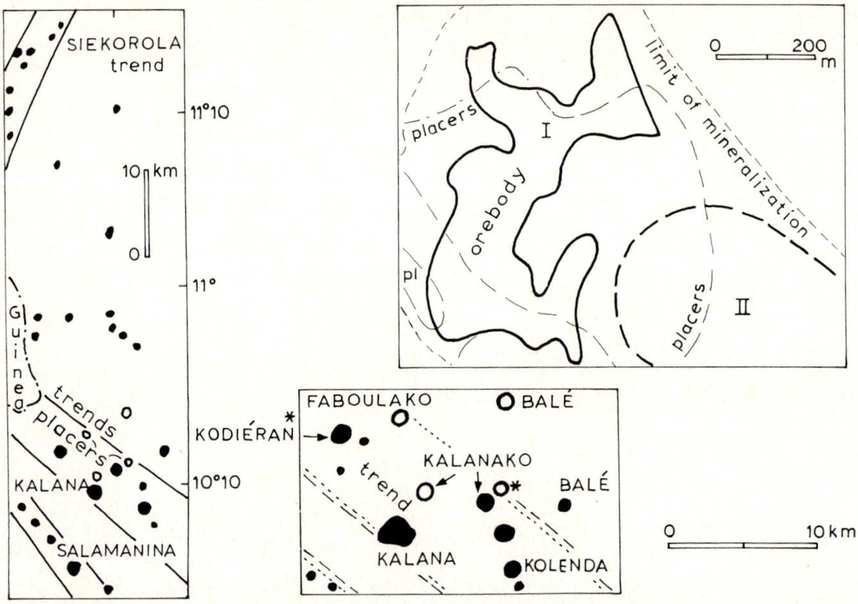

Fig. 22. The Kalana-Balé gold field (Mali).

coming, its capacity might treble to 150,000 t. A cement facility of 260,000 t has also been discussed for Gangouteri and a grinding plant for Tyenfala (SW). Some limestone is used by the building trades but marble production at Bafoulabé, nearby, has ranged from 1,000 ty to 500 m^3 ; 5-10,000 m^3 of granite were also quarried. So was clay at Diamou and ceramic kaolin at Sankaréné and Yanfolila (SSW).

Untapped resources

Bauxite reserves of 550 Mt grade 39-44% Al_2O_3 and resources attain 1.1-1.8 Bt mostly between Kéniéba and Bamako, on the Mandingue table land where PUK was active. Seven main areas can be distinguished over 1,000 km stretching east of the Guinea border :
I Sangalan and Fantofa (45% Al_2O_3, Satadougou county) extending from the Fello-Kéréwane plateau and Tembéré to the border, II Kéniéba-65 km west of Bamako 165 Mt, 45% Al_2O_3, 4% SiO_2, III South of Kéniéba 50 Mt 45% Al_2O_3, 2% SiO_2, IV Baléa west of Bamako, south of Kita, 330 Mt, 46% Al_2O_3, 2% SiO_2, V Koulouba near Bamako, VI M'Pébougou (SC, Ségou district) assays 50-60% Al_2O_3, 1.5-2% SiO_2, VII Ouenkoro, south on the Bandiagara scarp (SC).

Iron ore has to face the same problems in the peduncle (W) :
I Djidian-Kéniéba on the Falémé (SW), 2 Mt 65% Fe ; further east :

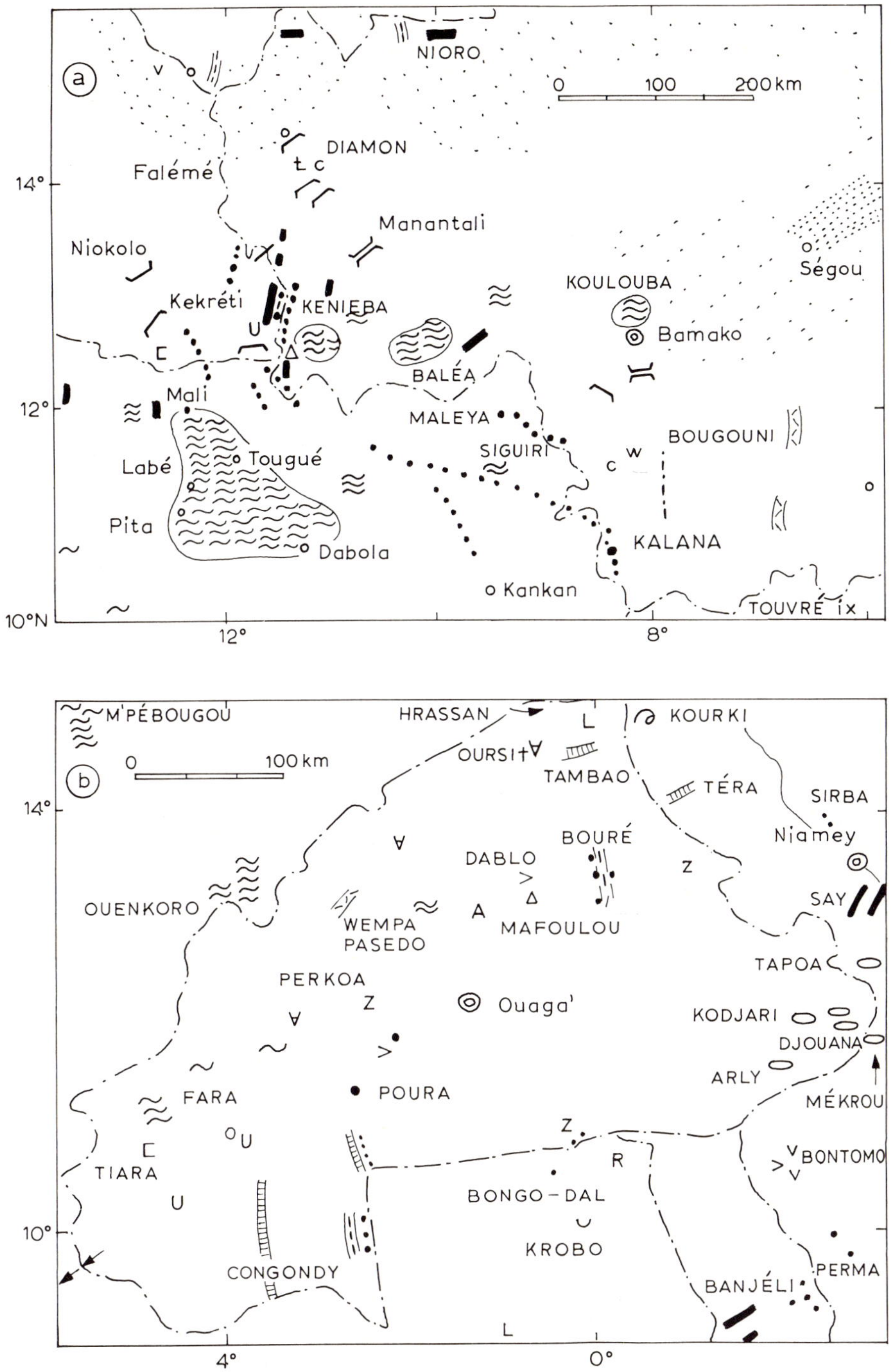

Fig. 22. (a) Western Mali, Upper Guinea and eastern Senegal. (b) Burkina and its neighbors. (Legend on page 13.)

II Makana east of Kita (Bafing valley) 150 Mt hematite 36-37% Fe + 500 Mt resources

III Kita 36 Mt ore 57% Fe, 25 Mt 56% and 31 Mt 55% Fe.

IV Gabé 60 Mt ore 52% Fe, 12% SiO_2

V Nioro (N peduncle) 10 Mt ore 63% Fe.

Based on Balé iron ores and Selingué power, Krupp studied for Somasid the feasibility of a 30,000 ty steel mill. Furthermore, Tikanasite Hill and Ofalibéné (E) near Ansongo hold 5 Mt ore grading 40% manganese and 3 Mt assaying 30%. At Tessalit (NNE) 150-200,000 t ore assay 7-10% zinc along with some lead. The Sinsinkourou pegmatite (SW) holds 100,000 t of spodumene (6.7% lithia) with three others containing 60,000 t near Bougouni (SW). Finally, other show are : Basemetals, tungsten: Balé placers (SW), (Bougouni-Sikasso), Kangaba, Southeast ; copper : Nioro (SW, Sirakoro, Lambatara), In Darset (ENE), south of Ansongo (SE); lead : Kayes (W) ; uranium, manganese : Agaoula, Tondibi Hill, Idelbel (Rarakoya), Tessalit, in Tamat (C), Gourma, Chag Gueret, Taoud, Hobori-Douentza ; thorium and zircon Tadhala ; tin and coltan : Movinio (Iforas, see U).

35 BURKINA FASO

The 120 km long Abidjan railway reaches Ouagadougou, the headquarters of CEAO. Plans call for a spur of 340 km to the Tambao manganese show (N). Gold and zinc require less infrastructure though. Meanwhile, remittances of half a million Mossi and Bobo, mainly from the Ivory Coast, support the budget.

GOALS : To urgently explore and develop groundwater resources. Secondly, to organize the working of domestically useful and thirdly, exportable minerals.

Table 51 Burkinabè resources.

Bauxite	Mt	7	Phosphate	Mt	20	Hydropower	Gw	12
Dolomite	Mt	16	Nickel	kt	100	Gold	t	30
Limestone	Mt	56	Cobalt	kt	15	Silver	t	100
Marble	km^3	60	Vanadium	kt	500	Zinc	Mt	1
Kaolin	Mt	2	Manganese	Mt	8	Copper	kt	320

Water, food, power, building materials and metals

There is less than 400 mm rain in normal years, but occasionally in the west rainfall is 1 m. Sandstones of Bobo Dioulasso therefore yield 90 m^3h. Along the northern border 11,000 km^2 of the Late Continental and the Niger basin provide an aquifer at a depth of 10-80 m. Natural recharge is here 430 Mm^3y, but 1-3 Mm^3y are mined. Most of the country is covered though by the Basement which totals only 3-4 Bm^3y. Weathering and fractured zones only yield here 0.2-2 m^3h in the dry season and rarely 25 m^3h in the wet period. Cattle died thus during the drought and people covered tens of kilometers to reach sometimes dried up waterholes. Finally, primary sandstones only cover 30,000 km^3, but have a 1.9 Bm^3y renewable resource. Nonetheless, a program of 2,000 bores and 600 wells seems ambitious because of pump maintenance and pipe making problems.

Electricity production has doubled in a decade to 110 Gwh - all thermic. The first dam to be built at Pama where the Kompienga falls 35 m (SE) could provide 14 Mw and AVV's Bagré could supply 10 Mw and irrigate 5,000 ha. Nambiel's potential is 60 Mw and Bui in Ghana could generate the 30 Mw now required. Near the Capital 50 cm of lignite were located under a cover of 3 m and the Bobo Dioulasso (W) and Banfora uranium occurrences have also been explored. Salt is recovered from brines. Phosphate deposits cluster in the southeast : a) at Kodjari where mean thickness 10 m, 25 Mm^3 or 60 Mt rock grading 27.5% P_2O_5 plus an additional 20 Mt of lowergrade resources. Current production should expand to more than 5,000 t for local use. b) Further east, Abobo-Djuana near Alub grades 20-30% and c) at Arli southwest of Kodjari 1.2 Mm^3 or 2.8 Mt assay 29% P_2O_5. The dolomites of Diounkan Samandeni and Dioungoko (9, 6 and 1, 35 million m^3)are near Bobo. There is limestone next to the Tiara marble deposit, but Tin Hrassan (N) and Beli (N) with 6 and 50 Mt are larger (and remoter). Cimat has lobbied, therefore, for a spur to Onaga and OMT seeks investors for a 50,000 ty clinker grinding plant at Ouaga'.

Other resources

Metals follow the Birrimian arc up to the Niger (N), so does gold (117 on Fig. 24) from a) the Gongondy copper district (S) 5-15 g/t Au and 10-30 g/t Ag to b) Poura. The mine has produced between 1959 and 1966, 1.4 t in 1964. Based on 23 t of gold (and silver) in assays of 11-15 g/t, a joint venture of Soremi 60%, IDB and Coframines 20% each, financed by EIB, IDB, CCCE and ADB has set targets of 2.5 to 3.7 t. c) Kwademen (C), d) the Guiro-Bayidiaga cluster (NE) might be developed by Somisa, a joint venture of BGM and San Martin Mining Co. of the U.S. e) Bani and Gangabe are in the same district and BRGM has studied f) Dionga and the Bouré Au, Cu, Zn deposits nearby. (See for details p 311 and 372 of MRA). Since the core of 10 Mt ore grades 23% zinc and 60 g/t silver, the Perkoa (W) project may become bankable. Further, lead has also been worked at Gon (NW). In 1940 300 tons of copper were recovered at Gongondy (S) and at Gaoua (EC) 40 Mt ore graded 0.8% Cu ; Dienema is nearby. Bauxite reserves of a) Kaya attain 6 Mt and b) the 1 Mt white and pink refractories of Wempapassedo (CN) are of excellent quality. c) The 400,000 t Kosso occurrences of ceramic grade (NW) and d) Fara with 3 Mt bauxite of abrasive quality have been occasionally worked. e) Other shows include the southwest Ndorola, Sabou and Lahirasso. Sovimpec has operated the Mafoulu antimony deposit (NC) since 1980 on a small scale. Still further, 70 Mt of laterite at Bongo (C) assay 1.5% nickel and 0.1-0.3% cobalt. Moreover, titanium - vanadium occurrences form the segment of a circle centered on the capital including Gountouwala (S), Gouba, Pigwa, Kolel near Oursi and Tin Edia (N). 50 Mt of resources assay 0.5-15% V_2O_5, 8-14% TiO_2, 40-55% Fe, 3-6% Al_2O_3, 3-13% SiO_2 and 2% MgO.

The principal manganese occurrences include a) Sakoura (SW), b) Tiéré (WNW, 121 on Fig. 24) including 3-5 Mt and 30% Mn, out of which 300,000 t reach 45%. Comiki (50% BGM) plan to produce here 25,000 ty for export and c) Rumanian Geomin was the last to study the 13.3 Mt of Poura assaying 50-55% Mn, 0.6-0.8% Fe, 2.6% Al_2O_3, 1.4% SiO_2, 0.15-5% P and 1.5% NaK_2O plus 13 Mt carbonates grading 48% Mn under the oxides. Lastly, other occurrences include :

Chromite : Koumé (N of Gorom-Gorom) ; diamonds and kimberlites : Barsologo (NE), Baorolége, Zabré (SE) ; gold : Dori, Yalogo, Bouroum ; kaolin : Foulou (SW) ; lead : Tougan (W) ; mica : the Niakado-Saria swarm ; molybdenum : Goron (CN) ; nickel : Dablo ; zinc : Tiébélé (Po, S).

36 NIGER

Uranium is the mainstay of the budget, providing a third of revenue and more than 80% of exports. Since 1971, when uranium mining started GDP has dramatically increased, roads and towns were built in the northwest in the area of Arlit. From here development efforts reverted to coal at Agadès and even further in the populated south to phosphates and cement.

GOALS : To develop water resources and to provide and distribute fertilizers. To find customers for uranium and to substitute imported oil by local coal and Nigerian hydropower. To pursue oil exploration.

Table 52 Niger resources, reserves and production.

Resources			Production			Rsrvs		Production			Rsrvs	
Lignite	Mt	3	Coal	kt	120	Mt	4	Uranium	kt	4	kt	230
Hydropower	Gw	2	Phosphate	kt	1	Mt	150	Molybdenum	t	40-100	kt	6
Limestone	Mt	4	Cement	kt	40			Scandium cap.	t	1	t	60
Lithium	kt	6	Gypsum	kt	3			Salt	kt	3		
			Sand.gravel	m^3	150,000			Tin	t	70		

Merchandise moves through Cotonou harbor, the 400 km Parakou railway and 1,600 km of road. The shareholders take of uranium produced and the EEC received 60% of the exports which declined from their 1981 peak of 5,200 t U_3O_8 and \$500M to \$230 M, but much less in franc terms (80-100 M CFA). Indeed, Cogéma's contracts protect this lifeline. The business arm of CEA also holds the key concessions and manages the mines which employ 4,000. Smaller operations keep 1,500 people busy and the state's Onarem seeks foreign partners for mineral development.

Uranium

The fossile Guezouman and Tarat deltas and the Tchirezrine streams deposited uranium which was further concentrated in the In Gall basin (W). Within the 30 m thick Tarat Madaouela beds there is some uraninite and coffinite (a silicate) in the sandstone, but the richer organic compounds of the shales are more difficult to extract. Although radioactive anomalies have been reported on both sides of the Aïr long ago, the Arlette, Ariège and Artois orebodies near <u>Arlit</u> have been contoured only in 1966 following Imreh's discovery of uranium minerals. Somaïr has already extracted 20,000 t uranium here and Arni near Tassa N'Taghalgue is to take over after the exhaustion of Artois . Further exploration might also delineate additional bodies next to Marianne and Marylin which are at a depth of 70 m at Madaouela. At <u>Akouta</u>, Cominak makes 2.5% profit, twice as much as Somaïr, and extracts 600,000 ty of ore through two inclines from 250 m depth. In the 30 to 50 m thick Guezouman Channels pays are 2 to 12 m thick.

Table 53 Uranium deposits in Niger (including projects⁚).

Deposit	Capacity t/y	Reserves kt	% U_3O_8	Onarem %	Cogema %	Other shareholders %	type mine
Akouta	2200	40	0.4	31	34	Ourd (25), Enusa (10)	u.g.
Arlit	2300	35	0.25	33	27	+ Imétal 19, Sgrem (18), Uran G + Agip 13	o.c.
Aoni⁚	1500	20	0.35	33	33	Ftcic (Kuwait) 100	o.c.
Madaouela		8					
Imarouren⁚	3000	70	0.2	30	35	Ourd	u.g.
W Afasto		50	0.4	37	37	Ourd (25%), 350 m deep	u.g.
Abkoroun	1400	13		50		50% Arabian Oil Co. (Sap)	o.c. u.g.

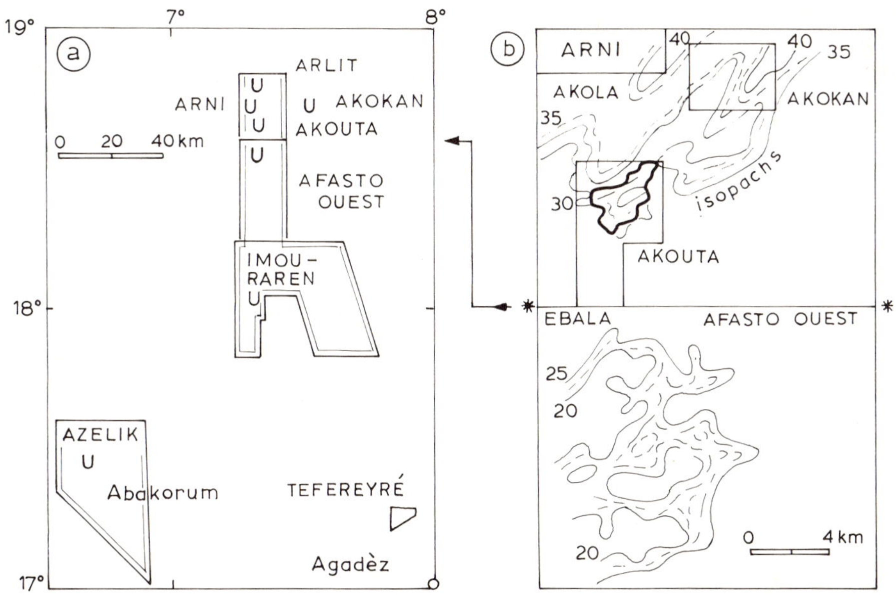

Fig. 23. (a) Uranium concessions in Niger. (b) Guezouman sandstones isopachs.

Similarly, the Afasto West orebody lies at 350 m depth. However, the largest deposit, Imarouren is covered by only 110 m of overburden, but ores of Abakoroun near Azelik, would require alkaline leaching. Finally, uranium also occurs west of the In Gall basin even in Algeria, the Japanese find of Irhavenzegirhan grades 0.19% U_3O_8 , but the Djado shows (NNE) are to date subeconomic.

At Arlit 9 t of overburden have to be blasted and stripped for each ton of ore, but the quarry required only 8% of investments against 20% for the camp, and 54% for the concentrator which has few walls since it never rains. But temperatures reach up to 45° in the shade. Ore is first ground to 800μ and solid-liquid separation is followed by soda precipitation of the end product, magnesium uranate containing 70% U. Each ton of ore requires 70 kg of sulfuric acid and 40-45 kwh.

Oil and coal

Electricity production may revert to its 1975/76 peak of 70 Gwh by 1990. Niger would also have to import then 1.3 Mb of petroleum products. More interest focused therefore on the Agadem trend (E), than on the Blima-Faya line. In 1975 a Texaco wildcat met some oil at Tin Touma at a depth of 1,500 m. The Sukor show has also been evaluated and hopefully exploration of the Djado (NE) and Tarak (E) basins will continue.

There are also 10-20 cm thick lignite seams at Tanout (S) and 30-90 cm and 50 cm thick lignite is buried at a depth of 40 m at Tahoua (SW) and 68 m at Fadama (Filingué) 120 km to the west. Chinyelga is in between and at Salkadamma (WNW of Tahoua) 2.6 Mt average 30% ash and 4000 cal/g. At Anou Araren (W) 1 Mt of recoverable coal out of 6 Mt measured and 3 Mt indicated reserves contains 51% ash and heats less, 3650 cal/g. On the other hand, the 5-8 m thick seams of Solomi 30 km north of Anou might be more attractive. Meanwhile, Sonichar (state 66%, IDB 34%) started operating the Araren colliery with a target capacity of 150,000 ty serving a 30 Mw generator for the uranium mines. Niger imports 40 Mw from Nigeria ; it has two hydropower sites, Kandadji I-II on the Niger and Dyodyouga on the Mékrou which it shares with Benin. Capacities and average capabilities of these sites are 125, 105 and 25 Mw, and 710, 750 and 80 Gwh/y.

Water and food

In addition to large irrigation schemes at Tarka, Goulbi and Maradi, government plans call for the drilling of 10,000 wells between 1983 and 1990. Meanwhile, the Niger valley supplies most of the water in the southwest. In addition to the underflow of the Wadis Dallol Bosso, and Azawak, the ancient sources of the middle Niger are also exploited, but hardly sloping alluvia provide inadequate aquifers. So does the Basement west of Niamey (SW) which hardly yields more than 25 ld. East of Niamey the Late Continental is locally under artesian pressure but the supply is only 3 mm/year. In the large south-central area underlain by the Intermediate Continental, groundwater depth averages only 40 m, but the Tégama sandstones lie at a depth of 600 m ! Yet in the parched uranium desert there is ample groundwater. The mills need 2-3m^3/t of ore and use the fossile-semifossile Tarat-Madaouela aquifer of the deposits at a depth of 40-50 m. Other water bearing sandstones include the Guezouman and the Viséan at 100 and 500 m of depth. There are even waterfalls in the Aïr (CW) and mineral water at Igouloulef and along the trails which lead to it from Tahoua (SW) and Zinder (S). Wells with sweet water dot the trail between the palmgrove of Djado (NE), Dirkou near Bilma (E) and N'Guigmi (SE), but quality deteriorates towards Lake Chad.

Table 54 Exploitable water resources of Niger.

Sediment	Area km^2	Feed Bm^3y	thick. mm	Resource Bm^3
Late Continental	96	1.2	13	40-90
Intermediate Continental	220	0.6	3	110-220
captive				28-60
Agadez sandstone	27			13-27
Lower Koramas	15	1.9	130	1.5-3.7
Pliocene free	104	0.3	3	27-57
Quaternary captive	26			6-13

Three areas of Sebkhet can be distinguished in the northeastern Kaouar district : Seguedine in the north, Blima in the south and Fachi on the edges of the Ténéré (C) where salt is recovered between April and October. Nomads sell far south salt evaporated over nine weeks at Teggid n'Tesemt and in the Gelele basins. Recorded sales have ranged around 3,000 t and other salt pans include Birini n'Kouni, Filugue and Gouré and Fofa (Salongari, S) in the south.

The hand operated Akker crushing plant near Takoa produces 5,000 ty of phosphates. Low grade 35-40 m thick beds of Tapoa (SW) include 1.25 Bt rock assaying 18-35% P_2O_5 out of which 500 Mt are commercial, with grades of 22.4% P_2O_5 in the upper layer and 27.8% and 34% in the two lower layers.

Other resources

There is limestone ang gypsum near Malbaza (S) and SCN (state 57%) could expand cement capacity from 40 to 220,000 ty.

Niger is the only African producer of molybdenum. More of this metal could be produced from uranium ores if dressing problems were solved. In 1978 0.7 t of scandium were also exported, but capacity is 1t. Moreover, copper shows in most of the uranium belt, e.g. at Kourki, so do in places cobalt, germanium, vanadium, barium, arsenic and nickel.

Since the fifties, recorded tin production doubled in the Aïr (CW) to 140 t of 70% concentrates in 1974 and customs now clear 50 to 80 ty to Nigeria. While effective output is probably higher, the paucity of water and labour hampers recovery in the Tarrouadji Mountains (CW) and in the adrars Elmeki and Guissat (C) where tungsten and columbite-tantalite accompany tin. In the Téra massif (SW), 18 pegmatites of Ayouron contain 300-350,000 tons of spodumene and grade 1.4-2% lithia. Reserves and LiO_2 assays of the three most important dikes are 70,000 t: 3%, 55,000 t : 2.8%, 42,000 t : 1.85%.

Some gold was panned in the Sirba (S) and in the Aïr (CW) at Tunan, Tilisdak and in the Oued Hgayague. 5 kg have been reported both in 1967 and 1968.

The 30 km wide ellipse of oölitic iron ore trends 150 km south from Libya, bulges to 100 km at 15°E-20°40 N and then bends north. Since the Niger has become navigable to Kainji dam in Nigeria, the 80 km^2 low grade iron ore plateaus 80 km south of Niamey have attracted some interest. Resources include : a) Dyabou 1.7 m thick, 280 Mt, 50% Fe ; b) Say 1.7-2.5 m thick, 270 Mt, 49% Fe, 19% SiO_2 and 3% P_2O_5. Out of these reserves UNDP has delineated low phosphorus zones grading 48-53% Fe. Lastly, other occurrences include :

Pisolitic bauxite : Gaya ; chromite : Makaloudji (SW) ; copper : oued Enneg, with Mo Liptako ; gypsum : Galmi, Keita, Madaoua, Bouza (S) ; lead and zinc : Firgoun ; manganese : Tatouss-Dabaga, north of Téra (SW) 31% ; molybdenum, Cu : Kourki, with RETh Pb N, Téra (SW) ; tin, titanium, iron : Iférouane (CW) ; titanium, iron : Sarua (SW), Liptako.

37 CHAD

Although some salt, soda, stone and gravel are quarried, oil is the best hope of wartorn République du Tchad. Sedigi might provide 2,000 bd and Mandouin flowed 1,600 bd of 25° API crude from 1,300 m depth. There are in effect two target areas : Borkou (N) and Erdi (NE) and around the Komé show near Mundu (SW).

Also in the south where much of the population lives, the Logone has hydropower potential. A prime site is Ganthioot Falls . Electricity generation has increased, meanwhile by 50% in the seventies and hovers since around 65 Gwh. Other sources of energy, geothermal steam in the north, and uranium shows at Ayamé in the Tibesti (N) are, however, too remote.

GOALS : Further exploration for hydrocarbons and detailed evaluation of groundwater and mineral resources as soon as conditions permit.

Table 55 Chad production and resources.

Peak production, past		Resources		Production		Resources
Gold	kg 200	Bauxite	Mt 10	Soda	kt 7	Mt 7
Wolframite	t 22	Hydropower	Gw 3	Salt	kt 4	

The internal basin of Lake Chad drains abundant fresh water but the desert which covers a million square kilometers has advanced further south. The drought affected this area too as well as a third of the population. There are oases though at Bir Anaga (W), Koro Toro (C), Oum Chalouba (E), Faya (CN), etc., but groundwater is more abundant in the Tibesti (N), e.g. on the Zouar trail at Bardaï ; it also flows from sources at Aouzou in the Libyan claimed strip. Still further, at Iriba-Guéréda (ENE), water is at a depth of 5-9 m and in the Precambrian areas of the east paleo-alluvia covered by younger sediments might hold water. So do underflows of wadis. Yet, east of Ndjaména, the contact of the Basement at a depth of 160 m yields little water. Further south, Cretaceous sediments are believed to underlie the valley of the Logone and to hold water. Farming is also limited to this arc covered by light and red brown soils which can be improved by fertilizer.

Mineral resources

Yearly evaporation and seepage of Lake Chad respectively equal 2 m and 20 cm. Salinity increases thus from 45 g/l at the Shari and Logone estuaries in the South to 90 in the middle and 300 in the north. The saturation point is reached at 205 g/l because the water of the lake also seeps north below wadis and depressions. Workers cut thus 20 kgs of high grade soda at a depth of 50 cm e.g. at Kalia. Cameliers carry the slabs 15 km south to Baga Sola whence they are shipped to Ndjaména and Nigeria. Exports have even reached the Central African Republic and capacity may be 25,000 ty. Magnesium carbonates constitute 10% of the material and there is also some potash. Diatomites also occur around the lake, particularly in Bornou. Clays and sand were worked around the towns, stone was quarried at Mani, gypsum near Ndjamena and limestone at Mayo-Kebbi, but most cement was supplied by the Figuil plant in Cameroon.

The Cameroon bauxite belt also extends to the Moundou area (SW). Five m thick layers include 3.5 Mt of bauxite grading 57% Al_2O_3, 8% Fe_2O_3, 4% TiO_2 and 3.7% SiO_2 on the Koro Plateau near Laï but the Koutou Kouma Mountains west of Koro hold only half a million tons.

In the same area CMOO extracted in 1939 200 kg of gold from the Mayo n'Daba stream near Kamboké, Pala county, but by 1940 only 2 kg have been reported.

Salt recovery averaged 6,000 ty in the Ennedi, Borkou and Tibesti (N) between 1955 and 58, and dipped to 2,000 in 1959 but current records are inadequate. There are extensive occurrences of tin group metals on the northeastern border (Fig. 18) and wolframite has been airlifted from the Yédri Mountains.

There are also traces of gold, copper and tin in the Tibesti and of copper, lead and zinc in the Ennedi, an area claimed by Libya, but there is not enough water for recovery.

4 WEST AFRICA

The Coast (Fig. 30, 25 and 24 to which (numbers)refer) has more water and minerals than the Sahel with which it forms ECOWAS. Most trade in this common market is carried out though between the French speaker's CEAO or in the orbit of oil-rich Nigeria, which represents three quarters of the GDP and 90% of the mineral output of West Africa. Moreover, Ghana also exports hydropower to its neighbors and Abidjan, Cotonou and Buchanan are outlets for products of the landlocked interior. Here are, indeed, the most advanced economies of black Africa.

40 GUINEA - BISSAU

There is a minimum of infrastructure in the Republica popular, but the area north of a Sare-Bacar line might benefit from OMVG's irrigation schemes. The northwestern coast is marshy with long inlets and elsewhere the Late Continental is the best aquifer.

> GOALS : Strengthening of administrative and planning structures. Improvement of the quality of water. Furtherence of oil exploration and establishment of a mineral map.

Black sands extend from Senegal south to Varela, BRGM located low grade phosphates at Farim (N) and the U.S.S.R. has shown some interest in 200 Mt of bauxite assaying 53-63% Al_2O_3 between the border, Béli, Madina de Boé (SE) and the Corubal.

Imports of 150,000 by of petroleum absorb the farmers export earnings. PEAT might spread south though to the Cacheu, Mansoa and Geba deltas and the Dôme Flore oilfield is offshore across the border.

Northwest of Bissau, the favorable Paleozoic layers are between 1,000 and 1,700 m depth, but the Elf consortium has not found oil yet, west of the Bisajos islands along the edges of the shelf. The state's Petraminas is a nonpaying partner. Royalties would moreover amount to 12.5-20% and income tax to 30-51%.

Power consumption declined from its 1973 peak of 18 Gwh. Yet, should it rise

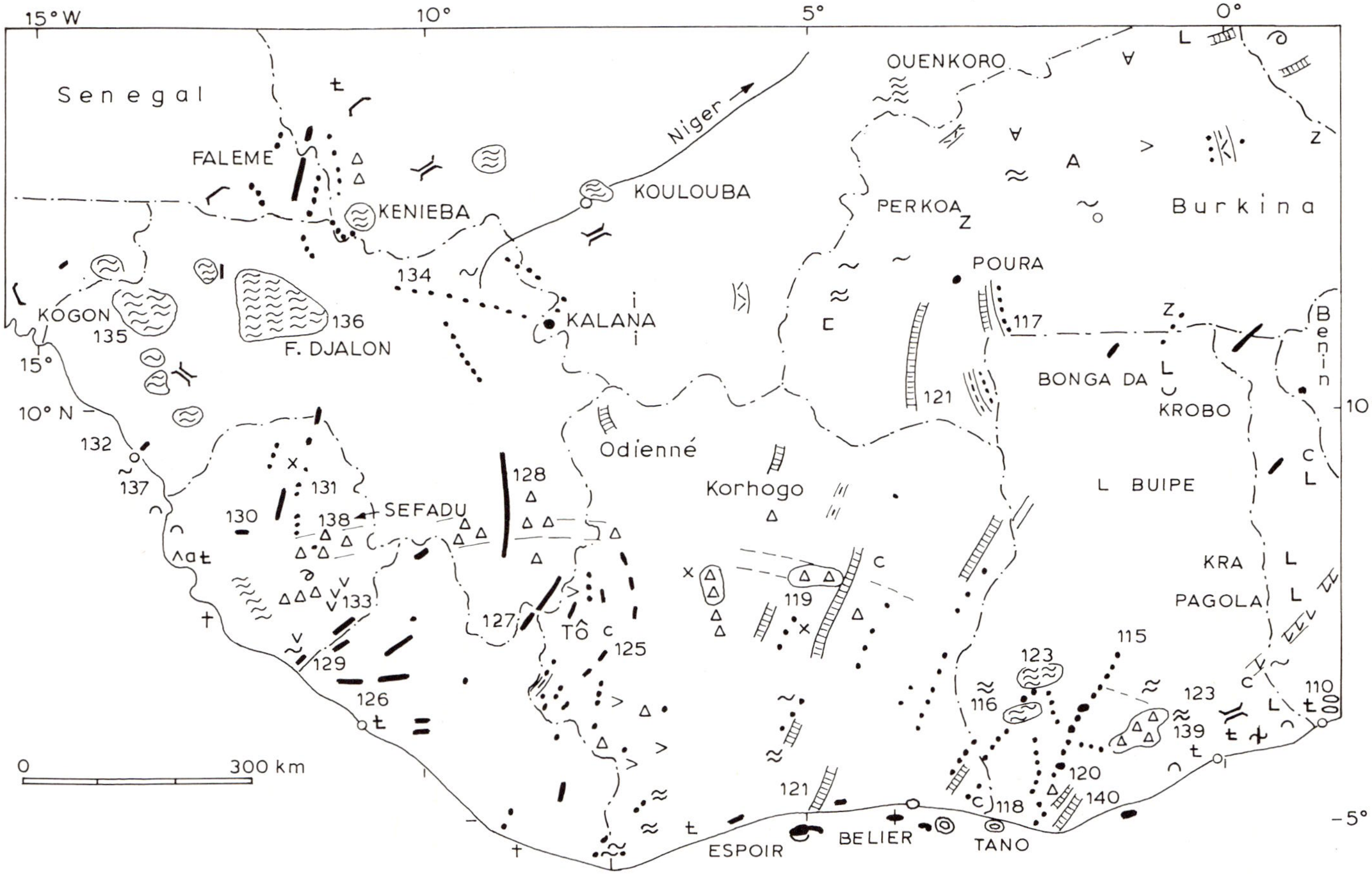

Fig. 24. West Africa and the southern Sahel. (Legend on page 13.)

to 25-50 Gwh in the nineties, peak demand would require 10 Mw. Moreover, at Kayanga on the Geba OMVG could develop 2 Mw across the northeastern border. Closer to the Bissau load center (W), the Corubal has a 240 Mw potential in the two Guineas and would generate 640 Gwh in a dry year. At Saltinho (SE), 170 km from Bissau, 9-11 m of head could be developed out of 19 m in order to avoid excessive flooding over the border, providing thereby up to 19 Mw and 105 Gwh out of the 66 Mw and 160 Gwh available.

Further downstream at Cusselinta 17 m of head and 17 Mw could generate 45 Gwh in the dry season.

41 GUINEA

The most important of the three Guineas holds a third of the world's bauxite. It had the makings of an advanced industry with its abundant hydropower when aluminum markets were favourable, but political factors delayed the progress of the République populaire révolutionnaire de Guinée by four decades. The investment climate has improved, diamonds are mined again and Guinea is to become the continent's second producer of iron ore.

GOALS : Political stability and managerial controls. To expand bauxite production, further develop alumina reduction and hydropower and to create an aluminum industry. To complete financing and marketing arrangements for Nimba, to extend the Buchanan railway and work the iron ores. To explore in detail the northeast for minerals and the coast for oil.

Table 56 Guinea resources and production.

Reserves			Production			Production		
Bauxite	Bt	7	Bauxite	Mt	13	Alumina	kt	700
Diamonds	Mct	5	Gem diamonds	kct	400	Industrials	kct	50
Hydropower	Gw	6	Hydropower	Gwh	80	Gallium (resources)	Mt	1.5
Iron	Bt	1.4	Clinker, capacity	kt	250	Limestone (resources)	Mt	50

Many of those who left returned but the population has not reached 7 M yet in an area the size of Pennsylvania. Bauxite production peaked with 14 Mt in 1980 and is increasing again since capacity is in excess of 20 Mt. Bauxite (alumina and diamonds) revenues are also the mainstay of the budget, they represent a third of GNP and almost 90% of recorded exports. Lately 40% of bauxite exports went to North America, a third to the EEC (mostly France, Germany and Spain) and 20% to the USSR, who took in the past most alumina. Forty percent are now shipped to the EEC which is also a target along with Japan and Nigeria for iron ore; most diamonds are shipped to the U.K. Alcan runs CBG and PUK Friguia whose shareholder is Olin (U.S.). U.S. Steel manage the iron ore project and the government holds 49% of the shares of the bauxite companies CBG and Friguia, all of the shares of OBK, 50% of Mifergui's iron ore and of Aredor's diamond venture. Mineral exploration has been carried out by BRGM, USSR and UN agencies, and the government also encourages small mining enterprises. Nonetheless any major venture would require the reinforcement and expansion of the Kankan railway ; 20 km will also be added at Nimba creating a new outlet. Supplies to Aredor are trucked from Conakry instead of Freetown and 5,000 km of formerly paved roads may require additions.

Principal customers were the British shareholders of the diamond company and for bauxite, the U.S. plus Canada 40-50%, the EEC (France, Germany, Spain) a third, the USSR 10-20%. Considering the rapid growth of aluminum demand, new opportunities arise for Guinea.

Bauxite

Some 5 Bt of resources can be added mostly on the Fouta Djalon to 7 Bt of reserves shown below including grades in % :

Area		Bt	Al_2O_3	SiO_2	Fe_2O_3	TiO_2
Dabola	C	1.5	42-44	19		
	C	0.3	45-48	0.8-1		
Pontiolo	C	0.2	47	2		
Tougué	C	0.5	45-46	2.6	20-24	2
Kindia	S	0.07	45	1-3	$\geq$ 20	2
Ayé Koyé		1.1	49.5	1.5-1.7		
NW		2.9	43-44	1.8-2		
Sangarédi		0.1	60	< 1	2.4	3-5
West		0.3	50			
Fria SW		0.1	40-43	1-2		

From the ENE to the coast (SW) the following 14 clusters can be distinguished
I. In northeastern Guinea grades range from 43 to 65% Al_2O_3, 2-3% SiO_2 and 5-30% Fe_2O_3.
a) Upper Niger highlands : the Bô-Oubé, Tamakaya and Kambarégna Mountains (65 km from the railhead), the western slopes of Télikan Kourou (30 km from the rails) and the Niandan Banié north of Kouroussa, in the Siguiri district (30 Mt), 600 km by rail from Conakry and nearer Dinguiraye (60 Mt) - b) Between Siguiri, Fatoya and Bougourou lower grade bauxite (plus gold occurrences).
II. 20,000 km^2 of bauxites and laterites cover the eastern slopes of the Fouta Djalon (136) 300 km from the coast, between the Mali-Labé line (W), the border(N) the railway (S), and Dabola and the Bafing (E).
a) samples of the bové of Horé between Sobori and Ndiré Yangueya, Lallabara, Lagui and northeast of Tougoué indicate 58-63% Al_2O_3, 2-4% SiO_2 and 2-15% Fe_2O_3, and at Pontiolo a larger area averages 47% Al_2O_3. There is hardly any overburden and 8 meters ore consist of cemented blocks, gravel and dark fine grained bauxite at the bottom. The same holds for b) the Dabola area including higher

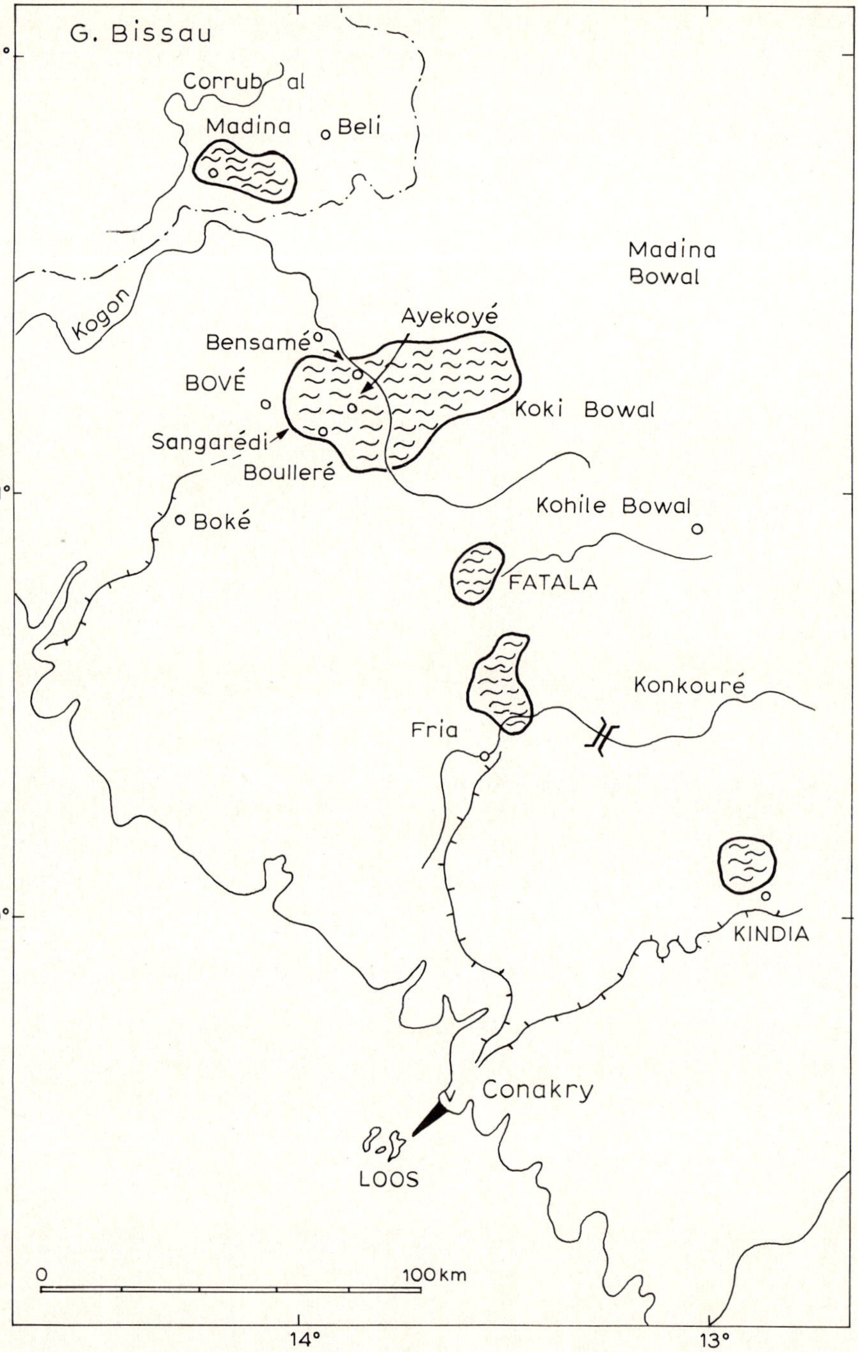

Fig. 24. (a) the bauxites of Guinea.

grades at Dabola proper, plus Siffray and Bissikrima c) 200 Mt at Pita (C) d) The Bassaris area at the sources of the Gambia south of the border and Gaoual (200 Mt) west of the Fouta Djalon.

III. The west. a) Near the Bissau border, in the center of the western bové a 2 m thick lateritic crust overlies 3 m of massive, high grade, bauxite followed by 5 m of gravelly high iron bauxite in the Ayé-Koyé-Bensamé area. Since they are nearer to the coast, these deposits might become minable first.
b) Sangarédi (135) nearby provided since 1963 9 Mty of highgrade bauxite (along with 1-2 Mty from Boké further south). Twenty m thick ore includes massive, pisolitic, white-brittle, conglomeratic and pebbly layers or lenses. Following blasting shovels load the material on railcars which move it over 130 km to Kansar harbor where it is crushed to < 1 cm. White abrasive material is calcined at 100°C. Kiln capacity has increased to 250,000 ty. Harvey aluminum manages CBG. Other shareholders include Alcan, Alcoa, PUK, VAW, Montecatini, etc.

IV. The south. a) The Fatala area of 60 x 10-25 km connects the bové with Fria. Its bauxites average 45% Al_2O_3, although large sectors grade over 50%.
b) Since 1960, PUK manages Friguia whose shareholders include Olin, Alusuisse, British Aluminum and VAW. Up to a meter overburden covers 6 m (4 to 14) bauxite at Fria and Kimbo. Capacity is 2 Mty and 1 Mty alumina, which could be expanded, is shipped via Conakry. c) The Debelé deposit at Kindia covers 2 x 5 km and averages 7.5 m thickness. OBK could ship up to 3 Mt raw bauxite crushed < 6 cm to the USSR. d) Bokoro, Farenja and Bassoya, northeast of Boffa (north of Conakry) and e) the Loos Islands (137) off Conakry have been worked out.

Iron ore

With 1.4 Bt of ore resources and a reserve base of 345 Mt Guinea probably boasts the highest grade iron ores of the continent. I) At Simandou (E) and Nimba (SE). Furthermore II) Yamboieli, Akoundiat, Lola and Kafou (WNW) hold two types of hematite grading only 35-40% Fe in the lower and 50-55% in the upper powdery ore. Finally, III) the capital was built on the Kaloum (132) iron laterites whence Compagnie Minière de Conakry and the state have exported 500-800,000 ty mostly to the Comecon, the U.K. and Egypt between 1953 and 1966 when production ceased. Moreover, an average of 1 Mty were still recovered from stockpiles between 1968 and 1970. Together with Mt Kakoulima where nickel also shows, resources attained though 1 Bt of ore assaying 0.06% P and 1.25% Cr. IV) Other lateritic deposits include Tomine (2 Mt with traces of chrome), Koumbia, Doumbiaghi, Cape Verga, Beyla and Foulamory.- Simandou (128, (E)): Ore resources attain over 80 km strike length 0.8-1 Bt grading 65% Fe, 0.05% P, 4% SiO_2 and 5.8% Al_2O_3 and they may reach 2.5 Bt ! Indeed, 350 m thick banded iron stone of the ridges broaden at Lamadougou Pass and Diodio. Thickness of this series exceeds 1,000 m at Fon peak 50 km to the south, but iron ores are narrower at Bétoubé and Sorayo between Simandou and Nimba (127) which includes 700 Mt ore assaying at the surface 67% Fe, 0.04% P and traces of chromium at Nzérékoré. Banded iron ores widen from the NNE over 10 km as far as Mt. Tô, and at the Goué River their width attains 2 km. They taper finally towards the border (SE). In addition to the Sempere and Château deposits, Pierre Richard contains 315 Mt of ore grading 67% iron.

Originally, production was to grow from 5 Mty to 15 Mty of natural sinterfeed grading 60% Fe, but the project has been scaled down to 8 Mty and $400 M. From an underground crusher the ore is to be transferred by a 7 km long conveyor to a third grinder which can reduce it to 6 mm.

In view of their proximity to the Liberian Nimba Lamco railhead, i.e. 265 km to Buchanan, this deposit will be worked by a consortium including Nigerian, U.S., Algerian, Libyan, French, Japanese, Rumanian and Yugoslav groups and Okura Trading of Japan.

Diamonds and gold

Diamonds were discovered in 1932 and Soguinex, a Selection Trust subsidiary recorded yearly outputs of 50,000 carats until this concession was cancelled. Officially, production ranged around 70-80,000 cty for two decades, but unrecorded washing may have supplied 150,000 carats of gems and 200,000 of industrials.

Soviet technicians provided assistance here between 1968 and 1973, but only in 1983 did the output reach the mark of 250,000 carats, 85% of which are gems averaging 0.53 ct. The target is 500,000 carats and the new mining company, Aredor (79% controlled by Bridge Oil (Australia)) has expanded at Kérouane from the Banankoro placers to pipes and more than 1 Mm^3y are to be worked out of 6.5 Mm^3 of placers. 30% of the stones weighed > 30 ct and the International Diamond Co., an affiliate, markets 70% of the stones. The government sells the remainder, plus all > 100 ct diamonds... In the past, more diamonds were recovered further east from the Macenta-Beyla field and the Bougban and Pougban drainages than from Banankoro, Férédon and Bonro, but diggers sell few to the official buyers. There were, further, diggings in the Jougbarako and Férougban. The first kimberlites were located here and in Fenaria with numerous low quality diamonds. Placers trend south, along the streams from Guékédou to Nzérékoré (SE, see 'iron'). On the other side of the divide, stones were panned in tributaries of the Moa and Milo up to 80 km north of Macenta and as far as Kankan, the railhead (E). Furthermore diamonds were found at Bounoundou west of Beyla, Bassam, Koulountou, Tominé and in the High Macéni.

Three gold belts can be distinguished in the northeast between Kalana in Mali, Bougourou (NE), Siguiri, and Mandia, where SAG, a subsidiary of the Chevaning Mining Co. set a target of 3 t in 1990. Other zones extend from the

Blue Lode, near the frontiers of Mali and Senegal, to Maleya and a third is near Mamou (Dabola). While little or no gold is reported in official records, Klaveness (Norway) also start operations.

Energy and other resources

There is up to 3.5 m rain on the coast and 1.5 m inland. Wells yield 10-30 m^3h near Conakry, exceptionally 70 m^3h. From the gravels of the Rio Nuñez artesian water rises at 2-50 m^3h, but in the east and northeast yields are better. Strongly weathered podzolized soils border here the laterites of the Fouta Djalon and red-brown soils.

Per capita energy consumption is a half of that of Sierra Leone and a quarter of the Ivory Coast. Since 1964, electricity generation has been flat at 500 Gwh (out of which 80 are hydraulic). The Konkouré alone has a potential of 1 Gw, but only 75 Mw have been developed at Souapiti which serves Fria. Other potential hydropower sites are shown below.

Site	River	Mw	Site	River	Mw	Remarks
Kaleta	Sarnou	116	Souapiti	Konkouré	350	
Kamarato	Baoulé	2	Koukoutouba	Bafing	85	680 Gwh
Foumion	Niandan	85	Bouréya	Bafing	130	85 Mw firm.

The uraniferous sediments studied by Cogéma in the north are subeconomic. In contrast, Davy Mc Kee with Saarberg Interplan, Nigerian and Moroccan support has indicated a radioactive orebody near Conakry. While Bridge Oil (see 'diamonds') has invested $15 M in oil exploration on the coast, an affiliate of Superior Oil has located favorable structures on the northern shelf. Meanwhile, petroleum imports hardly reach 2 Mb, but they would surely increase if the economy improves.

Soprociment's clinker grinding plant at Sinfonia near Conakry, has a capacity of 250,000 t and several other cement projects have been mooted : Kindia (200,000 t), Sougouita, 200 km north of Conakry (300,000 t), Lébékéré (N) and Siguiri on the Niger (NE) where there is abundant limestone.

Most bauxites average 200 g/t of gallium. Guinea probably holds, therefore, much of the world's lowgrade resources. Lastly, water percolating from the guano of Corail Island, in the Loos archipelago (S) developed phosphate grading 33% P_2O_5 and other occurrences include :

Blacksands : Coyah, Kansor, Conakry ; chromium : 1.5% in the iron ores ; copper, cobalt, silver : Firawa, Kointo ; pink granite : Morifonyé (SW) ; graphite : Lola (see iron) ; marble : Kindia (S) ; uranium and copper : Bohodou (SE) ; radioactive anomalies : Boua and Sessé.

42 SIERRA LEONE

Rutile, bauxite and iron ore represent 15% of GDP and three quarters of export earnings, to which might be added unrecorded removal of gems (and of gold) -

up to a half of total production. Nevertheless, mineral exports matched only a third of imports, led by oil.

GOALS : Development of hydropower, organization and control of diamond workings, incentives for their domestic sale. Increased mineral exports.

Sierra Leone is one of the world's two major exporters of rutile, which has some strategic importance as a source of the light metal, titanium. The government has also been successful in reopening the iron mines and it hopes to increase bauxite capacity.

Table 57 Sierra Leone production and reserves.

Resources			Production							
Hydropower	Gw	3	Petroleum	Mb	1.5	Diamonds gems	kct	600	Mct	3
Zircon	Mt	0.5-1.5	Cement cap	kt	200	"industrials	kct	200	Mct	23
Iron	Mt	30	Salt	kt	200	Rutile	kt	90	Mt	4
" planned produc.	Mt	1	Gold	kg	300	Bauxite	Mt	1	Mt	120

Alusuisse and Nord Resources of Dayton, Ohio, controlled the rutile and bauxite companies. The latter paid 50% tax and a $4 fee per ton ; the state's option for puchasing 47% of its shares runs until november 86. In contrast, the state controlled two thirds of Diminco, the diamond and gold producer. As usual, major export partners are the EEC (the U.K. and Germany) and the U.S.

Diamonds

Following their discovery in 1930, official recovery ranged around half a million carats between 1945 and 55, but a million carats were reported in 1942 and 1.8 Mct in 1961 when another million from various sources changed hands in Monrovia. EEC records average 900,000 ct in the seventies with sales declining in Liberia from 800,000 to 300,000. According to the statistics, the ratio of reported production and Monrovia sales decreased though, from 3:1 to 2:1 and their combined total declined too as shown below in million carats :

All diamonds	1965	1967	1968	1970	1973	1974	1972 gems	1981	1972 inds	1981
Sierra Leone	1.5	1.2	2.4	2	1.7	1.7	0.7	0.2	1.1	0.1
Liberia	0.5	0.6	0.7	0.8	0.8	0.6	0.5	0.1	0.4	0.2

Another comparison is explicit : 1961 SLST 720,000 carats, 1984 Diminco, its successor, 110,000 ct, licensed alluvial diggings 1961 1 Mct and 1984 30,000 carats...

Recovery from kimberlites doubled to 5-70,000 ct. In the Kono Reserve mining of pipes is said to diminish smuggling and Diminco plans to mine more underground. Conversely, little equipment is available at Yengema and lower grades are worked at Tongo and Ghahama. From Yengema kimberlites follow over 15 km a northeasterly line via Sefadu and Koidu (138). The majority of the diamonds has been found in the Bafi drainage, but stones have been carried 150 km downstream by its collector, the Sewa from Jaiama Nima, Yema as far as Sumbuya where the tides reach ; they also spread over the land between Tongo and the Woa as far as Kissiwuri (E). - The placers of Moinde, Meya and Woyie locally produced as much as 350ct/m^3, including a 770 ct gem, and recently three 100 ct stones turned up.

Rutile, bauxite, iron ore

Between 1966 and 1971, Sherbro Minerals Ltd dredged 120,000 tons of rutile from the Mogwembo at Gbangbama in the Moyamba and Benthe districts. Sierra Rutile Ltd of Canada surveyed dressing from 1976 on and was on 3 shifts with a capacity of 100,000 ty. The concentrates are screened on the dredge, tabled and passed through meter magnets in the dry mill to obtain 95-96% TiO_2. Ilmenite ferrotitanate and zircon are stockpiled. Sherbro reserves of 160 Mt averaged 1.75% rutile ; the top soil and mud grades 2.5% TiO_2 and sand and gravel on the bedrock 3.5%. Additional 10 Mt of rutile are a) nearby, b) in the Jong (E), c) Sembelhun (W), and d) Lanti (S) where concentrates grade 37% rutile, 28% ilmenite, 20% zircon, 1.4% monazite plus some corundum.
Other occurrences include :
Blacksands : Tokeh to York, Whale River estuary (S), Kambia district, Port Loko (N) ; asbestos : Masari (N) ; graphite : Gbangbana Hills, e.g. Moselo.

The 250 km long bauxite belt extends far southeast from Hastings (CW) and includes the 30 km long Mokanji Hills, largely exhausted, and 3 km of the Gbong Hills. Out of 15 Mt of reserves Sieromco (Alusuisse)'s output doubled since 1963 at Gondama. Ore is shipped 25 km to Moyamba and by lighter on the Bogon to the drying plant at Nitti Harbor on the Sherbro. Further, heterogeneous lenses could supply 100 Mt of washed bauxite grading 47% Al_2O_3, 21 Fe_2O_3, 2 TiO_2 and 4% SiO_2 at Yenkiso (N). While there are only 250,000 tons near Waia (N), all in all resources reach 230 Mt.

SDLC operated opencast Marampa (130) from 1933 to 1975. Out of reserves of 100 Mt grading 38% iron output averaged 2.5 Mty during the last years, with 1.5 Mt shipped to the EEC. Austromineral managed until 1985 the remaining reserve base of 90 Mt assaying only 30% Fe for Miomco but did not reach its target of 1 Mt. Other ore resources (Mt) and grades (%) include :

	Mt	Fe	Al_2O_3	SiO_2	P	Deposit	Mt	Fe	Al_2O_3	SiO_2
Numbara, Simbili		60	15	3	tr.	Bagha (N)	400	41.6	6.6	1.2
Pampana, Waka						Gt. Scarcies (NW)	15	30		
Total Tonkolili	600	54	8.5							

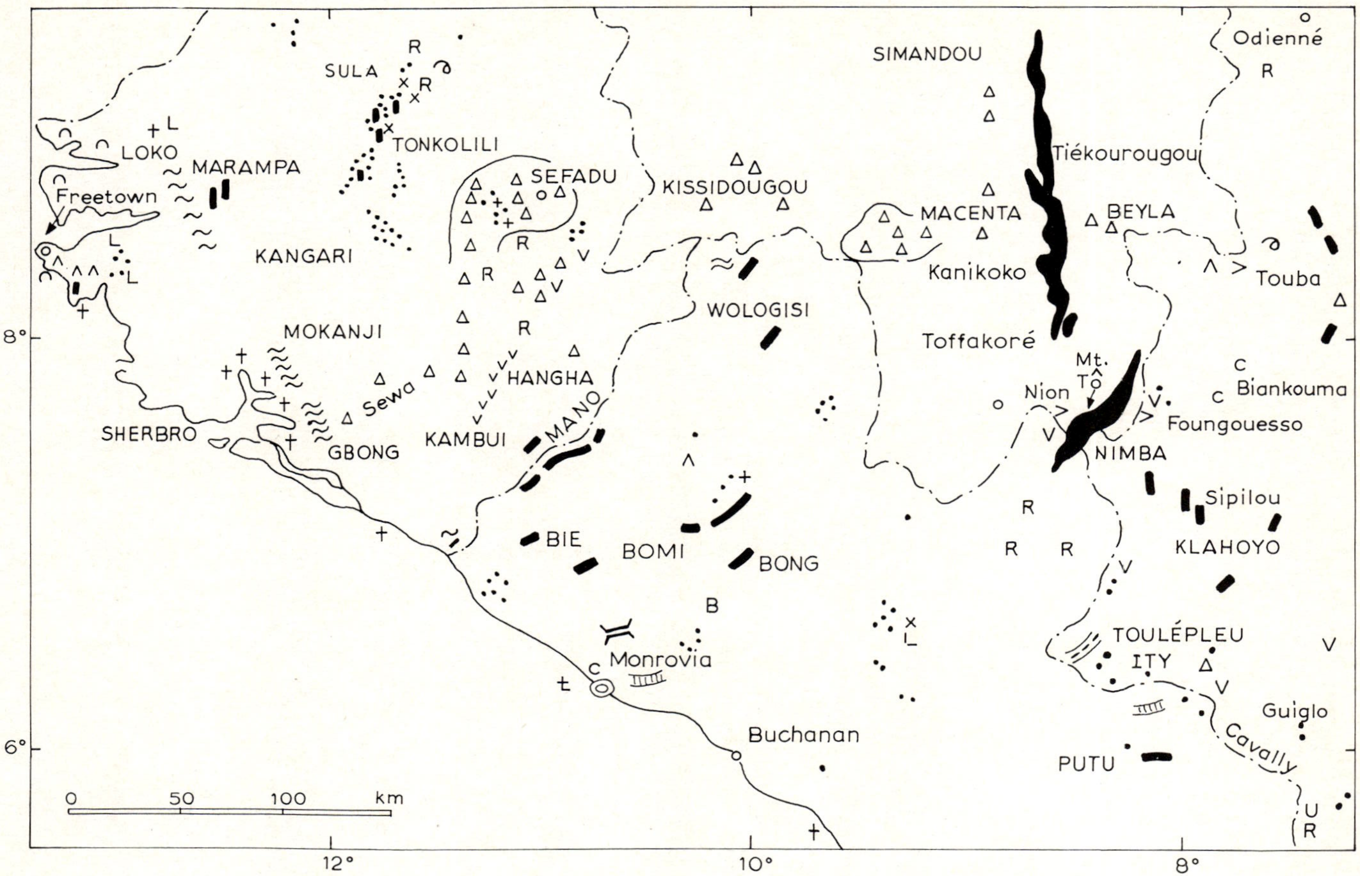

Fig. 25. Sierra Leone, SE Guinea, Liberia and western Ivory Coast. (Legend on page 13.)

The sedimentary basin covers the coast and the west, but oil exploration has hardly started. There is, further, lignite at Newton (W) and plans call for a hydropower plant at Bumbuna on the Seli (N). Meanwhile electricity requirements are met by thermal generators whose production trebled in the sixties and is creeping upwards from 200 to 250 Gwhy.

The nepheline syenite of Bagbe is of ceramic grade and the Sierra Leone Brick and Ceramics Factory is still operating at Wellington. Ciments d'Origny and CCC control Seracem's Freetown cement plant where grinding capacity has been expanded to 200,000 t out of which 50,000 are targeted for export.

Between the border and Freetown, salt is traditionally recovered in the deltas of Samu chiefdom. Plans call for increasing output from 200 to 250 kt... .

Between 1930 and 1956 when industrial production was interrupted, 10 t of gold were washed from placers of Sula, Kangari, Gori, Loko, the Nimini and Kambui Hills. Output peaked with 1 t in 1936, dipped to 50-100 kgy and is rising again. Meanwhile, Liberia exports five times as much. Since the Bank of Sierra Leone increased prices of placer gold there is reputedly less smuggling. Some 150 buying licenses were awarded, but there are other dealers too. Diminco is the principal producer, however DDX diamond Distributors and Explorers of the U.S. now holds the rights to Nimini Hills along with EVL. The Burkman group operates in the Bô district and others have explored the Dalakuru, Yrisen and Baomahun veins in Diang, Kafe and Valunia chiefdoms. Bin Rafah prospected Matotoka near Tonkolili and UNDP was active in the Sula Mts. and Kangari Hills.

Between 1929 and 1949 miners extracted 170 kg of platinum from the Colony Peninsula (NW). Similarly, mining of up to 20,000 ty of chromite ceased at Hangha (133, ESE) in 1963 when lumps were exhausted. Between 1954 and 1956 some tinstone and coltan were also panned in Valunia and molybdenite from the Wankatana. Lastly, other occurrences include :

Chromite : Bender Yawie (Gori Hills), Yanahun ; copper : Kalakuru (CN) ; corundum : Kono, Koinadugu, with garnet : diamond placers ; molybdenum : Mapoko (N), Nyandehun, Petema, Wangi (N), with base metals at Kalakurù in Koinadugu district ; nickel : Kambui (N) ; niobium and tantalum Sula, Kangari Mts, Issia, with uranium : Bagbe (SE).

43 LIBERIA

Iron ore has contributed almost half of the GDP and exports as well as 20% of the budget, second only to rubber. Production peaked with 27 Mt in 1974, yet the heydays are gone : highgrade deposits have been worked out and the two remaining companies accumulated losses. At the same time, competition increased. Although less iron will be mined in the next decade, Guinean ore will transit through Buchanan harbor.

GOALS : Encouragement of hydrocarbon and diamond exploration and of the Guinean Nimba project.

Commodity	Resvs	Product	Prod.	Power	Ins-talled	Pot.	Commodity	Prod.	Reserves
Barite	Mt 1.5	Clinker	kt 60	Hydro	Mw 300	Gw 6	Iron ore	Mt 10	Mt 250

Most mining labor is in the interior ; their number has exceded 30,000. In order to attract more business a favorable investment code was enacted including exemption from some duties and taxes. In effect, government subsidises mining. It owns 50% of BMC from whom it receives 50% tax when profit is made (but no dividends) and 62% of LVJ. Foreign shareholders are also the principal customers. The EEC takes most iron ore exports, Germany alone a 30-40%, and Italy 20-25%. The Benelux and the U.S. import less.

Iron ore

I. On the Mano (129) river six hills (W) originally held 50 to 200 Mt of ore reserves grading 54% iron plus 200 Mt assaying 40-50%, but remaining reserves are said to average only 47% Fe. Exports started in 1962 via Monrovia, some mixed with Bomi ore. By 1981 NIOC's output dipped from 4 to less than 1 Mt, mostly fines, even though the former managers expanded capacity to 3.25 Mt.
II. The BIE ridge extends eastward over 40 km. Reportedly, 1.2 Bt of magnetite and hematite would grade here 54.4%.
III. On Bomi Hill and West hills, further east, LMC stripped 3-6 Mty between 1951 and 1977. High grade lump has been worked out and the remaining 65 Mt (out of an initial 100 Mt) grade 45.3% Fe, 0.05-1% S and 1% P.
IV. At Bong-Zaweah, 80 km by rail from Monrovia, Delimco received a 75 year concession in 1958. This subsidiary of German Steel producers led by Thyssen, controls 34% of the shares of BMC, Italian Finsider 16%. BMC does not pay, however, export taxes, royalties and duty on imports. It is undoubtedly the healthier company, even though its ore is the poorest, as shown in the Fe, SiO_2 and P assays of five categories :
a) hard itabirite : 38.2, 41.6 and 0.03%, b) slightly altered itabirite : 37.4, 40.4% and trace, c) altered itabirite 44.32 and 0.05%, d) laterite 42.24 and 0.06% and e) boulders 46.25 and 0.06%. As to reserves, there were originally 370 Mt grading 38% Fe, 7-8% SiO_2, 0.03% S and 0.03% P with a magnetite/ hematite ratio of 61/39.

In 1980 215 Mt were left at Bong Mine and ± 134 Mt on Bong Peak - with a capacity of 30,000 td, 7-8 Mty concentrates are exported, averaging 66.5% Fe and 6.3% SiO_2, including the throughput of two 6,000 t pelletizing plants, i.e. 2.4 Mty assaying 63.5%.

V. The enriched Bassa Hills and Finlay Mts orebodies are accessible by road from Harbel through Owensgrove.

VI. In the Wologisi-Bea Mts Lisco held the largest resources, 1.3 Bt ore grading 45% Fe. Unfortunately, the rail link to Bong would be 250 km long.

VII. In the Liberian third of Nimba (127) resources originally reached 1.6 Bt, excluding the underlying itabirite which grades only 32% Fe. By 1980 reserves of 65% Fe included 50 Mt on Mt Nimba, 30 Mt on Gbahm and 25 Mt on Tokadeh. There were also twice 200 Mt assaying 50-55% a) on Tokadeh, b) at Yeulliton, Gangra and Beeton combined, which would have to be enriched. At full capacity LVJ stripped at Yekepa 13 Mt waste and shipped 13.5 Mt ore over 260 km to Lower Buchanan. This subsidiary of Lamco is managed by Swedish Grängesberg and the reader is referred for further details to pp 86 and 255-7 of MRA.

VIII. On Mount Jedeh, the highest peak of the Putu Range (SE, MRA page 260) 520 Mt of ore consist of up to 10 m of high grade magnetite interlayered with itabirite grading 35% Fe.

Other resources

Monrovia is probably the rainiest capital of Africa and water problems are limited to quality. Degraded yellow and red soils cover the Basement and the rubber plantations grow on the rich soils of the coast.

Oil shows at Monrovia and exploration continues off Grand Cape Mount county. Meanwhile, Liberia imports 4 Mb petroleum. Mines have half of the 300 Mw generating capacity out of which LEC supplies a fifth from the Mt Coffee Dam on St Paul River. During the drought their flow has been reduced to a fraction and plans of the Mano River Union call for the development of its potential. Requirements will decline though when the mines close.

The Norwegian run Liberian Cement Corporation has a 150,000 ty clinker grinding capacity at Bushrod Island near the capital, which is an active diamond reexporting center (see Sierra Leone chapter). Yet, customs clearances have declined from their 1961 peak of 1 M carats to 240,000 ct and $11 M in 1984. Some stones might come, nevertheless, from diggings in the Lofa drainage (NW), Kakata (W) south of Nimba, Tapeta (E), Camp Alfa or even as far as Ghana. There is, furthermore, a 100-125 by 50-75 m pipe at Mano Godua, but the kimberlite fissure fillings in the west are barren (personal communication of Dr. Eugene H. Shannon, Director, LGS).

Similarly, gold easily crosses frontiers and 30-100 kg of exports are usually reported. Further, Bentley International Trading Co. planned to dredge 30 kg gold per month in Grand Geteh county. In addition to the Cavally, the Dubo can be dredged and the reader is referred for further details to pp 87 and 310 of MRA.

Out of six barite veins in the 50 km^2 Gibi area (NW), Nr. 2 contains 70% of 1-2 Mt reserves. Similarly the kyanite deposit near Buchanan (C) may have some importance. Lastly, other occurrences include :

Bauxite : Webo (see iron), Wanigishi (N), Careysburg dist (W), Voinjama (NW) ; chrome : Nzérékore (N) ; clay : Monrovia ; columbite - tantalite, lead, zinc : Tawalata, Kokoya (C) ; ilmenite : Greenville (S), Cavally, Robertsport (W) ; iron ore : Webo Hills (SE), Kpo Range, Mts Kitoma and Wantike, Wangisi and Kpandemai Mts.; manganese Wautiken (SE), Mt Dorthrow (E) 31%, Kingsville (W, 23%) ; platinum creek, Freeman creek, Gbarbor with gold ; uranium : Bong and Lofa counties.

44 IVORY COAST

Côte d'Ivoire's success story is based on years of stability, good management and farming. Hydropower is abundant though (when the rains come) and there is oil and gas. While the development of large low grade nickel (and iron) resources will have to wait, small scale mining of diamonds and gold revives. Through Sodemi and Petroci, the government promotes one of the best run exploration programs and it has been able to attract foreign capital for prospecting and industrial development.

> GOALS : Maintenance of a favorable investment climate. Development of water resources in the arid north and processing industries in the south.

Table 58 Ivory Coast resources and production.

Production			Production Resrvs.					Resources		ore	metal
Cement	Mt	1	Hydropower	Twh	1.8	Gw	0.8	Iron rsvs	Bt	1.2	0.5
Petroleum	Mt	12	Oil	Mb	9	Bb	3	Iron rsrc	Bt	1.7	0.7
Diamonds[P]	Mct	1	Gas			Tcf	4	Nickel	Mt	270	2.8
Feldspar resour-	kt	5	Sulfur cap.			kt	10	Bauxite	Mt	13	0.7
Clay ces	Mm^3	3.5	Gold			t	30	Manganese	Mt	3	0.9

Abidjan, the 'Paris' of West Africa, is the headquarters of the ADB. All roads, and the 1,200 km 'Niger' railway lead there.

Water and energy

Water is scarce in the north, the area of the Senufo, bordering Burkina Faso and the Korhogo area has been hit by the drought. There are only secondary aquifers at the contact of the weathered Basement in the west, but rains are abundant. Likewise, the eastern schists are not much better endowed. Luckily, the best aquifers are the recent and late Continental basins which provide 200 m^3h at Teichville (S), where clays protect water from underlying salt. Although tropical soils cover the southwest, most of the country belongs to the grasslands whose grey-brown soil can be upgraded. Accordingly, the Ivory Coast is a key exporter of coffee. Yet only 10% of the land is cultivated. Bituminous sands show at Ebocco and Eboinda (SE) at Beoou asphalt is at a depth of 3,600 m and at Groguida near Grand Lahou (S) oil and gas seep from 1,500 and 2,000-2,600 m. The 450 Mb Bélier (SE) field of SIOP (Esso 65, Shell 21, Petroci 15%) flows 6,000 bd from a depth of 1,900 m in 60 m of water. Even better, Espoir (Phillips 57, Agip 25, Sedco 10, Petroci 10%) yields in 70 m of water off Jaqueville 17,000 bd from 2-3,300 m depth, to which should be added Kumbo nearby. At Foxtrot 3.5 Tcf gas would require a 62 km long 16" pipeline to Vridi. Moreover other gas resources total 2-3 Tcf and EECI (60%) and SIR (40%) plan ammonia production.

The Société Multinationale de Bitume has a 10,000 bd plant and Gulf and Petroci control the 40,000 bd SIR refinery which is to process 5-6 Mby of Nigerian crude. In effect, Abidjan wishes to emulate Rotterdam, or rather Mombasa , as a refiner.

Even though the boom has slowed down, electricity needs are expected to increase from 2.5 to 4 Twh in the early nineties. By 1987 450 Gwh will be fuelled by oil. Meanwhile 88% of nameplate capacity ought to be generated by hydropower, at Ayamé 50 Mw and 200 Gwh, Koussou 175 and 450, Taabo 210 and 960, and at Buyo respectively 165 and 900.

The Missouli project which would regulate the Marahoué upstream from Taabo has been shelved. Nonetheless, the 1.6 Twh Soubré could provide are indispensable and the N'Diélesso and Malamalassa sites on the Comoë could be harnessed in the nineties with 730 and 600 Gwh.

Clays, cement, diamonds, gold

According to the ECA, clay resources have included (in 1000 m^3) : Gounioubé near Abidjan 1,400, ceramic grade, Moussu 1,200 brickclay, Katida (NEC) 200 colored , ceramic Bafina (W) 30 colored pottery, Cavally (WSW) 300 white, good quality, San Pedro (SW) 130 brickclay.

Cement deliveries declined from the 1.3 Mt of 1980 to half of rated capacity: a) Société des Ciments d'Abidjan (State 40%, Omnium Tropical 24%, SCOA 12%) : 800,000 t, b) Société Ivoirienne de Ciment et de Matériaux (Origny 72%) also at Abidjan : 800,000 t and c) and the clinker grinding plant of Socim (a joint venture of the state and Omnium) at San Pedro (SW) : 300,000 t.

Between 1948 and 1960, 1.5 M carats of diamonds have been recorded and diggings of the Upper Sassandra (W), the Bandama and Comoë and of Seguéla (WNW) supplied 50,000 to 400,000 ty with current purchases of only 20-30,000 cty. At Tortiya (CN) output reached 200,000 carats in the sixties with an industrial/

gem ratio of 1.5. Between 1971 and 74, 300,000 cty were reported, the Watson Co.'s Bobi shaft of 1977 is rehabilitated, Sodemi reasserted control over the diggings in 1983 and Comimat reopened Tortiya in 1982. Output passed thus again the mark of 300,000 ct in 1979 and reportedly 100,000 ct in 1983.

Gold washing continued since 1895 at Toulépleu (W), Soubré (SW) and at Seguéla (WNW) :

SMI : Ity 5.5 t, 8.4 g/t, Comici : Nero and Tabou (SSW) eluvia 70 kg, 1 g/t.
JC syndicate (Eden Roc, Canada and Sodémi 51%) : Aboisso, Ahunta, Akrassi,
Aféma 1 t ; Ayema 2.4 g/t, Aruri - 1.6 t, 16 g/t and 1 t, 0.7 g/m^3. Issia :
Babadougou (Kodian) Sodemi : Bagadiou volcanics (NW) 6 g/t, 0.46% Pb, 0.16% Zn.
Veins Goma, Benou, Agovia. Placers Bandama (SC), Marahoué 1-4 g/m^3, Lobo (WSW).

For further details, the reader is referred to pp 89 and 310-311 of the 'Mineral Resources of Africa'.

Metals

As shown in table 59, laterites grading 1.3-2% nickel still have to be delineated. While there are 5 magnetite and 2 minette deposits, the shipment of Comiferci's (Sumitomo 40%, British steel 20%) iron ore to San Pedro harbor (SSW) would have required a 380 km long watersuspension pipe or rail.

In the sixties up to 80,000 ty manganese were also produced near Grand Lahou (S) and a total of 115 t of columbite-tantalite near Bouaké (C, 119).

Table 59 Principal Ivorian ore deposits with grades in %.

Nickel ore	Mt	%	Manganese ore	Mt	%	Iron ore	Mt	%
Sipilou W	230	1.3	Gd. Lahou (121)S	1.5	44	Klahoyo W	1000	36
Moyango W	16	2	Ziémogoula NW	1.2	47	Mt. Totr W	230	40
Founfouesso W	15	2	Lagnokaha N	1.1	35	Mt. Tia W	160	25
Biankouma W	10	1.3	Dassoumblé N	0.4	20	Mt. Gao W	150	40
Ilmenite			Bauxite	Mt	Al_2O_3	Mt. SegayeW	40	48
Addah S	0.2		Bénéné	10	53	Monogaga SW	150	40
Gd. Lahou-Fresco	0.2		Guéto NW	1.4	55	Sassandra SW	50	42
Southeast	0.1		Elinzué	1.3	44			

Other occurrences include :

Bauxite : Ngouinou, Yaouré, Orumbo ; chrome Doliémé, Wémélhoré (NW) ; cobalt, Cu : Mt Trou, Boundoukou (E), Dabéakola, Guiglo ; Cobalt, Cu, W, U : Boundiali, Gagnou, copper, Pb, Zn San Pedro, copper Tibeïta, Kong, Feldspar Diéou (CW), Manganese Koishi-Ndawa (E), molybdenum Monogaga, Guehielbi, Farake, nickel Mbayakoro, Bouafte, Kansankourou, Iyola, nickel, Co, Cr Biankouma (W), Katiola, Marabadrassa (W), niobium, Ta, Be, Li Baola, Tourré, Agboville, Korhogo-Odienné (N), niobium, Ta, U, Th Tabou (E), rutile Youna (W), pegmatites (Feldspar) Alépé (SE), Anombrako, tungsten Man (W), Katiola, Dimbokro, Hana-Lobo.

45 GHANA

Infrastructure, rolling stock and mines are being rehabilitated and the Economic Recovery Program provides better incentives for cocoa, as well as gold, diamonds and aluminum (Fig. 26) which supplied 10% of GDP, 10-15,000 jobs and 40% of foreign exchange to which should be added hydropower exports.

GOALS : Pursuance of the ERP : A new start for mining, an integrated bauxite-aluminum industry. Further development of water resources in the north.

Table 60 Ghana production and reserves.

Production					Production	Resvs				Production	Resvs	
Aluminum	kt	40	Hydropower	Twh	5	Gw	2	Gold		15→ 20^{p}	t	150
Cement	kt	300	Bauxite	kt	50	Mt	550	Silver	t	1	t	10
Petroleum	Mb	5	Oil	kb	300	Mb	7	Diamonds gems	kct	150	Mct	15
Salt	kt	50	Gas			Bcf	4	" industrials	kct	800	Mct	60
Steel	kt	15	Iron resources			Mt	600	Manganese	kt	250	Mt	5

The state' share of companies is : gold AGC 55% ; SGMC 100% ; bauxite GBC 55% ; diamonds GCD 55%, Oda 100% ; oil 43% ; petroleum 100% ; manganese 100% ; cement 75%.

Kaiser Aluminum controls Valco (with Reynolds 10%) ; other managers and minority shareholders are : Lonrho (AGC), Alcan (GBC), BP 's CAST (GCD), Texas City Refining and South Norway Cement Export Co. Company tax is 46%, dividend withholding tax 5% and profits should reappear before 1990.

Gold, diamonds, etc.

Ashanti gold already reached Europe in the Middle Ages ; recorded output averaged 20 ty in the fifties, peaked with 33 t in 1961 and bottomed with 8 t in 1983, costs and illicit washing have increased rapidly, but with better policing, more supplies, better recovery through the carbon in pulp process, as well as credit from IFC, CCCE and Lonrho, AGC increases production to 12 ton. Furthermore, Canadians strengthen SGMC's management and a 60 ty target has been set for 1995.

Statistics of unreported do not exist, but twenty years ago Ariston provided 4 t of gold, Prestea 1.5 (0.6), Bibiani (W), 2.5 (O) and Tarkwa 4 (0.8 t), where a target of 2 t has been set. The latest figures are in brackets and to these should be added 5 kg from Konongo and 100 kg from the Dunkwa dredge. Both here and at Prestea reserves declined along with maintenance. Yet AGC's underground output dropped less - from 10 ty to 7 ty in 1983. Nonetheless profit has been wiped out.

Tailings and many deposits dot the five trends of hardrock (banket) deposits and six river systems :

Primary deposits

Trend I Kumasi - Obuasi - Prestea (115)

a) Konongo, Awere, Odumuse, Akyenase (20 km, N)

b) Oboum, Obouso, Adakwai, Begoso, Wiawso

c) Obuasi, Sansu, Kan Su, Akapdi Su (8 km)

d) Akropong, Bokitsi, Twoofoo, Atasi

e) Prestea, Anfargah, Ekotokoo, Tuappim, Ampassam (S)

Trends II-V

II Effuenta-Tarkwa,Abbontiakoon (118), Abosso, Akontansi (35 km, S)

III Sariehu, Mamnao, Bepa Ysi (SC)

IV Kumasi-Mansu Kwanta-Mape (160 km, poorer, west of A)

V Bibiani (116, W)

Placers

VI Ankobra with Fura (C)

VII Birim with Pusu-Pusu (CE)

VIII Ofin with Dunkwa (W)

IX Tano, Nantrang (near W border)

X Pra, Jimi, Niwi

XI Nangodi (NE)

Arsenic, the tracer of gold, is abundant at Beposo, Bibiani and Nanwa. For every ton of gold 20-50 kg silver can be recovered and output again approached the one ton mark.

Diamonds are widespread between Akwatia and Aboabo in the 1000 km^2 Birim basin (139, E), including the Atiankama Subinsa, Subin, Supong, Abansa and Esuboni drainages. Stones are small though, often grey, sometimes green. Apparently, the ratio of gems to industrials is 1:11 and clandestine mining affects GCD's profits. Over 50 km downstream of Akwatia, 29 Mm^3 of gravel graded 2.4 ct/m^3 including 15-20% gems in the Oda elbow. One meter of overburden has to be stripped and 15 m^3 bucket draglines are to remove a million m^{3}y. Antwerp merchants contributed funds ; they are to buy through Swiss Inadco 40% of the gems. The small Bonsa field (S) will be rehabilitated too with Pintotum and Nimasin. All in all diggers provided 30% of the recorded output which reached 1 Mct in 1937, 2 in 1951, 3 in 1957, 2.5 in 1972, bottomed out with 300,000 ct in 1983 and might match the 1 Mct of 1981 in 1986. Finally, saphires show at Bonsa, chrysoberyl at Nantia and Beshea and along with tourmaline in most diamond placers.

Water, food, energy, aluminum

There are two aquifers on the coast. Cretaceous-Tertiary sandstones and gravel provide three quarters of the supply, including the cities and in most villages 10-15 m deep wells are sufficient for a family. Three quarters of these are covered by government water supply and needs will double by the year 2000 to 170 Mm^3.

When the rains failed grassland soils of the north suffered most since soils in the remaining two thirds of the country are more moist. They sustain farming when fertilizer is available but only 12% of the land is cultivated.

Salt is evaporated along the coast between Elmina and Keta and recovered at Dabiya (N) along with lowgrade barite.

Saltpond (SW), operated by Primary Fuel of Houston, flowed 1-2000 bd and 36 Mcfd. At Tano 800 bd of 29° API oil flowed from 1900 m depth in 90 m of water. Other shows include Keta lagoon and Agip's Axim. Meanwhile the Tema refinery (5000 bd) imported oil and the bill increased tenfold in ten years.

Hydroelectricity passed the mark of 3 Twhy in 1972 and 5 Twhy in 1980. Nameplate capacity is 760 Mw at Akosombo and 160 Mw downstream at Kpong. However, during the drought inflow from the Volta reached only a fifth of fifty years' average and in 1983 all potlines of the Tema smelter shut down. But by 1986/7 a capacity of 200,000 ty might be reached - at a higher cost since VRA increased kwh prices sixfold. Valco also pays a toll-charge of 68% on the LME aluminum price. The company imports alumina from Jamaica but the Kibi project could provide 800,000 ty from Atewa bauxites. Indeed, various resources totaled 1.7 Bt.

Table 61 Ghana bauxite reserves, resources (Mt) and their grade in Al_2O_3 (%).

Deposit	Mt	%	Nyinyahin (SCW)	Mt	%	Nyinyahin (SCW)	Mt	%
Atewa (SE)	260	43.5	South Aya	30	47	Chichiwere	100	44
Ejuanema (SEC)	1	48	Abantiakrom	20	50	Mpesaso	70	50
Affo (SW)	20	50	Asafo (W)	25	52	Aya Bepo	50	50

Low grade resources : Kanainyeribo, Supiri Beppo, Bekakhirri, Afumba-Nfatahur : 50 Mt ; 15 km NE : Angwinyare, Sumanchichi Hill.

Firstly, the Bascol zone of the Atewa chain contains 120 Mt bauxite grading 42.5-45% Al_2O_3 on Atiwiredu Hill and Asiakwa. Secondly, GBC's (Alcan 45%) exports via Takoradi from the Awaso mine at Ichinoso (Affo) plummeted. Hopefully, the 400,000 ty target will be reached though. Thirdly, Chemokomplex (Hungary) has studied 30 km of the Chichiwere Tente Hills, the longest deposit.

Other resources

The Nauli (SW) limestone is suitable for cement. Lime can be burnt from 2.7 Mt at Asuchuali-Bator (SEC) ; Alcusa (SE, 0.6 Mt) is also well located. At Buipe Busumi (NC) 6 Mt are cement rock out of 100 Mt of magnesian limestone and at Benga Da (NE) 15 Mt overlie dolomite. Meanwhile, in the south Ghacem's (25% S Norway Cement Export Co) grinding plants rated 600,000 ty at Takoradi (SW) and 800,000 ty at Tema (SE). Nonetheless Ghana will import half of its requirements until rehabilitation is finished in 1987.

Other industrial and building materials include :

Andalusite : Abodum ; asbestos : Anum and Akuse with talc, Zanidaw and Cape Three Points with magnesite ; clays brick Accra railway, Subin Valley, Takoradi, Ankaful, Cape Coast, Ho, Dunkwa, Winneba, Inchaban gypsiferous, Lake Bosumtwi, Temale, Bui (N) ; kaolin : Saltpond (S, 400,000 t) ; pigment : Jamise, Huni, Foso, Intabrawso ; pottery : Kwahu, Eddubura, Dukludja (E) ; corundum : Neibwale, Birim ; feldspar : Princes Town, Winneba, Mangoase, Cape Coast ; Etuakrom and Saltpond with mica, Garnet Ho and Mankwadzi in rocks ; Osudoko, Adaklu Ho (SE) placers and with diamonds ; gypsum : Pokoasi highgrade, Krobo (NE) ; kyanite : Jang, Gurupe (N), Kulapawn ; lime : Amedua-Tefle ; limestone : Kangan-Twendani-Edu (S), Akubi, Longoro Juju Hill, Wenchi, Dabaya (N), Jimam (N) ; mica : Yipala (N), Ampia (S), Ajumaco (S) with talc ; nepheline Kpong ; silica glass sand : Petepong, Tarkwa, Aboso, Eduapriem, Abontiakon ; pure quartz : Zuarungu ; quartzite : Nyafomang (SE) ; sandstone Ajena, Elmina, Komenda (S) ; Bolgatanga (N) ; staurolite with diamonds ; talc : Birim.

Nsuta (SW) has produced much of the world's battery grade manganese. Production has peaked with 1 Mt in 1952, rose to 400,000 in 1975 and declined to 160,000 t in 1973 and in 1982.

Simultaneously, reserves of high grade oxide ore of GNM have also diminished. The new plant can supply, however 300,000 t of MnO_2 nodules from 500,000 t of carbonate feed and exports should increase. Ultimately, the oil price will determine loss or profit. At Nsuta 22 Mt carbonate grades 31-20% Mn ; little is left at Yakan, but other resources include : Three Hills 2.5 Mt 30%, Odumasi (SC) 1.7 Mt 20%, Kalimbi (NW) 6 mt 14%, and Kapili (N) 4 Mt 24%. At Shieni (ENE) 1.3 Bt of iron ore grade 35-50% Fe, 18-37% SiO_2 and 0.33 P. At Open-Mansi (SW) 50 Mt average 52% Fe and at Pudu (CN) 5 Mt might be upgraded to 52.5% Fe. The Tema (SE) smelter uses meanwhile scrap and produces 5-15,000 t steel with a rated capacity of 30,000 t, and other occurrences include :

Beryl : Cape Coast-Winneba (S), Barkum (W), Senya Bareku ; chromium : Anum ; columbite-tantalite : Makoba lagoon, Akim Oda (S), Assuboad ; lithium : Ejisu (SC) ; monazite : Kpong (SE).

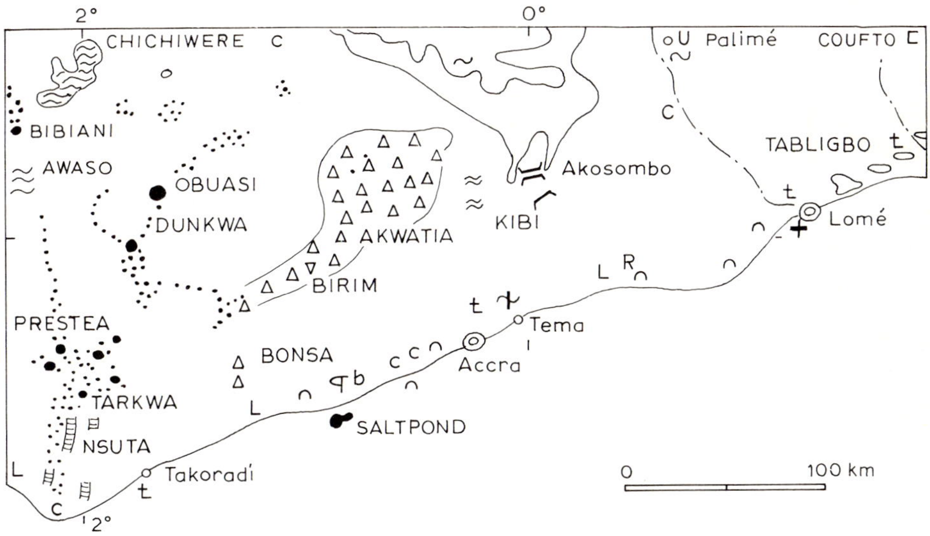

Fig. 26. Southern Ghana and Togo. (Legend on page 13.)

The République togolaise extends 600 km inland and Cotomib's phosphates spread from Adéta near the coast beyond Momé (Fig. 27).

OTP, the world's fourth exporter markets the concentrates which contribute half of the exports, 15% of GDP and a large part of government revenue.

> GOALS : Production of attapulgite, phosphoric acid and later fertilizers. Offshore oil exploration, diversification into building materials and development of water resources in the north.

Togo exported 100 t of asphalt in 1979, but it imports oil, hydropower and steel scrap since iron ore was 450 km from Lomé.

The government owns and promotes the mineral industries which employ 4,000 people ; it also provides generous benefits for investors. The EEC was the principal trading partner, but with shipments to Austria, Turkey and India France's traditional half share of phosphates has declined since there was even some trade within Ecowas.

Table 62 Togolese production and resources.

Resources			Production					Production	Resources
Fuller's earth	Mt	30	Fertilizer*	kt	330	Phosphate P_2O_5	kt	800	Mt 360
Limestone	Mt	205	Petroleum	Mb	(3)	Phosphate resvs			Mt 80
Dolomite	Mt	150	Salt	kt	200	Hydropower	Gwh	20	Mw 500
Iron	Mt	220	Chalk	t	300	Clinker	kt	700	
Bauxite	Mt	1	Steel	kt	5	Cement	kt	300	
			Rolled steel	kt	10	Marble	kt	1	Mt 500

Phosphates, clays, cement

From their beginnings in 1961 shipments increased steadily to 2.5 Mt concentrate in 1974 and hovered just under 3 Mty, i.e. 0.8-1 Mt P_2O_5 until 1980. The Kpemé concentrator has a capacity of 3.5 Mty, on the coast 30 km from Hahotoé and Kpogamé (110) where 4 and 2 M rock can be quarried. Indeed, two tons of rock grading 65.5% BPL are required to produce a ton of 79-80% BPL concentrate. Overburden may reach 30 m and its ratio to rock ranges from 3.2 to 5.6:1 (Krauss, 1983), but only one out of three phosphate beds is worked now. First, 25-50% of clay is washed out with seawater from the ground rock. Hydrocyclons and a second washing and drying provide the highgrade concentrate carried by a conveyor belt to ships anchored at Porto Seguro. The recovery plant produces 8% iron and aluminum oxide and slimes grading 31-32% P_2O_5. Engineering work has also started on SET's 1000 td phosphoric acid and fertilizer plant.

There are five deposits : Momé is followed by Dagbati, where phosphorites are up to 4.5 m thick ; this deposit might be worked before the nineties. Nearby, the largest, Hahotoé-Akoupamé extends over 11 x 1 km and the phosphate series is up to 8.5 m thick. Lastly, at Kpogamé and Avéta (SW) phosphorites are up to 10.5 m thick.

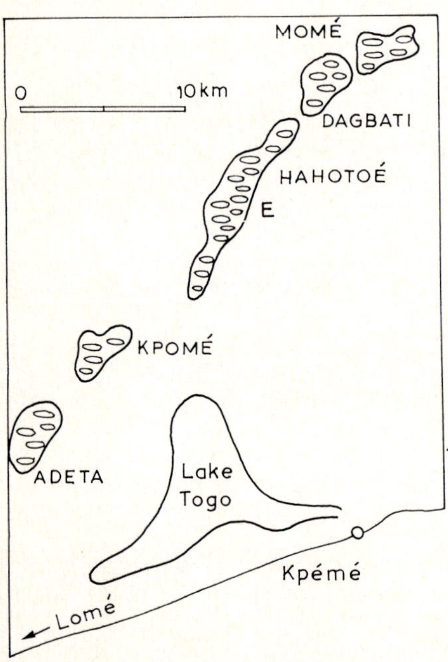

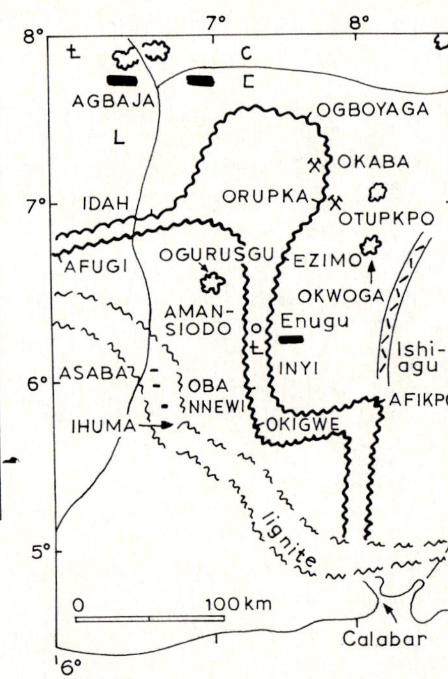

Fig. 27. (Left) Togo phosphates and attapulgite.

Fig. 28. (Right) Coal in Nigeria. (Legend on page 13.)

There are traces of uranium in the phosphates which are underlain by bentonitic clays and Fuller's earth (attapulgite - montmorillonite) and limestone. The largest deposit of the latter is at Avéta (200 Mt), followed by Kpogamé (4-10 Mt). The Cimao plant, a Togo, Ivory Coast and Ghana joint venture at Tabligbo near Mono (not far from Hahotoé) had a capacity of 1.2 Mty and projects called for shipping 300,000 t of clinker for grinding over 80 km to Ciments du Togo (a partnership of Swiss Lambert Fr. et Co. and the state) at Lomé, rated at 800,000 ty. To continue, sand, gravel, high grade clays, bricks and tiles are worked near the towns and Salinto recovers salt from Anécho lagoon.

Other resources

The quality of water increases towards the coast where clays and sands yield 3-5 m^3h/m and sandstones 5-40 m^3h/m. At a depth of 4-500 m the Late Continental offers here the best potential with 6 m^3h. The Cretaceous yields only 2.3 m^3h and the Paleocene only 0.4 m^3h which add up to 75 Mm^3y.

There is less water in the north and typical yields are in m^3h according to the seasons :

Region	wet	dry		wet	dry	formation	wet	dry
Dapango	10-30	0-20	Buem	30-55	10	Atacorian	20-30	0-40
Oti basin	40-55	10-55	Kandé	20-40	0	Dahomeyan	20-70	0-20

The 20,000 bd Lomé refinery has been shut down, Diesels generate 100 Mw and three times as much hydropower is imported from Ghana.

A 100 km long rail line reaches Palimé, near Ghana where copper shows. The Mt. Agou bauxites, nearby, grade 46-53% Al_2O_3, 18-30% Fe_2O_3, 1-1.5% SiO_2 and 1.2% TiO_2.

In the center, on Mt. Ahito, PUK delineated 30,000 t ore grading 25-32% Cr_2O_3 plus additional resources on Mt. Djéti nearby. Magnesite is in the vicinity and titanium further northeast between Agbandi and the railhead where limestone crops out.

Dolomites are quarried at Kra (CS) and along with marble at Pagola (C, 150 Mt). Gold shows at Kéran, Bungba and Momoin around Mt. Togo and 50 km NE at Agbandi.

The Lomé furnace had a 20,000 ty capacity, and the rolling mill rated 40,000 ty corrugated roofing. Far away, Mt. Djolé, at Banjéli, holds 42 Mt of iron in hematites grading 42-54% Fe with additional resources at Kabré nearby. However, Bidjabé with 600 Mt ore assaying only 32-33% Fe, 20% SiO_2 and 0.23% P is bigger. In the north the Sassané-Mango and Oti iron-manganese belt extends into Niger. Gold shows at Kara and next to the Perma placers of Benin, and clay at Kpodji

(NE). Finally, SMCC in which the government has also invested works the 500 Mt Gouawoulou marble deposit.

47 BENIN

Palmoil is the principal export of the République populaire du Bénin and Sémé oil alleviates the import bill. The principal trading partner was the EEC, mainly France, followed by the U.S.

GOALS : To find and distribute more water in the north. To continue the search for oil and other commodities.

Table 63 Production and resources of Benin.

Resources		Production		Production		Resources	
Marble	Mt 6	Cement	kt 300	Oil	Mb 3	Mb	100
Limestone	Mt 50	Gravel	kt 20	Hydropower	Gwh 20	Gw	2
Phosphate	Mt 1.5	Salt	t 200	Iron		Mt	160
Gold	t 3			Chromite		kt	10

The government controls the cement industry but the oilfield is operated by Dr Vittorio Fabbri's Pan American group and UNDF, Soviet technicians and BRGM prospect the north.

Unocal discovered in 1968 23° API low sulfur crude at a depth of 1900 m at Sémé in 40 m of water. Saga Petroleum has increased flow with water injection to 10,000 bd and Pan American has set a target of 25,000 bd. Oil also shows 30 km inland from the western coast and lignite lenses are at depths of 60-100 and 80-200 m at Grand Popo and Porto Novo (S). Twenty Gwh of hydropower are generated, six times as much as a decade ago, and a joint scheme with Togo calls for Nangbéto. Conversely, thermal power has declined from its 1972 peak of 90 Gwh. Installed capacity is 25 Mw, mostly at Cotonou and another 25 Mw are available from Ghana. The Mékrou project has not materialized however, for lack of funds.

Other resources

The Arlhan limestones are covered by only 0.5-1.8 m overburden and thick beds average 80% as far as Abeokuta, Nigeria. Onigblo and Mecise, near the Pobé railhead each had 18 Mt reserves and SCO (governments of Benin and Nigeria 49%, and 41% and F. D. Smidth 10%) might reach an ultimate capacity of 200,000 ty cement. On the other hand, the state's two clinker grinding plants, SCB at

Cotonou and Sonaci at Porto Novo have each a 200,000 ty capacity.

Groundwater constitutes a problem even along the coastal cities. The Maestrichtian provides 45-100 m^3h, mostly north of the Lama depression. The Late Continental has also been tapped, but the Kandi sandstones supply only 4 m^3h in the wet and 1 m^3h in the dry season. The problem is further compounded in the hard rocks of the north and dry soils of the savannah extend as far south as Abomey, the old capital of the sultans of Dahomey. There is marble at Coufto, nearby and at Diado further north. One outcrop out of four holds 6 Mt of dolomitic marble.

400 ty salt are recovered at Porto Novo (S) and the phosphate belt crosses the south from Lama to Toffo and Pohé. At Athiémé only traces remain, but Mékrou holds only 4.5 Mt of strippable reserves. There is no mining either at Borgou and in the far north, the larger Tapoa deposit of Niger overlaps the border.

During World War II workers recovered 300 kgy of gold from the upper Perma placers (NNW), the chromite lenses of Buem (Bontomo, N) hold only 6,000 and 9,500 t ore grading 37% Cr_2O_3 (Cr/Fe 2) plus traces of nickel and the iron ores (NNE) include

	Mt	Fe %	SiO_2	Al_2O_3	P	S
near Kandi Loimbou-Loumbou	250	50	16	0.5	0.8	0.4
550 km from coast Madekali	60	58	3	2	%	%

Other occurrences comprise :

Bauxite : Kandji (NE), especially Bodjécali, chromite Moti, Matéri, clay : Coufto (SE), copper, glass sands Houéyogbé, Sémè, gold Konadogo, M'Binah, ilmenite Coufto (Abomey), kaolin Kétou, Sakaté (SE), monazite Duémé, with kyanite Okpara, rutile Birni, Djougou (N), Kilibo, Pehunco (N), Savalou (C), Tchaourou, Toui (S), Tin group Dawa, Parakou-Ouaria.

48 NIGERIA

The gross national product rivals that of South Africa ; at peak output, the value of minerals surpassed it and funded an unparallelled development program for a population which according to statistics matches that of 33 African countries combined. The Federal Republic also has problems, mostly financial and managerial. Debt service is prohibitive, yet government revenues and exports have been halved by restrictions. Since Nigeria is a wayward member of OPEC, the share of oil also varies from a third to a fifth of GDP. Admittedly, illegal exports and demand have been scaled down. Still, the strategic question can be summed up thus : oil revenues now or later ?

> GOALS : Discipline and leadership. Agressive marketing of LNG and energy projects and development of water, gas and coal resources. Downstream, broadening of the industrial base from processing into manufacturing.

Table 64 Nigerian production and reserves.

	Production		Reserves			Production		Reserves	
Cement	Mt	4			Petroleum	Mt	60		
Limestone	Mt	1			Oil	Mb	4-500	Bb	17
Marble	kt	3			Gas produced	Bcf	700	Tcf	50
Shale	kt	150			Gas marketed	Bcf	10	reinjected	2
Stone	1000 m^3	400			Hydropower	Twh	4	Gw	10
Kaolin	t	700			Coal	kt	100	Mt	90
Clays	kt	30			Lignite			Mt	50
Feldspar	kt	5			Iron ore			Mt	380
Fertilizer	kt	7			Steel	kt	400 P		
Salt	kt	3			Tin	kt	1	kt	10
Lead	t	200	kt	100	Columbite	t	70	t	500
Zinc	t	100	kt	80	Tantalite	t	10	t	100
Gold			t	6	Monazite			kt	13

Some 60,000 people work in the mineral industries which are fully or majority owned by the Federation. NNPC is, therefore, the country's largest corporation, and the other oil companies report a much smaller turnover.

NNPC share %		in refineries		NMC share %		NMC share %
80	Shell	80	Pt Harcourt	60	ATMN	100 NCC coal
60	Mobil, Gulf	100	Warri	60	Bisichi-Santar	100 NEPA power
60	Agip, Phillips	100	Kaduna	40	Makeri smelter	100 NSDA steel
				100	WAPCO cement	100 Delta steel

After two appraisal wells per field and the expensing of 12.5% royalty on oil and 10% on gas 66% of tax is paid on earnings until the amortization of preproduction costs, when it rises to 83% ! Government tends to control allowable liftings and buys back crude occasionally. Since NNPC delays investments oil companies favor neighboring countries, even allies of Comecon. Conversely, NNPC develops fields itself and state companies promote useful projects and the merger of the tin mines.

As to exports, half reaches the EEC (mainly France, Germany and the U.K.) and a quarter the U.S. from the Bonny, Forcados and Escravos oil terminals and

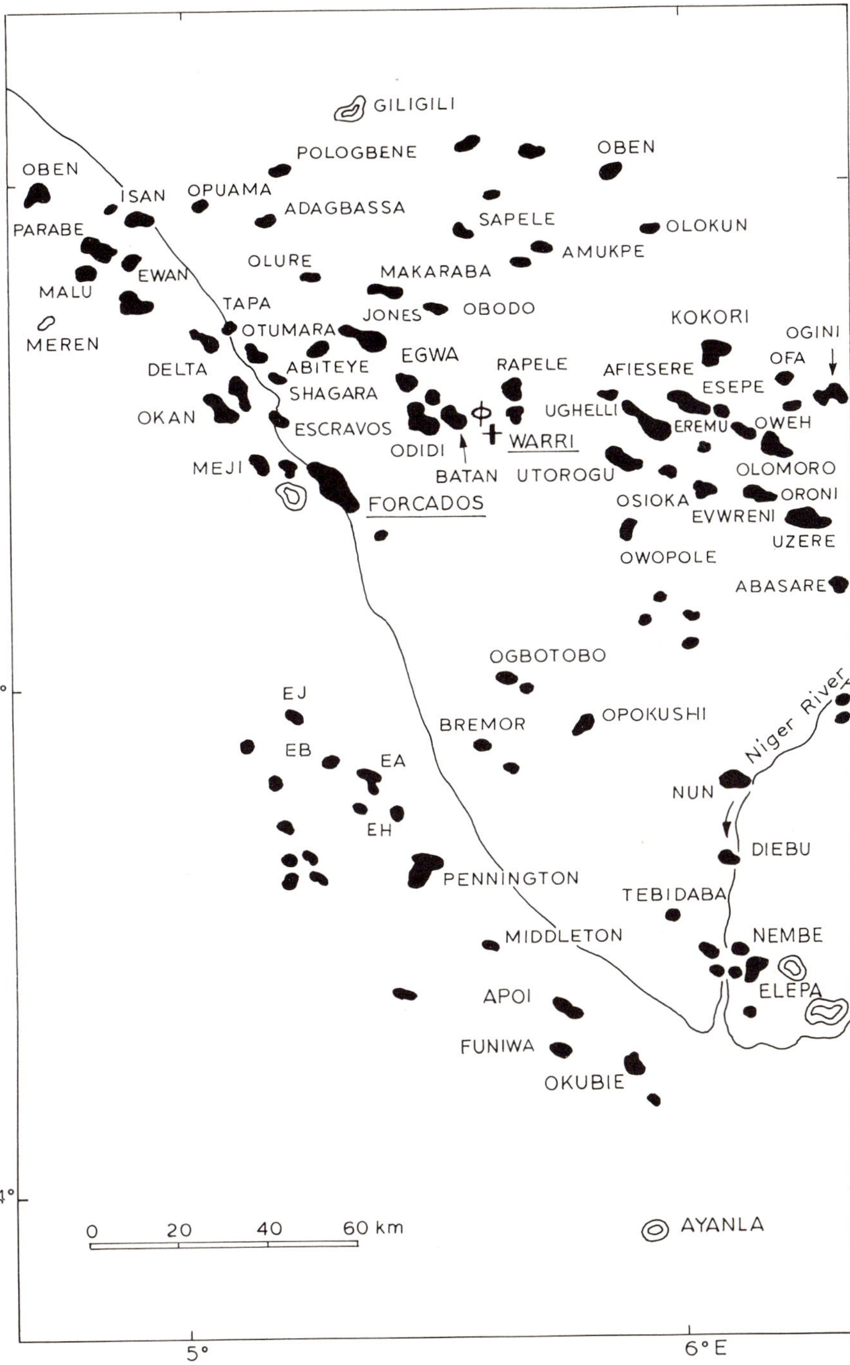

Fig. 29. Oil and gas in the Niger Delta (west).

Pt. Harcourt, Warri (and Calabar) harbor.

Oil reserves and production

In the early 1900s, Germans saw oil seep at Okitipupa (SW). After two decades, Shell-d'Arcy found Oloibiri in Tertiary beds in 1956. Production started at Bomu and Imo in 1958, between 1976 and 78 seven giants were identified offshore and in 1981 Mobil located the largest, Edop, on land. More than 100 fields operate out of 300 discovered and the ultimate number of commercial fields may reach 500.

Production peaked with 800 Mb in 1974 and 78. Although capacity has now declined to 600 Mb, by 1990 cumulative output will not be far from 20 Bb. While proved recoverable reserves ranged from 15 to 20 Bb in recent years, total reserves including the undiscovered, have been put at 33 Bb. Since traps are of the rollover type, mostly limited by growth faults, fields average only 50 Mb and 1,600 bd per well, i.e. as low as 500 bd per pay , whose depth ranges from 2,300 to 4,000 m. Yet, on the eastern shelf a quarter of the fields flow from only 1,500-2,000 m.

Demand permitting up to 1,500 wells can flow in the Delta, a third of which is in shallow waters. However, the favorite Agbada reservoir sands hardly extend beyond 40 km from the shore or 200 m of depth. While only one field out of ten is operated offshore, their numbers are to increase with tax incentives.

The proportion of 'giant' fields - one out of forty - is one of the lowest in the world.

The less important among multiple pays become more interesting therefore, as well as deeper traps. Constant exploration is required thus. Luckily, Nigeria has one of the most productive wildcat/find ratios. Yet the western extension offshore, the Anambra platform to the north and the Bornu-Chad basin (NE) are less enticing.

Fields and operators

Out of a maximum height of 100 m, only an average of 15 m are productive ; to put it otherwise : 70% out of thicknesses of less than 15 m and only 5% out of a column of more than 45 m. Because of the variety of reservoir sands, source rock properties are poor. In some fields porosity reaches 40% at 1,500 m and then declines to 15% at 4,000 m. Because of the high sand/shale ratio water drive is not efficient either, only 50 wells are on artificial lift, and in some fields, less than a third of in situ reserves can be recovered. Consequently,

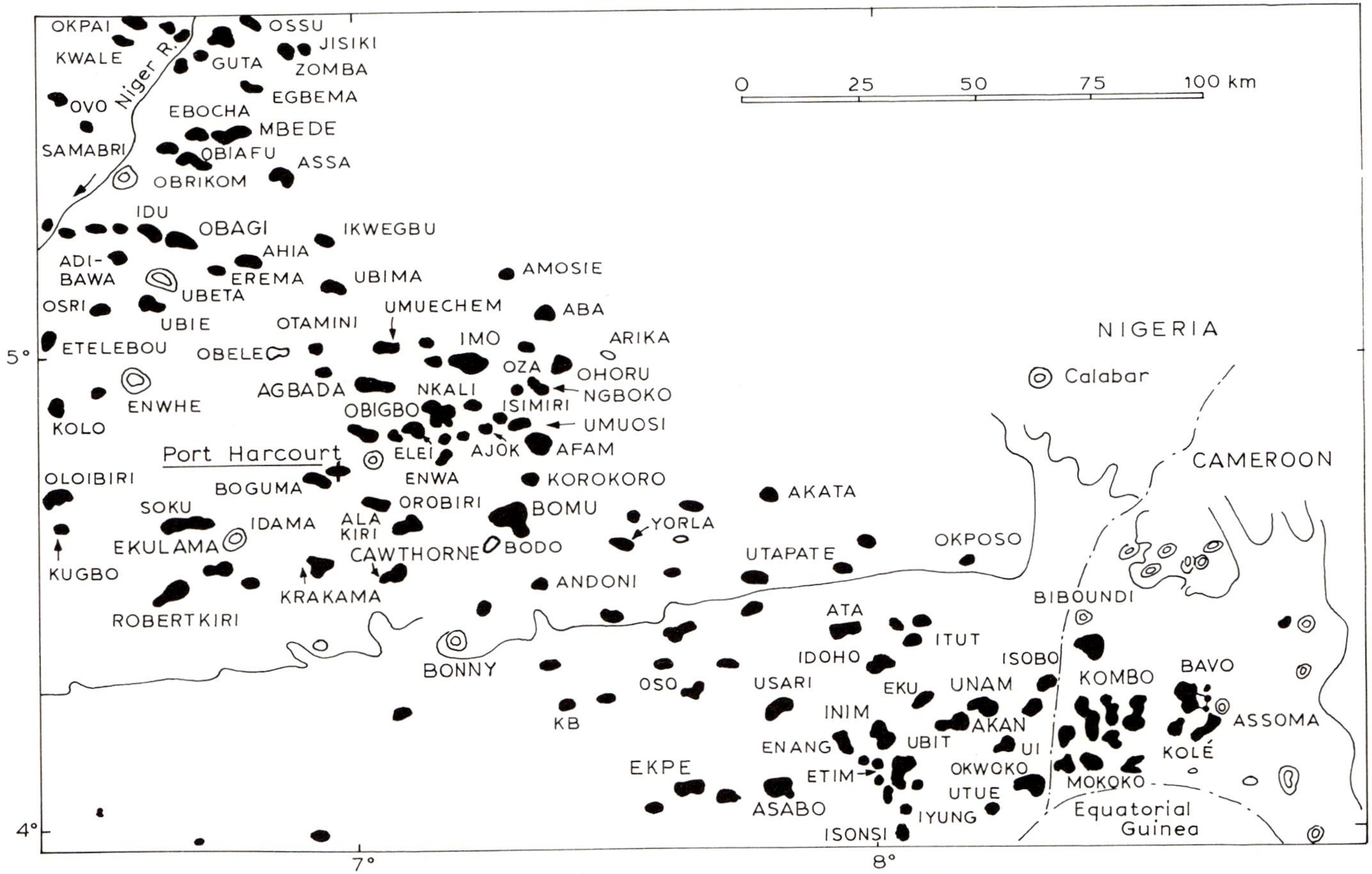

Fig. 30. Oil and gas in the Niger Delta (east) and the Rio del Rey basin.

oil properties vary widely - from 16° API to the light 50° API. Heavy - low pour point - crudes are found higher than lighter, paraffinic, oil. Medium crude with 26° API is predominantly naphtenic and popular 'Bonny light' has 37° API. The average is though 34.6° API, 5% paraffin wax and 0.15 weight % sulfur.

The largest fields are (in Mb) :

Onshore : Edop 1000 bd, Jones Creek 800, Imo River and Yokri 700 each, Nembe Creek 650, Olomoro 600, Bomu, Kokori, Mbede and Obagi 500 each.

Offshore : Meren 650, Delta South, Eken and Okan 500 Mb each.

The same fields (Edop excluded) also supplied the largest cumulative output and the biggest current producers are in 1000 barrels per day : Obagi 60, Gulf Okan 30, Meren 30, Chevron North Apoï 35, Forcados Shell Midwest : Yokri 50, Benisede 40, and Opukushi 35, Shell East Nembe Creek 55, Imo River 35, Ekulama 30 and at Mobil Ubit 30. Others include :

20-25,000 bd Shell Midwest : Node Oben, Jones Creek, Otumara, Olomoro, Odidi, Shell East : Cawthorne, Kolo Creek. Mobil : Enang.

15-20,000 bd Agip : M'Bede, Obama, Tebidaba, Gulf : Delta South, Makaraba, Mobil : Asabo, Ekpe, Iyak, Shell Midwest : Afiesere, Egwa, Kokori ; East : Adibawa, Etelebou, Agbada, Obigbo, Oguta.

Shell's share of production is more than 50% ; so is their number (Nr) of fields:

Area	share %	Nr	Area	share %	Nr	Operator	share %	Nr	Operator	%	Nr
Shell Midwest			Shell East			Gulf	14	15	Texaco-Socal	4	5
Forcados	16	13	Pt Harcourt	10	14	Mobil	15	15	Ashland,		
Ughelli	9	15	SE Pt. Harct.	6	5	Agip-Phillips	10	22	Impex	1	4
Node	2	4	Other	13	11	Elf	9	6	NNPC, Deminex		
									Tenneco		

Gas, distribution and refining

The Delta is even more of a gas than oil province. Since gas was not the prime target of exploration, gas reserves have not always been registered when found. Currently, 70% of these are onshore, the proportion of gas to oil increases though down dip and seaward, where much exploration remains to be done. Variations in estimates are not suprising thus from a total of a) 35-40 Tcf proved (Oil and Gas Journal) to b) 46 Tcf (Petroleum Information International) c) 75 Tcf and d) 85 Tcf proven plus 65 Tcf undiscovered recoverable reserves (IBRD, 1983). Average reserves per field are small indeed : 170 Bcf associated and 210 Bcf of non-associated gas and only 17% of the latter are in fields of more than 1 Tcf. Proportions of these gases also vary from b) 15/60 and c) 50/50. To be sure, three quarters of the associated gas are caps which might disperse through the basin as more oil is produced.

The produced gas/oil ratio has been rising constantly though from 800 to 1300 cf/b, but it has not reached yet the EUR reserve ratio of 2700 cf/b. More than 90% of associated gas has been flared ; in effect, a third could not be economically recovered because of gathering costs and the low depletion cost of non-associated gas in the ground. More gas is reinjected though since 25 ¢/Mcfd penalize two thirds of flaring.

There are now only a few small gaspipelines at Oben and NEPA's Sapele plant, at the Aladja steel mill, between Warri and the Ajaokuta steel complex, but the earliest pipes of Shell at Port Harcourt and Aba have to be replaced. If the necessary investments were made, gas consumption might increase to 3.50 Bcf or 170,000 boe, equivalent to current oil usage and requirements might reach by 1990 440 Mcfd at Lagos, 150 at Ajaokuta, 300 at Aladja-Ughelli and 370 each at Warri-Sapele and Port Harcourt-Afam.

Two thirds of the total would be used for power generation and the $6 B required for the Bonny scheme might be found by 1990. Meanwhile, Saipem and Snamprogietti build a 36" 350 km pipe from Escravos to Egbin near Lagos with a 600 Mcfd initial capacity. In effect, most of the 3.5 Twh power is oil and gas generated and the production of methanol and fertilizer has been considered.

Before the year 2000, 500,000 ty of butane and propane could also be recovered from LNG of the Warri-Lagos pipe with an investment of current $1.5 B. While apparent usage of low priced LPG seems to increase rapidly, better policing of the borders has reduced requirements of subsidized oil by 100,000 bd. Some observers, therefore, put current needs at only 150,000 bd, which is met by existing refineries. Port Harcourt, the oldest, has a capacity of 60,000 bd and Warri and Kaduna are rated at 100,000 bd each. Before 1990, demand will outstrip nevertheless current capacities and a new 150,000 bd refinery at Alesa-Eleme near Pt. Harcourt has priority including a 40,000 bd of motor gasoline.

Coal, hydropower, uranium

Sub-bituminous coal production started eighty years ago, doubled to half a million tons by 1944 and almost reached 1 Mt in 1960. Calorific values range from 5,000 to 6,000 cal/g or from 5,500 to 6,500 airdried.

All production comes from the pear-shaped a) Enugu basin which extends from Dekina (SC) to Odokpono and thence 150 km to Okigwe (S). From north to south reserves of $>$ 1 m thick seams are in million tons : Ogboyoga 110, Okaba 70, (Otupkpo, E), Orukpa 60, (Okwoga, E), Ezimo 50, Enugu 50 (Amansiodo W, Ogorugu NW), Inyi 10 (Awlaw). The bracketed deposits are located near the preceding fields. The basin also has several outliers : I the Afikpo-Calabar (SE) elbow II the Idah-Afugi (SC) strip III Koton-Karifi and Jamata on the Niger, further north and IV Lafia, north of the Benue.

Productions (and capacities) after rehabilitation are underground : Onyeama and Okpara 25,000 ty each (150 to 400,000 ty) and opencast : Owukpa 25,000 and Okaba 15,000 (300,000 ty each). These could supply the Nkalagu cement plant and the Yander and Ashaka projects (NE). Coal shows though there too b) along a 100 km long arc centered on Gombe station.

To continue, the 20-40 km wide lignite belt connects Orlu (SE), Umu, Ezeala, Umuahia, Ihoma, Nnewi Oba (S) and then across the Niger to Ogwashi-Asaba-Mgbiligba and Adiase-Uti. But for other occurrences the reader is referred to pp 98 and 532 MRA.

Uranium shows in the Kaffo Valley and between Gombe (see 'coal') and the Benue, in the Waza waters, and phosphates assay 0.004-0.007 % U_3O_8 at Ososhun, 0.0065 at Shoyinka (Ifo) and 0.008% at Apon.

The Niger has a regulated outflow of 860 m^3d at Kainji. The 65 m high concrete dam stores 11 Bm^3 water and generates up to 4 Twh.

Table 65 Capabilities (MW) and averages (Gwh) of Nigerian hydropower sites.

Niger	Mw	Twh	Site	River	Mw	Twh
Kainji W	760	2.2	Shiroro CW	Kaduna	300	1.3
Jebba P W	500	2.2	Zungeru^p			
Lokoja^p SC	2000	8.5	Makurdi^p	Benue	1000	4.2
Onitsha S	750	3.3	Yola CE	Benue	600	4.7
Others : Afikpo 180 Mw, Atan 180, Gembu 130, Garin Dali 130, Richa 70, Sarkin 50,			Katsina N	Katsina	260	1.1
			Beli SE	Taraba	240	1
			Ikom SE	Cross R.	400	1.7
Guadjikiri, Kurru 40 each, Kombo, Gwaram, Kasimbila, Ifon 30 Mw each.						

Water, food, diatomite

Water supply is inadequate in the arid, but densely populated north and center. Some water is imbibed though in the Eocene sandstones of Sokoto (N) and in the sandy interbeds of the Chad clays (NE) which lie at a depth of 100-250 m. Worse, at Maiduguri the hypothetical water level at the contact of the granite is at 1050 m of depth. Nonetheless, some artesian wells flow here, in Bornu, 0.5-20 m^3h. In the Niger basin, sandstones provide good aquifers between Kontagora, Mokwa and Bida, but clean water is scarce even in such areas of heavy rainfall as Adamawas (E), Benin, Owerri and Onitsha, where the water table is at a depth of 200 m since impervious sediments cover large areas of the south.

Still, a third of the land is cultivated, a maximum in Africa and a tenth is irrigable. To continue, leached laterite soils only cover the coast and the red-brown soils of the savannah can be improved by fertilizer.

Unfortunately the Ajegunle-Ifo Junction beds at Ososhun (SW) are thin. Some phosphate also shows at Akinsinde, but numerous farmers need fertilizer and a 1500 td urea and 1000 td NPK plant has been considered at Onne, using 24 Bcf gas from the Pt. Harcourt area.

Brine springs are abundant along the Benue and salt recovery is scheduled to increase from a few thousand to 200,000 ty at Ijoko (Ogun) and Oyhereki near the Sapele oilfield. A salt dome is at Makurdi (CE) and in Bornu (NE) high grade diatomite covers a 250 m wide basin. At Abakire layers are 2.5 m thick ; diatomites also crop out at Badice, Garin Maina and Bugzu as far as Maiduguri (NE) and lie at a depth of 40-60 m at Gujba, Bolokolo and Sabon Kassa.

Building and ceramic materials

Asbestos Shemi, basalt Jos, Bin, Longonda plateau, clay Benin, Onitsha, Oweri , fire clay in Lower Coal Measures of Enugu with shale, kaolin Ropp (C), light clay Jos, pottery clay Ikorodu (SW), Abuchi, Okigwi, residual clay Kwali (Abuja) white clay Arkella NW, Dakingari, Garkidda, Kutigi, dolomite : Osara, graphite : Birnin Gwari (N), in the northeast : Maio, Butabe, Hossare Nuwa, Gayo, Jauro-Jalo, gypsum Chad formation (Bornu, NE), Sokoto (N), Benue (E), Onitsha (S), limestone Abeokuta-Ijebu, Nkalagu-Idumale (SC), Sakura (SC), Ukpilla (SC) with marble, Ewekoro (SW 20 Mt), Gombe-Numan (EC), Sokoto (N 100 Mt), marble Jakura (SC), Lokoja (SC), Mfamosing (SE), mica Egbe, glass sand Enugu escarpment, stone Ifon, Ondo, talc Ilorin, Oyo, Zaria, Niger (Abuchi).

Granite quarries can produce 3-400,000 m^3 aggregate, works 30 M bricks ; Jakura ships limestone to the Ajaokuta steel mill and Mfanonsing to Warri. Finally 8 cement plants work at half of their nominal capacity of 4 Mty ; as shown in table 66 :

Table 66 Nigerian cement companies, shareholders, plants, kilns, fuels and capacities (1000 t).

Company	share%	Plant	kiln	fuel cap.	Company	Location	kiln	cap.
WAPCO	NNSC 60	Evakoro	SC 2sd	o 700	Cement of North	Kalaimbana		600
WAPCO	Portld. 40	Shagamu	SC 2w	o 900	Ashaka Ct.Co. NE	Bauchi	2d	680
Nigerian Cemt[a]		Nkalagu	SC 5w	c 600	Benue Ct. Co b)SEC	Gboko (Yandev)		900
Ukpilla CmF.		Auchi	SC 2d	o 700	Calabar Ct.Co, SE,	CrossR. state,banks		400
			o(il),	c(oal)				

Shareholders : a) Anambra and Imo states, b) Benue State 35%, Cementia 12%. The Federal Government is also involved in the Nkalagu and Calabar companies and owns 43% of Benue Cement. With most capacity concentrated in the southwest, it also plans a plant at Makurdi (900,000 ty) and at Maiduguri (NE, 700,000 ty).

Steel, iron ore, metals, barite

At Itakpe (SC) 600 Mt ore assay 39% Fe, 0.9% P and 0.08% S and near the coal and limestone deposits of Enugu (S) 200 Mt of laterite grade only 32% Fe. Exploration continues on Muro Hill (C) and AOM hopes that at Okene (SC) removal of 800 and 300 kt waste from east and west will provide 200 and 300 kty ore. Meanwhile, the Aladja plant of Delta Steel Co. imports iron ore and Nigeria is part of the Guinean Nimba consortium. It has access to Warri harbor and uses 70 Mcfd of natural gas. With a capacity of 1.4 Mty this high cost facility will satisfy part of the 4 Mt demand and the Ajaokuta mill is rated at 1 Mty. It has its own light section and provides billets for the Jos (C), Oshogbo and Katsina (N) mills which will roll together 400,000 ty. Katsina has even expanded to a capacity of 300,000 ty.

Some bauxite shows at Oshogbo (SW) and barite has been crushed and sized at Azara (C). Up to 2 m thick veins occur at Lefin (Ogoja), at Dumgel (Adamawa, E) and this drilling aid is abundant in the Ishiagu-Gwonda (SC) zinc-lead belt. Arofu (CE) and Zurak (E) are smaller here than Abakaliki (SC) where 1 Mt of ore assay 10% Pb and 8% Zn, 70% of which has been qualified as reserves. Other estimates claim 700,000 t of lead and zinc. Closed in 1976, another shaft has been sunk nearby at Amari. Silver shows at Zurak and it accompanies fluorite at Arena and Akwana. On the other hand gold is thinly spread over the 150 km long Kontangora belt (SW) e.g. in the old workings of Iperindo, near Ilesha (6 t of reserves) next to the Itagunmodi jig and at Beni and Laka where NMC has delineated 950 kg of reserves. All in all 50-60 kg have been reported between 1963 and 1974 and exploration continues in Sokoto (N).

Tin and columbite - tantalite, etc.

From a peak of 10,000 t and 7,000 t during the war, tin and columbite concentrates plummeted, even though additional resources of 600,000 t and 500,000 t (!?) reportedly exist. Yet low cost deposits have been worked out and little labor is available in the Jos plateau (CN), where the Nok culture flourished 500 to 100 B.C. ATMN now provides only half of the concentrates, three quarters of which is supplied by contractors, but the Dorowa dredge was still operating. The minors, Bisichi-Jantar, Ex-Lands and Kaduna Prospectors are also part of the rationalization scheme of NTMC (an offshoot of NMC) which might slow the decline.

Some 15 m of basalt has to be first removed now and drilling at Ngell was not

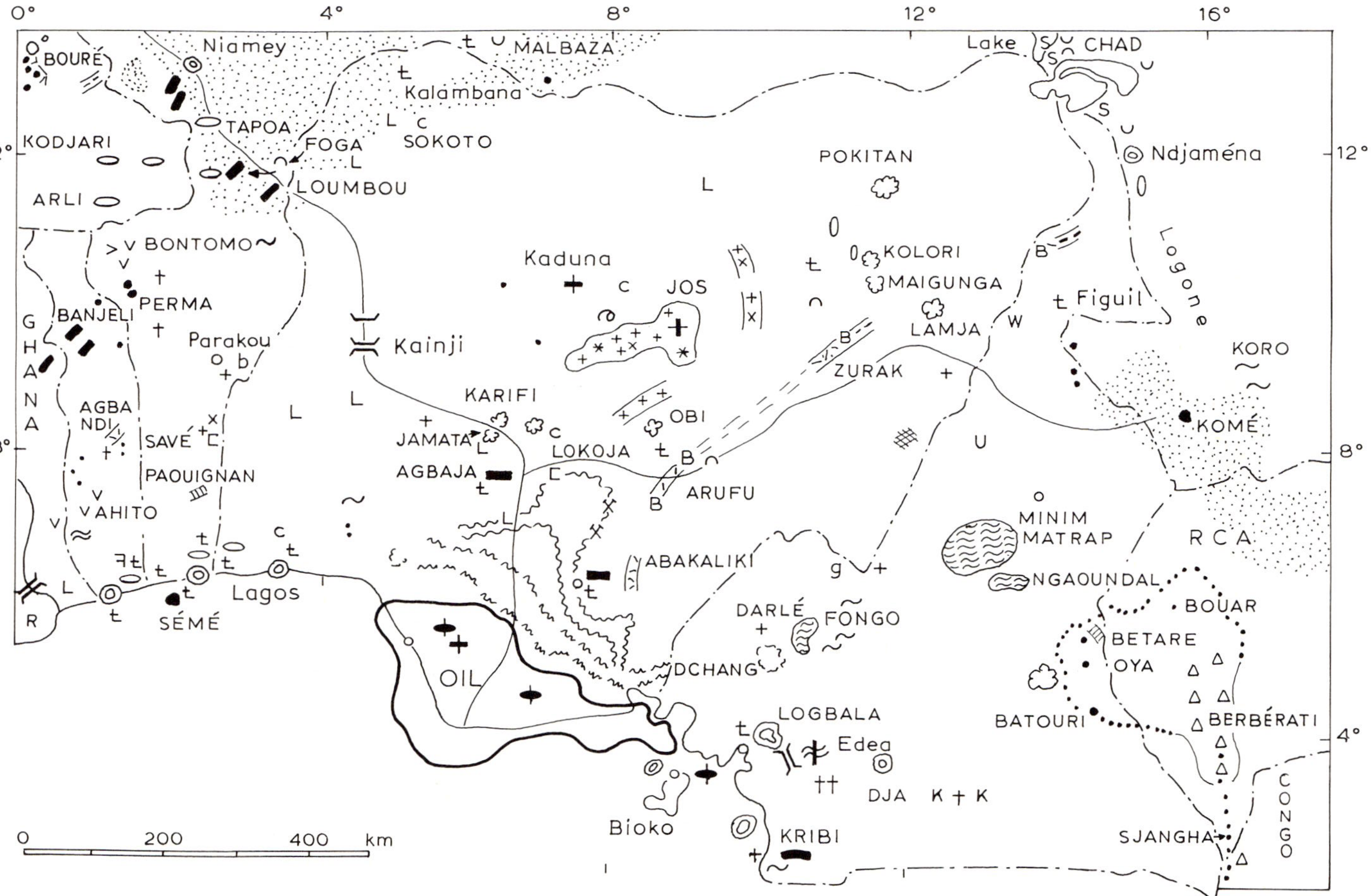

Fig. 31. Nigeria and its neighbors. (Legend on page 13.)

encouraging. Remains, meanwhile, NMC and Gold and Base Metals Ltd's Lirui n'Kanb dyke at Ririwai (NC) with 2,000 and 4,400 t proved tin and zinc in assays of 0.82 and 1.7% plus as many indicated reserves. Another project, TAMP concentrates on Gurum, Rafin, Jaki and Banko. In Benue, Niger, Oyo and Ondo states laborers occasionally wash some cassiterite and ATMN produces a ton or two of wolframite. Tantalum shows up in the Makeri slags and reserves of 8,000 t are said to be computed on this basis. The smelter operates at part of its 18,000 ty capacity and uses imported lead for welding supplies and even the diecasting of motors has been considered.

Lastly, zircon, monazite, garnet and tourmaline show on the Plateau, but better beryl shines at Kabba and molybdenite occurs on Kigon Hill (C).

49 CAMEROON

The République Unie has low debt charges by African standards since rapidly rising oil production matches three quarters of the budget and half of exports. It is the motor of the rapid industrial development along with gas and aluminum. Yet, in the distant but underpopulated north bauxite is more abundant than water.

GOALS : Further development of oil, hydropower and groundwater resources. Utilization of flared gas for LNG, fertilizer and possibly steel reduction.

Table 67 Cameronese production and reserves.

Reserves			Production			Production			Reserves	
Bauxite	Bt	2	Aluminum	kt	80	Petroleum cap.	Mb	15		
Kyanite	kt	50	Cement	kt	600	Oil EUR	Mb	45	Mb	500
Ilmenite	kt	20	Limestone	kt	60	Gas			Tcf	5
Tin production	t	10	Marble	t	600	Hydropower	Twh	2	Gw	23
Gold output	kg	20	Pozzolana	kt	80	Coal, workable			Mt	10

SNH owns 60% of Sorepca, 50% of Tepcam, 20% of Pecten and 66% of Sonara ; SNI controls 48% of Alucam, 43% Amencam and 35% of Total Cameroon and of Sefercam and the state companies also seek partners for development. Although oil companies bear most development costs and heavy taxes, the investment climate is excellent. The largest companies are SNH, Elf Cameroun and Alucam and the major export market is the EEC, above all France, Germany and Italy ; the U.S. is an excellent customer for oil.

Oil and gas

Sorepca's (Elf 40%) dozen fields in I the Rio del Rey basin (Fig. 28, SW) contribute 80 % of production which is to peak with 160,000 bd in 1986. Ekoundou and the Kombo group, discovered between 1972 and 78 flow 50,000 and 30,000 bd of 24-29-36 and 35-37° API crude respectively from depth of 1550-1950-2150 and 1800-2150 m. Smaller fields around Kombo include Bavo, Boa, Inoua and Kita. A 36,000 bd pipeline will serve Lokele and Pecten's (80% Shell) Abana and Mokoko. Further, Modale Marine is Pecten's latest find.

The geology of the Agbada sands is more complicated and shale diapirs (summits) constitute the trap of the Asoma, Biboundi and Kole fields.

II. Tepcam's (Total and Mobil) Moundi field flows 7,000 bd of 40° API oil instead of 30,000 from 1,500 m depth 90 km offshore Limbé in 50 m of water. EUR reserves of the southern structure might reach with gas injection 30 Mb. Other deposits include Sanaga Sud (1.3 Tcf gas), Jabane, Natandi, Hiari and Obanekan. Further, the Sonara (Total 10%, Elf, Mobil 8% each) refinery at Port Limboh is being uprated to 43,000 bd and serves Chad and the RCA.

III. In the Douala basin sediments reach 7,000 m thickness ; reserves are of Turonian-Santonian ages and salt 'diapirs' appear. The first Cameronese field, Logbaba (14 Bcf) supplies gas here to Douala since the fifties. IV. The offshore Kribi gas field (S) is just within the limit of the 500-1400 m thick salt layers, typical of Gabon. Its 3.5 Tcf reserves of gas could supply the Lobabe LNG plant which Sogazcam might build in the nineties. This joint venture of Elf, Mobil, Pecten, Total and SNH has already initiated negotiations with a consortium of Ruhrgas, Gaz de France and Snam (ENI).

Hydropower, iron, aluminum

The Mamelles (Nipple) Range and Mewongo Hill between Kribi (SW) and the border contains 130 Mt of iron ore grading 57% Fe in laterites and 37% in hard rock. Séfercam (BRGM 31%, GFM Inc. 24%, Krupp 10%) seeks to develop the deposit which is only 20 km inland, i.e. near the gas field and the government would join partners in a 11,000 ty rebar mill.

Alucam's (52% PUK) capacity at Edea will be increased to 160,000 t based on imported alumina. Fluctuations have to be attributed to the dry season flow of the Sanaga hydropower plant which has a capacity of 140 Mw. The Song Loulou plant should increase capacity to 380 Mw by 1990, for a total of 680 Mw. Other dams and projects include Mbakaou 2.6 Bm^3, Bamendjin 1.8 Bm^3 and Mapé 3.1 Bm^3. The Lagdo could generate 72 Mw, Njock-Mpoumé on the Nyong 350 Mw, but harnessing of the Djerem and Noun, two Sanaga tributaries, may have to wait. All in all production doubled in twelve years out of a potential of 144 Twh and requirements may reach 3 Twh by 1990.

Moreover 500 Mt of coal accumulated east of Letta (SE) ; they would be difficult to work, like the thin lignite seams of Dchang (SW) in the extension of the Nigerian belt. Hot springs in the Adamawas are not of immediate interest either. Likewise, uranium shows near Poli in the Adamawas, and with molybdenum at Ekomedion. Other occurrences recall the Gabonese Franceville ores.

There is also a 1.2 Mt bauxite occurrence I at Rocher du Loup (SW).
II More important, the 2000 km^2 of lateritic plateaux in the eastern Adamawas near Minim and Matrap (NC) are 1000 km from Douala by rail, at an altitude of 1,200 m. They are 8-9 m thick and 1.4 Bt of reserves plus 600 Mt of resources grade 43% Al_2O_3 , 3.4 SiO_2 , 23% Fe_2O_3, 4.5% TiO_2, 0.5% MgO, 25% water and less than 0.2% K_2O, 0.2% Cr_2O_3, 0.01-0.02% V, 0.03-0.07% Mn and 0.05-0.15% P_2O_5.
III At Ngaoundal 11 m thick layers hold 90 Mt grading 41.7% Al_2O_3 and 1.4% SiO_2.
IV At Fongo-Tongo near Nkongsamba station (SW) Alucam estimates that access roads to an integrated facility via Melong, Dschang and Bafoussam would cost \$40 M and a hydropower plant at Noum \$130 M. 36 Mt of bauxite assaying 45% Al_2O_3 and 4% SiO_2 can be screened out of 46 Mt. To the northeast, 4 Mt of ore in the southern Bambouto area average 46 % Al_2O_3 and 4.3% SiO_2. East of Dschang bauxite shows between Bafang and Bafoussam and there are also traces at Fokemezoum-Fokoné.

Other resources

Sabem, a joint venture of Milchem Inc., Houston, and of the state, imports barite from Morocco and Egypt. The grading plant has a capacity of 50,000 ty and SCE's closed Bonaberi (SW) fertilizer plant is rated 90,000 ty. Even aluminum sheet for roofing has been produced since 1962.

Soft white trachyte of Mt Cameroon is a good building material too and the Bonaberi cement plant (SW) imports clinker and gypsum for a capacity of 900,000 t. Cimencam, a joint venture of Italian Finsider (31%), French Lafarge (15%) and the state also has an oil fired dry kiln at Figuil (N) ; the grinding units have been uprated to 100,000 t and capacity can be increased to 570,000 t to facilitate exports to Chad.

Water supply is more of a problem in the north too. There is next to no groundwater in the Mandara Mts and in the plain of the Logone and Shari. Some Artesian water (2-5 m^3h) rises from the 250-300 m isobath near the Basement in the Chad basin (N) and Cretaceous sandstone reservoirs yield 50 m^3h in the Benue basin (NW). Finally at the foot of the Adamawas (W), surface alluvia drain water towards Lake Chad. The most gold (750 kg) was washed in 1942 in the Lomkader region, near Batouri. About 9 tons have been recovered to date in the southeast, mostly between Meigonga and Bétare Oya. Some gold was carried further downstream in the Sanaga drainage and 1.5 kg have been reported at Faro, near Ngaoundéré (NC).

Workers wash up to 20 ty of tin from Mayo Darlé, notably from tailings. The placers yielded 320 t in 1939, including Taparé, and a target of 500 ty tinstone has been set again at Wum Banyo.

There are reportedly 180,000 tons of rutile in the Akonolinga. From this tributary of the Nyong, near Edea and Banyo east of Kribi (SW) recovery peaked with 3,300 t in 1944. According to BRGM estimates 140,000 t of sands grade 5% ilmenite near Kribi, 300,000 t assay 4.6% near Douala and other shows include : Blacksands Ngambe, Mbalmayo, Akonolig (W), brickclay Nkolbissou, Djapoma, Edea, Mabenga, clay Douala, Yaoundé, Méfou, graphite Bétare, kyanite Dja, Niyba (Edea), Yoo, nickel Akouafim, rare earths Mt. Elephants (SW), tungsten Nahé near Garoua (N), with bismuth Ribao, with lead and zinc Goutchoumi (N), Uranium Ngoyavang, Djoun-Oveng.

5 CENTRAL AFRICA

Until deforestation started jungle covered most of this subregion which is bigger than the U.S.A. It is further distinguished by the world's largest hydropower potential, its biggest copper and cobalt deposits, important diamond workings - licit and illicit - and by great differences of income, ranging from rich Gabon to impoverished RCA. Although one of the better endowed parts of the continent, it is also far behind South Africa and the Maghreb in processing.

51 CENTRAL AFRICAN REPUBLIC

Boats sail from Bangui down the Zaïre/Congo towards the sea, 1500 km away with cotton and coffee. Their revenue outranks diamonds which ought to provide a third of exchange earnings. In order to promote official sales the RCA lowered the export tax on gems.

On the other hand, diggings provide some livelihood for 40,000 families.

> GOALS : To control the diamond and gold trade, to find partners for the uranium and cement ventures. To provide more water in the drought stricken northeast, to continue exploration for oil and to establish a mineral map.

Table 68 Production and resources of the Central African Republic.

Resources		Production		Resc.	Production	
Iron resv.	Mt 2	Hydropower	Gwh 65	Gw 11	Diamonds gem	kct 500
Iron resc.	vast	Uranium^{+}	t 1.5	kt 17	" " industrials	kct 100
Limestone	Mt 50	Phosphate		Mt 2	Gold	kg 150

Plans call for regulating the M'Bali since 10-30,000 km^2 of the Shari (N) basin are flooded 3 to 4 months each year. In contrast, water is scarce beyond the Bongo Mts (NE). Elsewhere shallow wells tap the water table and the moderately developed soils of the southern corner are wet. The use of fertilizer and lime could of course improve the leached soils of the tropics.

Resources

While diamond exports rarely exceded 100,000 cty since their discovery in 1914, half a million were cleared through customs each year in the early sixties and late seventies, just to decline to some 300,000 cty later on. Two thirds are purchased by 200 ambulant buyers and many have been cut by CND, a joint venture of Americans and the state.

In the Berbérati district (W), diamonds occur in the right hand tributaries of the Mambéré and in the terraces of such rivers as the Batouri Bolé and Ngoélé (Fig. 18a, 31). On the other hand, in the Kotto drainage of the Bria district (E), at Ouadda, Yalinga, Boungou and Ouandjia diamonds are transparent or green brown.

More than 5 tons of gold have been produced, mostly before the war, but only a few kilograms per year have been reported since then. Then, after 1978, recorded output increased to 30-50 kgy at : I Pouloubou and Roandji (E), II Dar el Koutti on the Kotto (E), III Bouar-Baboua (W) and the Upper Sangha diamond fields (SW). Further, exploration continues at Yaloke and Abba.

The Fatima-Bobassa limestones extend over kilometers south of Bangui. Ceramic clays are near the capital and the EEC has financed a cement survey. Tocages imports half a million barrels of petroleum, but hopefully oil exploration will start again in the northwest near the Chadian shows. Bangui is, meanwhile, served by the Boali hydropower plants with capacities of 9 + 10 Mw (80 + 150 Gwh) and the Kotto, Safari and Lobayé could generate 360, 300 and 260 Gwh.

More interest focused on 6.3 Mt of phosphates scattered in 80 m deep and 70-150 m wide depressions of Bakouma (S) which grade 14, 17 and 27% P_2O_5 and 0.4, 0.22 and 0.25% uranium in three layers. Primary ore averaging 0.25% U_3O_8 includes most of the 17,000 t reserves of uranium. URCA (1/3 Alusuisse, 1/3 state, 1/6 CFMU, 1/6 Cogéma) has operated a pilot plant and other indices include : Chromium : Bossango, copper Biaro, Cu, Ni, Cr, Co Kanga-Bandoro (C), manganese Tayoro (Goutekeri,W) , mica south of Ippy (SE), tin Bocaranga, Bambari with gold, iron ore East of Ouadda (NC), M'Berri between Rafai' and Yalinga (E), Roandji (C), Kaga Yama, Kaga Bagongolo, Bogoin near Deamora north of Bangui. Bogoin proper represents only half a million tons of iron ore and a larger area 3.6 Mt.

52 GABON

Fortunate République Gabonaise (Fig. 32-35) is a middle income country. Luckily oil matches half of GDP since debt service is high. Yet only 7,000 work in the mineral industry. Elf Gabon, the largest company, employs 1,400 ; 3,800 men produce a third of the world's manganese which covers a third of U.S. imports. Although France is the major trading partner, exports even reach the Far East and the U.S.S.R. Extension of the 700 km long Transgabonais reroutes manganese to Libreville. Hopefully iron ore will find customers too by 1990 when a long delayed branch reaches the northeast.

Table 69 Gabon reserves and production.

Reserves			Production			Production			Reserves
Iron	Mt	700	Petroleum	Mb	8	Oil	Mb	90	Mb 600
Barite	Mt	1.2				Gas produced	Bcf	70	Bcf 600
Talc	Mt	100	Cement	kt	180	Gas marketed	Bcf	5	
Copper	kt	250	Gold	kg	100	Hydropower	Mw	160	Gw 18
Ferro manganese,	kt	80	Scandium cap.	t	1	Uranium ox.	kt	1	kt 40
Silico project	kt	50	Mn pellets	kt	110	Manganese ore			Mt 400

GOALS : Marketing of more gas and development of more hydropower. Broadening of the industrial base and extension of the rails towards the heavy ore deposits.

Government is increasing its share in the principal oil companies, Somifer and the gold diggings from 41 to 45 and 49%, to 50-70% in the refineries. It only owns, however, 20% of the manganese, 30% of the uranium and 35% of the barite companies. French Elf and CEA dominate the oil and uranium companies and U.S. Steel holds 36% of the shares of the manganese company. All businesses make profit, and private interests even have a share in the power company SEEG. Government further fosters investment with 5 and 10 year exemptions on profit and real estate taxes and reduced import duties.

Oil and gas

are in three areas : a) the Pointe Wèze show (NW) b) Port Gentil (W) and c) Mayoumba (S). The decline of reserves has been halted and production will hopefully reach again its 1974 peak of 80 Mb. Out of 100 fields or shows, most of them offshore, 20 are usually in production through 300 wells, but 120 have al-

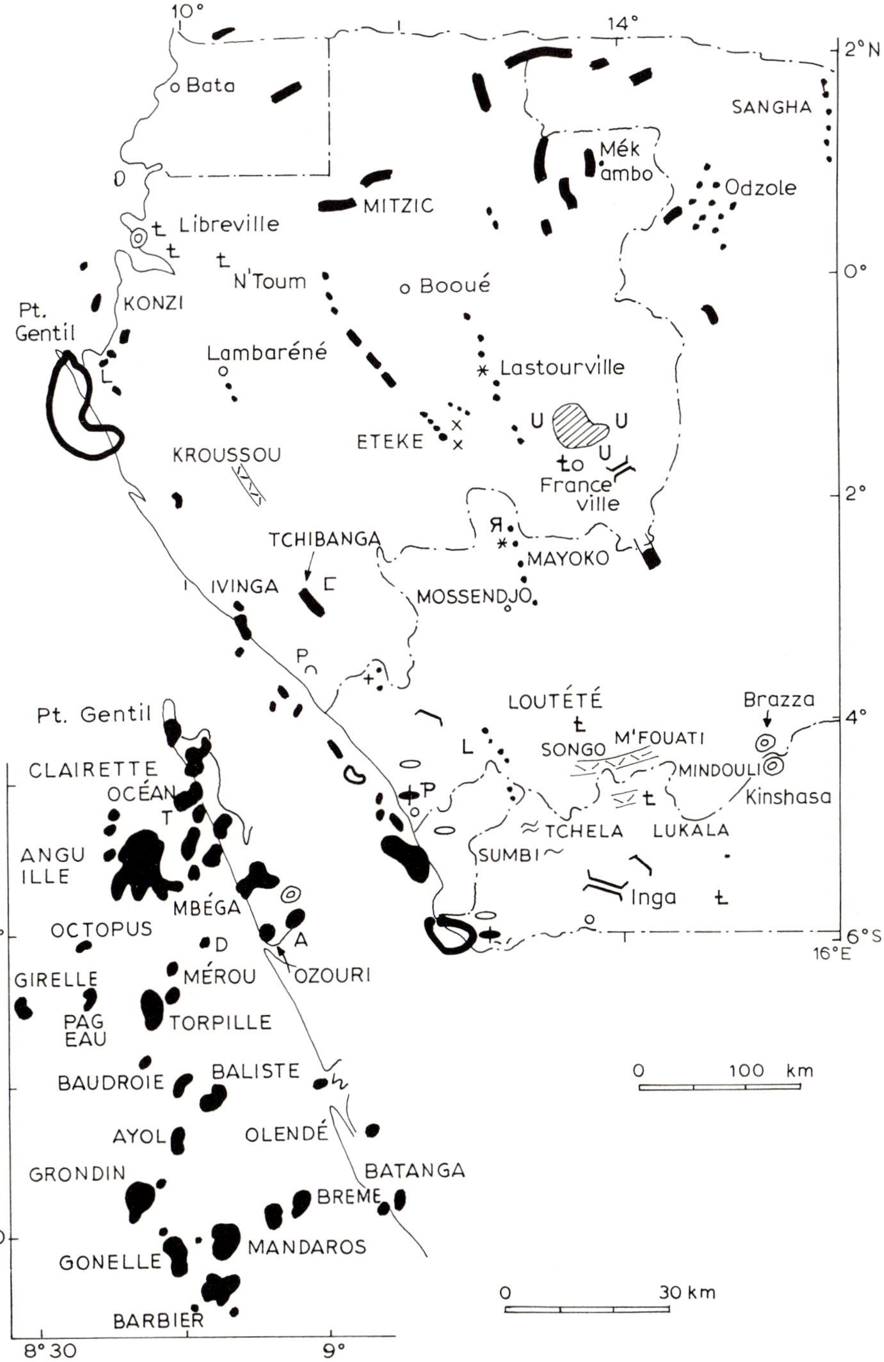

Fig. 32. (Upper right) Gabon, Congo and west Zaïre. (Legend on page 13.)

Fig. 33. (Lower left) The Port Gentil oilfields.

ready been shut in. Two thirds of the Port Gentil reservoirs are in the Batanga sands which surround the Ogooué delta. Out of a gross thickness of 200 m, pays average 25-30 m. 'Post-Salt' marine shales of Turonian and Senonian age are the source rock and salt mountains (anticlines) the usual trap. Porosity ranges from 15% (N'Dombo sandstone) to 30% (Gamba sands). Mandji crude is heavy, low in sulfur and of intermediate paraffin content, as shown below.

Field	depth m	1000bd	API°	Field	depth	1000bd	API°
Ogouendjo	1800-2000	25-40	37	Grondin	2250	13	16
4 Anguilles	2200-2760	26	36.5	Barbier	1800	13	16
Mandaros	1470	14	20	Brême	1500	12	36

Elf (SNEA) and Elf Gabon produce 130,000 bd and control the 20" 70 km long sea pipeline. Their discoveries include a) Olendé and M'Poulounié 3500 bd, Dorée Dôme onshore, b) Barbier (S) and Konzi 5600 bd (50% Petrogab, 15% Elf Gabon, 35% SNEA), Baudroie Marine 8000 bd, Grande Anguille 6000 bd in 1986, Ablette Ouest, Eyena offshore and c) M'Bya, M'Wengui, Kala and Paka (S).
Others comprise :
Mitsubishi and Elf-Gabon at Baudroie, Shell, Elf and Gulf : Lucina, Wed Gabon : Ayol (50% Japanese National Oil Co), Tenneco leader at Obando (30,000 bd, 36° API) and Octopus (6000 bd). Amoco alone at Gombe (6000 bd), developed Ogouendjo in 30 m water with Wintershall, Preussag and Deutscher Schachtbau (35%) and BP (15%). Burmah Oil operator at Longhi.

Only 8% of the gas is used and the Port Gentil refineries operate at part of their capacity. Sogara (51% state, 19% Elf Gabon, 19% GCP) is rated at 33,000 bd and COGER (70% Elf Gabon, 30% state) can export 27,000 bd.

Uranium and power

The 100 Mw share of SEEG's thermal station's should decline though, since more hydropower will be harnessed :

River M'Bei	Mw	River	Mw
Tchimbélé	70	Ogooué petite Poubara	20
Kinguélé down	30	Ogooué gde Poubara proj.	320
Kinguélé upstr.	40	Ngouiné ch. Impératrice	420
Other sites : Liboka, Lebombi, Mafoula-Matato, Akréni, Angouma.			

Fortunately, the grande Poubara project could be developed with a first 100-110 Mw and 300 Gwhy step.

Uranium is widespread along the edges of the Moanda manganese plateau (Fig. 33) in the 'Fl a' Franceville sandstone. Comuf's principal deposits follow over

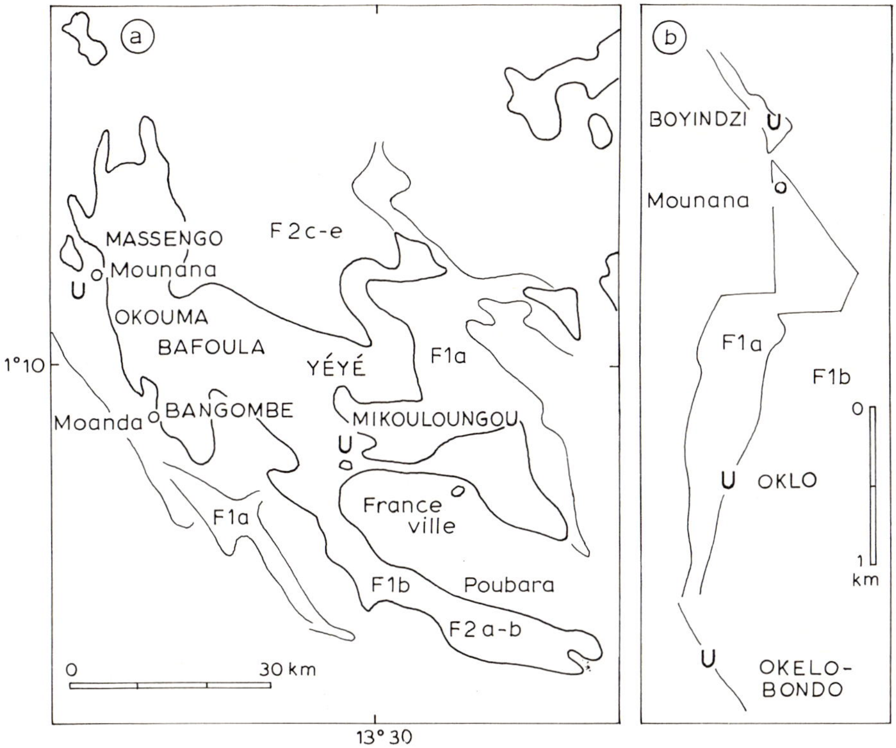

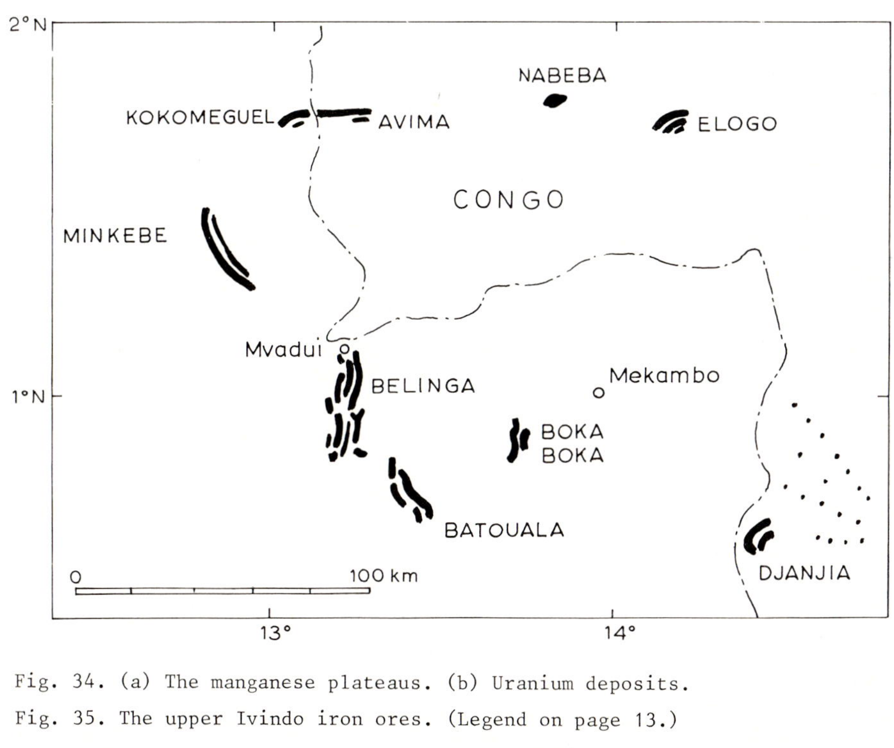

Fig. 34. (a) The manganese plateaus. (b) Uranium deposits.

Fig. 35. The upper Ivindo iron ores. (Legend on page 13.)

10 km precisely such an edge. Their original reserves were in 1000 t (with grades in %) :

Oklo 18 (.42%), Okelobondo 8 (.44%), Mikoloungou 10 (.35%), Mounana 6 (.48%) Boyindzi 3 (.44%).

Some 20,000 t have been already extracted and Mounana has been depleted. Still > 40,000 t resources are left and Mikoloungou, further east, has not been worked yet.

Further northeast, the Taiwan Power Co. explored the Booué-Lastourville-Mounána area, other partners prospected southeast of Franceville and Kaya-Kaya (S). A joint venture of Cogéma (49%), of Korean Power Co. (41%) and the state, has already located a deposit at Nord Leyon, north of the Comuf zone. Others prospected on the upper Ogooué towards Akiéni and Kouala-Moutou (SE).

Comuf leased its permit from CEA whose subsidiary Cogéma held 19% of its shares, and Imétal, its affiliate, 39%. Comuf's two underground mines can supply 450,000 tons grading 0.35% U to the concentrator which upgrades ore from 5 to 35-55%. The yellow cake plant produces 70-75% uranium oxide ; recovery is 95% and potential throughput almost 2 Mt. Sales depend though on the nuclear fuels market. The same holds for scandium : in 1978 600 kg were produced, but vanadium is not recovered.

Manganese and iron ore

Manganese ore (and metal) reserves of the southeastern plateaus attained in Mt : Bangombe 90 (40), Ikouma 110 (56), other Moanda area 200 (100) plus another billion tons of resources. At Bangombe a 4.5 m thick layer covers 50 km^2, but Okouma spreads only over 25 km^2 . It is also more irregular, like its extension, Bafoula. At Yéyé-Maila only boulders are left and the ore of Massengo is more siliceous. Other occurrences are scattered a) at Okoudja 100 km northeast of Moanda, b) on the Kanda (C), a sub-tributary of the Ogooué and c) east of Kimongo.

Moanda is the world's largest manganese mine. Production peaked with 2.3 Mt in 1975 ; it is dependent on the steel market, since shipments of chemical and battery grade manganese tend to reach only 150,000 t. Ore is moved by a chain conveyor to Moanda and was shipped thence by an 80 km long ropeway to M'Binda and a 270 km long narrow gauge railway to the Congo-Pointe Noire line.

Comilog's shareholders include Coframines 17%, Imétal 13, Samaf 7 and Elkem 6%, and Sogaferro's : Comilog 10%, Norwegian and Japanese interests 20 and 15, Belgium and Italy 10% each. However, the smelting of up to 150,000 ty of ferro- and silicomanganese out of 300,000 t ore depends on 1.25 Gwh from the greater Poubara scheme. Meanwhile, Comilog and the state acquired control of the Société de Ferromanganèse de Paris-Outreau.

Plans call for shipping 15 Mty of thrice homogenized, graded iron ore of 6/30 mm and agglomerated fines of 0/6 mm by 1990 from South Bakōta-Mombo, Kombo, Babiel and North Bakota hills at Belinga (NE). Indeed, shareholders of Somifer, (currently Comilog 17%, BRGM 6, GEM 10, Bergbau AG, Voest 4 each), had to wait thirty years for the completion of the railway, which will absorb together with Owendo harbor 50% of the $1 B investment (in 1982 currency). According to Socomine, the managers, reserves averaging 64% Fe, 3.4% SiO_2, 2.6% Al_2O_3 reach 345 Mt with 0.06% P (cut-off 0.07%) and 416 Mt with 0.07% P (max. 0.08%). To these should be added 110 and 210 Mt Batouala and Boka Boka, but other ridges have not been delineated yet.

Banded ironstones are also found at Ngama. However Mebaga Hill (N) only averages 50-55% iron, Mela (NW) 45% in 900,000 t ore per meter of depth and Milingui Hill (S) 43.5% in 80 Mt ore.

Other resources

Salt and potash occur above the hydrocarbons of the coast, e.g. at Lake Azingo. Inland, there are only patches of sandstones in the Basement and, therefore, few continuous aquifers. In contrast, surface water is abundant and rapidly breaks down the humus of tropical soil.

There are excellent limestones at Achouka and the Société des Ciments d'Owendo (Lafarge-Coppée 70%, state 20%) has grinding and cement capacities of 270,000 and 140,000 tons. Cimenga (state 45%, Lafarge-Coppée 15%, Owendo 15% and Elf 24%) has a clinker capacity of 350,000 t at N'Toum (NW) and 150,000 t at Franceville (SE).

Gold has been washed in the M'Bigou, Koula-Moutou, Lastourville and Lambaréné districts where Dr Schweizer lived. There is primary gold at Ovala and Domdo Moti and in 1962 Eteke (C) produced 500 kg. Sogarem (Comuf 35%, BRGM 20%) seeks to revive these workings and other occurrences include :
Base metals with uranium, Kroussou, Offobou (SE), bauxite Makongonio (SE), copper N'toum (W), Makokou, talc and marble N'Dendi (S), tin and columbite-tantalite Dkjolé (Ndjolé), Makokou (NE) ; with gold Etéké, Kouala Moutou.

53 CONGO

The République populaire's wealth comes from the oilfields off the shelf which provide at capacity more income than Zaïre's whose diamonds and gold also transit through Brazzaville (Fig. 32, 37).

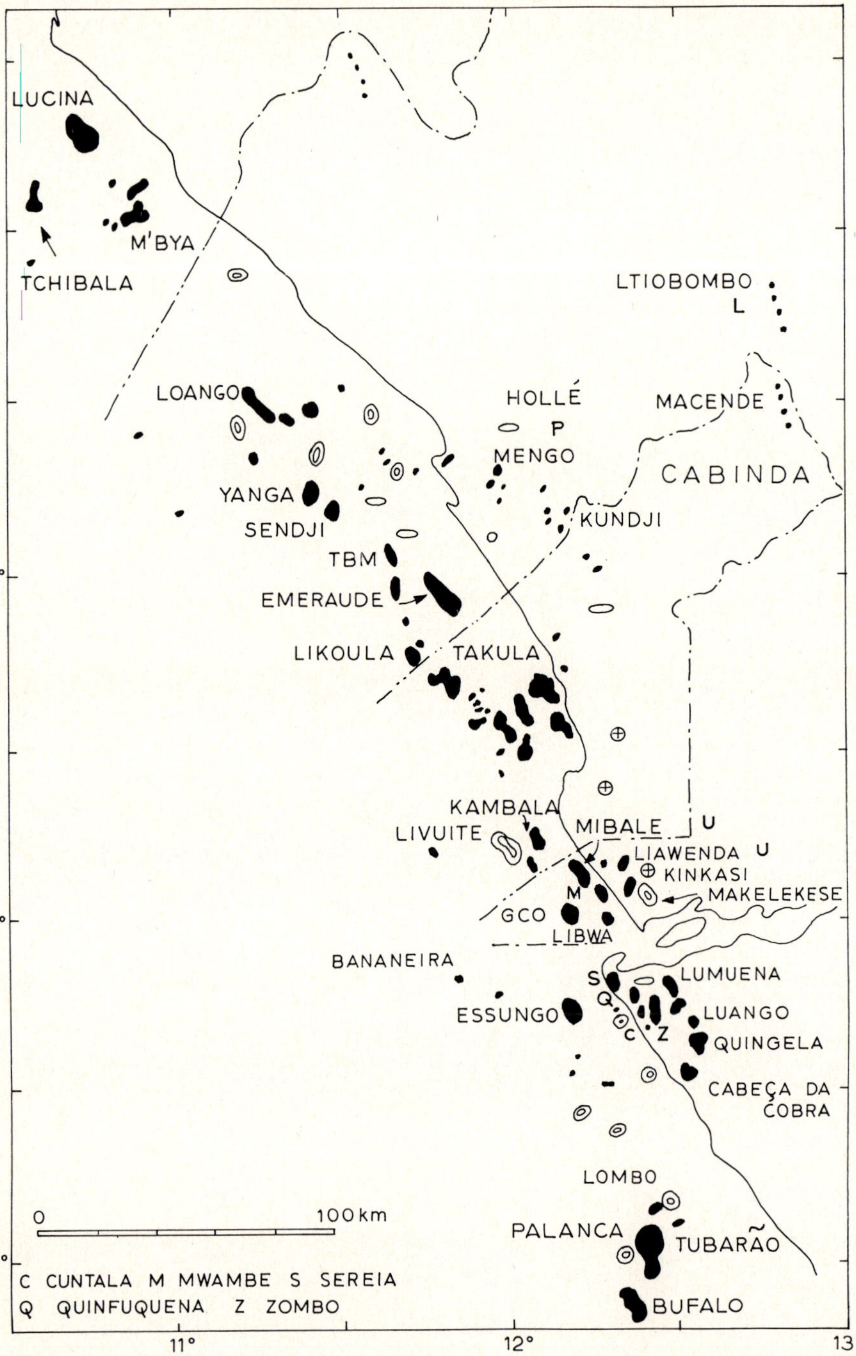

Fig. 36. Oilfields from south Gabon to Angola.

GOALS : Incentives for further oil and mineral exploration on the coast and up country. Utilization of gas, processing of minerals and rehabilitation of the potash mine.

Oil sales increased fivefold in five years to match 70% of revenue and 90% of exports, a third of which covers debt service. While Congo has close ties with the U.S.S.R., its major trading partner is the EEC, above all France. French Elf, along with Agip, are the principal oil operators. Both earned profit on a turnover of $1 B. They employed 1000 and 300 people respectively. Prudent Hydro-Congo only owns 20% of the shares of the local subsidiaries and 60% of the refinery. Hydropower plants and the cement works are state owned. So is the struggling base metal group Sonamif, assisted by Soviet specialists, but they employ only 3,000 people.

Table 70 Congo production (also peaks and past), reserves, resources.

Concentrates produced to date	kt	Peak	kt	current kt	Resources kt total	Yanga			Production		Reserves	
Lead	150	M'Fouati	7	4	500	40	Oil		Mb	40	Mb	500
Zinc	70	Hapilo	5	3	400	40	Gas, gross		Bcf	40	Tcf	2
Copper	25	Mindouli	1	0.1	400	165	Petroleum		Mb	6		
Gold			kg	100	8		Hydropower		Gwh	170	Gw	9
Reserves : potash Mt 20, Iron Mt 60							Cement		kt	40		

Energy

Oil seeps along the edges of the coastal basin, but hydrocarbons have migrated from their source rock, black marls, which lie under the salt layer, into the green pre-salt silts,sands and limestones of Turonian-Cenomanian age. There are half a dozen major fields out of fifteen (Fig. 34) and their average size is bigger than in Gabon.

	1000 bd	depth m	API°		1000 bd	depth m	API°
Yanga	40	1200	27.5	Likouala	18	1200	33
Emeraude	25	350	23	Sendji	18	1200	27.5
				Loango	20	900	27

To date Emeraude, 20 km offshore, has supplied half of Congo output, a heavy oil with 0.8% sulfur. There is 20% associated gas but recovery rates remain low. They reach only 28% at Yanga out of 480 Mb of reserves in 100 m of water. Projects call for increasing output here to 100,000 bd and at Sendji nearby to 130,000 bd. Except for Likouala, all Elf fields are on artificial lift. Agip, who own 65% Loango, installed another platform and four fields are on gas injection. ARC (Agip) also prospects at Mandingo in the north, but in the other fields the Elf/Agip proportion is reversed with Elf Congo leading with 65%.

Tchibouéla, the largest will come on stream in 1987. Elf is to get 45% and Hydro-Congo 30% of the oil. Elf (85%) also evaluated N'Kassa and Tchendo ; total capacity at Bouygue's Mengo, on shore, of Kundji, Paloukou and Libondo should not exceed 1,000 bd. East of Likouala oil shows at 1,650 m depth, Louvessi Profond tested 2,700 bd and 5 Mcfd of gas and deep Litchendjili flowed 9 Mcfd from 2,200 m depth.

As usual, little gas is used and both pipelines serve the Djono terminal. The Pointe Noire refinery, rated 2,100 bd, even exports 2 Mb of fueloil.

Power consumption doubled in ten years to 200 Gwh, mostly hydropower. The Kouilou alone could generate 800 Mw. At Imboulou on the Lefin, China builds a 100 Mw plant which can be expanded to 300 Mw and the Bouenza dam will be followed by ten micro-hydros. There are, moreover, traces of uranium in the Hollé phosphates, the Pointe Noire phosphorites (20-150 ppm) and the M'Fouati ores.

Other resources

There are free aquifers in the northeast. While the Basement offers only limited possibilities, surface runoff is abundant elsewhere too ; it also contributes to the weathering of hard rocks.

The soils are of the red tropical type with few exchangeable bases.

North of the Kouilou (SW) and further, irrigation of the Nyanga valley is to expand. Along the coast 300 m thick salt layers lie at a depth of 250-600 m and at Hollé the thickness of three potash levels ranges from 1.5 to 5 m. Before flooding SEPH had a solution mining capacity of 2 Mty ore and 400,000 t contained potash out of 200 Mt of resources.

There is some gypsum-related anhydrite above the column of salt and phosphate rock grades 20-25% at Tchivoula nearby. Débris are unevenly distributed along the Gabon-Congo coast, but 1 mm phoscorite grains accumulate in loose sediments near Pointe Noire in only 30-40 m of water ; they assay 0.6-0.8% Al and 1.5-2.5% Fe.

Limestones are of course widespread ; they extend inland to Dolisie and Sibiti and Cidoulou's Madingou plant could add another oil fired kiln to its current capacity of 80,000 ty of cement.

The Upper Sangha goldfield straddles the border (N), but the Mayombe (S) supplied most recorded output which peaked with 1t in 1942.

Workers also washed 50 ty of tin before 1969 and 10 t of columbite-tantalite between 1973 and 75, 5 t of wolframite in 1961 and 200 t of corundum at Mayoko Mossendjo and Moufoumbi between 1944 and 47.

Although production of base metals plummeted in the Nyanga Valley (S), Sonamit and Soviet technicians have set ambitious targets based on 4.6 Mt ore at

Yanga-Koubanza and 900 kt Boko Songho (Djangile) and Moubiri. While 100 kt of tailings averaged 4-10% at Mindouli, 40 kt of black soils assaying 4.5% copper at M'Passa required the Torco process.

There are 30 Mt ore at Mayoko (SW), with 56% iron, 9% SiO_2, 2% Al_2O_3, but a 170 km long rail link would be required to strip 100 Mt ore at Zanaga grading only 43% Fe, but 20% SiO_2. Together with Avima, Nabebo and Elogo (N), extensions of the Gabon trend, and even remoter Abolo, resources add up to 0.9 Bt ore grading 40% Fe. Lastly, chromite and nickel show at Souanké and Sembé, manganese at Kimongo and bauxite at Konyi (SW), 30 km from the Gabon railway.

54 ANGOLA

Rising revenues cover two thirds of GDP, most of the budget and 80% of foreign exchange of Africa's fourth oil exporter. They also service foreign debt which has increased five fold in five years. Hopefully other production will also return to its preindependence levels.

Table 71 Angola capacity, production, plans and reserves.

Production			Production					Production	Reserves
Cement	kt	220	Petroleum	Mb	17	Oil	Mb	80 → 200^p	Bb 1.8
Gypsum	kt	20	Gas marketed	Bcf	8	Gas gross	Bcf	90	Tcf 2.2
Kaolin	t	200	Gas reinjected	Bcf	40	Hydropower	Twh	1.2	Gw 10
Resources			Asphalt	kt	25	Diamonds gems	Mct	1.4	22
			Salt	kt	50	" industrials	Mct	0.3	8
Bauxite	Mt	10	Manganese cap.	kt	17	Phosphate	kt	10	Mt 2
Coal	Mt	10	Steel	kt	10	Iron ore	Mt	1^p	Mt 60
Lignite	Mt	30				Copper resources			kt120

The oil terminals and harbors (Fig. 36, 37, 38) operate, and fleets of trucks, Cuban convoys and aircraft serve the interior.

> GOALS : Reestablishment of communications, rehabilitation of plants and mines. Better water supply in the south, utilization of gas, control of diamond trade. Mineral exploration.

Laws permit additional deductions and Western investment and Comecon technicians are welcome. Sonangol owns 51% of the oil operators; it is the largest corporation along with Gulf (Socal). Endiama owns 77% of Diamang, Sibéka and Société Générale de Belgique 17.5%, De Beers 3%, but profits have been wiped out. Taxes are low or negotiable and the principal export markets were the U.S., and EEC (Spain, Portugal, Germany, U.K.).

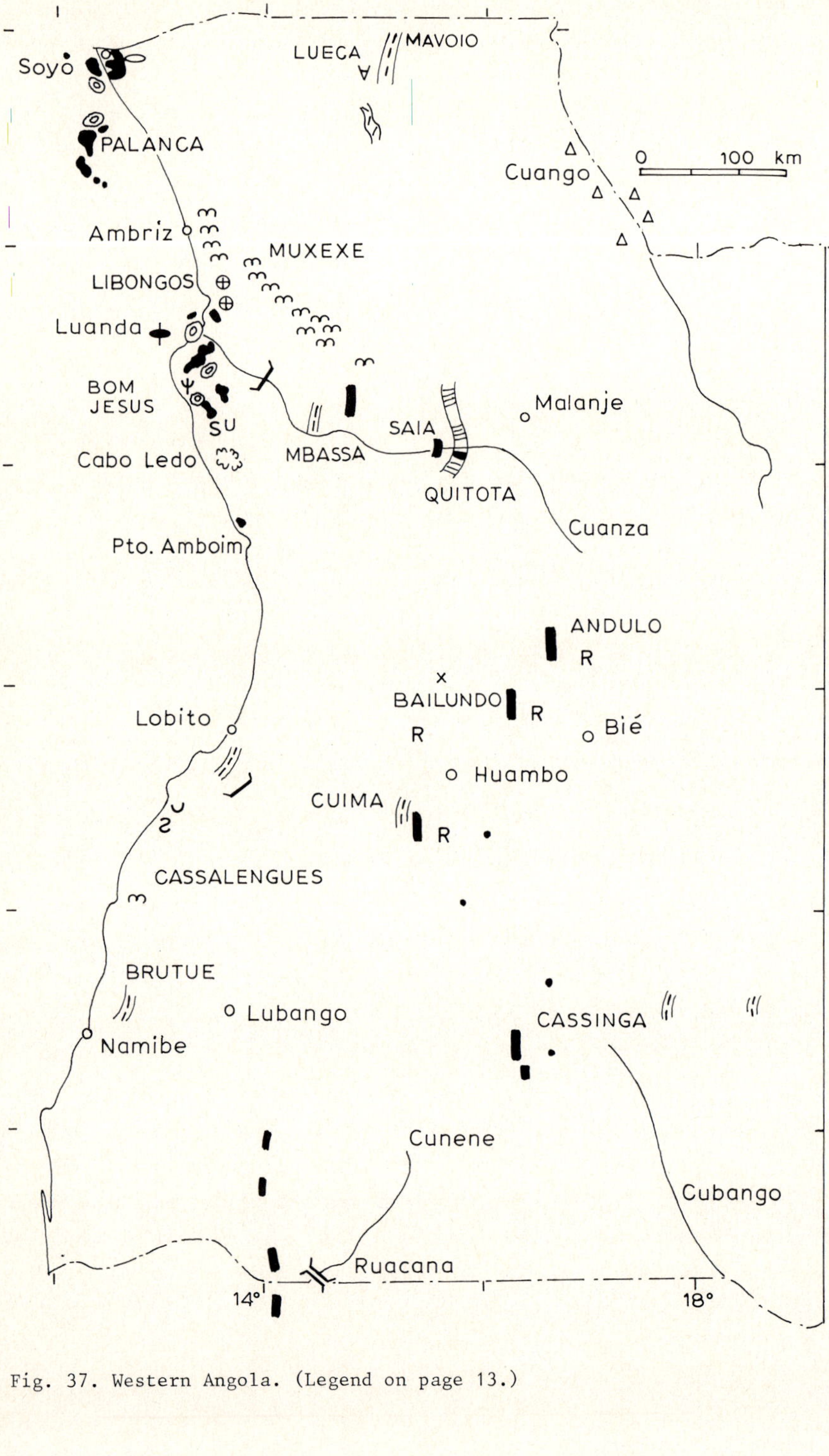

Fig. 37. Western Angola. (Legend on page 13.)

Oil

Output was in excess of 60 Mb in 1974 and rose again from 1982 on. Indeed, Takula is the biggest producer south of the Sahara. Another 30 Mby will be developed in 1988/9 at a cost of $800 M, and the pattern and shares of operators in % might reach :

Cabinda 300,000 bd, Gulf (Socal) 49 ; Congo Soyo 35,000 bd, Petrangol 33, Texaco 16 ; Bloc 2 20,000 bd, Texaco 40, Braspetrol 17.5, Total 17.5. Bloc 3 30-90,000 bd Elf 50, Mobil 25, Agip 15 ; Bloc 4, ? Braspetrol 30, Petrofina 30, Sonangol 30, BP 10 ; Kwanza 10,000 bd, Petrangol 49.

Table 72 **Major Cabinda fields.**

Field	1000 bd	API°	depth m
Takula	350^p	31.5	1200-1950
Malongo	40	22,20,25	450-2600
Kungulo	20	32	1200
Limba	15	31.6	900-2600
Kambala,&	10	31-33	3-3300

Oil migrated from bituminous shales of Cabinda into 50-180 m thick sandstones and limestones, mostly above but also below the salt layer. The higher reservoirs hold heavier, undersaturated oil, gas caps cover the lower level, but with injection the old fields are expected to last longer : half of the 300 wells are already on artificial lift. Petrangol's largest, N'Pangala and Quingila supply 15,000 bd to which Quinfucena, Quengela and Kitona add another 12,000 bd. The Belgian company also has secondary recovery in the Soyo basin.
- Texaco's Lombo East flows 5,000 bd of 41° API crude, Tubaro up to 3,000 bd from 2800 m depth and Pintangueira, 47 km from the coast, 4,600 bd. Essungo and Cuntala tested 10,000 bd in the northern part of bloc 2. There are also new fields further south : Sulele 24 km offshore ; Mavanga 25 km, in 34 m of water and Maleva 2 km to the north - i.e. 8,000 bd combined. Elf's discoveries in Bloc 3 are also coming on stream : Palenca (30,000 bd), Pacassa (40,000 bd) and Bufalo (7,000 bd). Lastly, exploration now extends south of the Kwanza.

There are also plans for the extraction of LNG and for fractionators ; LPG would require, however, pipelines or storage vessels. Sonangol has its service center on Kwanda Island, 5 km from Soyo, and its refinery at Luanda has a 33,000 bd capacity. The industry employs 2,500 people.

Other energy and coastal resources

Bitumen impregnates the eastern edges of the oil basin over 900 km, these limestones have been worked at Dondo, Libongos and Husso Norte. Undwi, Iembe, and

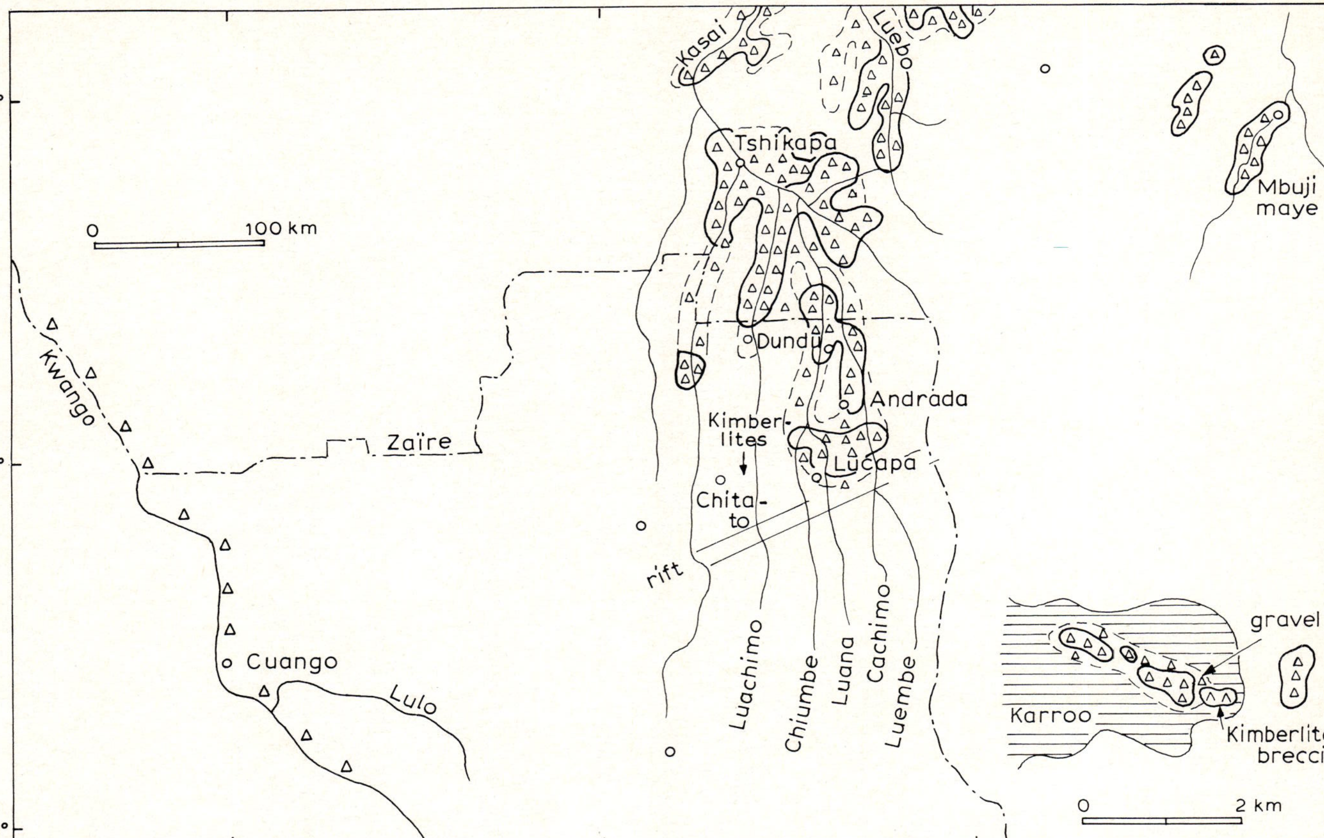

Fig. 38. Diamonds in Zaïre and Angola.

Mussera, further north, remain, however, inactive. There is also asphaltite at Calucala and Quilungo and peat at Bom Jésus. At Cabo Ledo, lignite resources attain 500 Mt, but coal resources of Menongue (SC) only reach 100 Mt.

As usual, phosphates carry traces of uranium which also shows south of Huambo (CW), in the Tchivira and Bonga carbonatites (SW) and in the Namib desert (SSW). Gas is a cheap source of power. While needs are difficult to predict, electricity supply doubled in ten years. While three quarters of the 1.6 Twh generated are hydraulic, the largest plant is on the Cunene nearby in Namibia. The oldest, Cambambe, can generate 56 Gwh. Also on the Kwanza, the Mabubas project could have provided 4 Twhy but it was scaled down to 22Mw, while the Biopio (W) and Matala (28 Mw, CW) plants can generate only 40 and 70 Gwh. The Kapanda project of Teknopromexport will provide 450 Mw capacity if/when completed.

Limestone is of course widespread in the coastal basin ; 3 oil fired wet kilns of Cimangola at Cimaco (Luanda) had a capacity of 650,000 t and Encimé's kiln at Lobito rated 150,000 t but they require rehabilitation like the power-plants and 28 saltworks which also need purification according to Cuban specialists. Rocksalt and potash reach 300 m thickness in the Kwanza basin, e.g. at Cuvo, but most salt is won at Benguela (W). Nearby from Ouche, abundant gypsum has been exported, sulfur is widely scattered in it, e.g. at Dombe Grande (NW).

Bulgaria assisted Fosfang at the 15,000 ty Kindonacaxa plant (N). Phosphates even extend into Cabinda, where Yugoslav Energoprojekt explored 10 km^2. They are near the surface at Coco (near Luango) and at a depth of 80 m at Cacongo (near Mongo Tando) : 6 Mt of rock grade 32% and 50 Mt of resources 18% P_2O_5. Recovery of half a million tons has been discussed here.

Diamonds

Following the discovery of diamonds by Belgian prospectors in 1912 along the Zaïre border, production reached the mark of half a million carats by 1950 and doubled by 1960. Recovery peaked with more than 2.4 million carats in 1970 and recorded sales of Diamang have varied since between 1 and 1.5 Mct. Estimates of illicite mining and exports range up to a half of reported gems. Effective sales might, therefore, exceed 1.5 Mct of gems and half a million ct of industrials.

Deposits extend over a 120 by 80 km area from Lucapa to Zaïre in the north, to which should be added 90 kms of the Cuango (Kwango) valley further west, famous for its "potholes" where 2-5 ct are found by m^3 : the Dunge gulley alone contained 400 m^3 of gravel. Rivers have to be diverted of course, to reach these deeps. Normally placers and terraces are removed by scrapers and draglines. Reported yields of Andrade and Lucanga reach 0.2-0.3 ct/m^3. The Andrade and Cuango divisions provide 60% of sales (with a third of reserves at Cafunfo); Lucapa supplies the remainder. Only weathered kimberlites have been worked at Colonda (Andrade), Catoca and Camajico, and hard rock remains in reserve at Camafuca, Catoca, Camajico, Camaxia, Cemetue, Caixea, and Camazombo. As usual, gravel is washed, screened and separated by dense media and greased belts. The 8-10 Mw Luaximo plant provides power.

According to statistics, gems average 8-10 ct. Although 40% of the stones are colored, they are not opaque.

Iron and manganese

Ore reserves and additional resources of 2.1 Bt and 1.2 Bt grading only 40% iron include, from north to south :

I . In the Lucala manganese belt : a) Dala-Tando near Champambe dam 1 Bt of hematite 32% Fe, b) M'bassa, 150 km E of Luanda 10 Mt titanomagnetite 58% Fe, c) Kassala-Kitungo 70 Mt reserves, 230 Mt resources, 33% Fe, 0.4% TiO_2. II. Chilessa, Sarakete, Essale, Changa and Bailundo, 70 km north of Huambo. III. Cuima 160 km south of Huambo 10 Mt 60% Fe. IV. Kassinga 520 km from Namibe 20 Mt detrital 44% Fe, 100 Mt opencast, plus 1 Bt resources (?). V. Iron pans of Kuanhama in the Kalahari. VI. Picolo, Otchinjao, Chibia, Chitado on the Kunene and VII. Niela and Manhinga (E).

The capacity of M'bassa reached a million tons of concentrates in the sixties. The socialist government is also interested in upgrading Kassala (I) ores and then moving them over 230 km as a slurry to a 2 Mty, gas fuelled pelletizing plant at Luanda.

Between 1969 and 1974 production averaged 3.5 Mt and Austrian Voest has rehabilitated Ferrangol's Kassinga (IV) mines. Ore has been shipped from Mussesas by conveyor and from Indungore and Osse by truck to the Jamba plant designed to produce 1.1 Mty concentrate grading 62.5% Fe from 2 Mty ore. Tchamuete ores, further south, will follow later on.

The Malanje belt (NW) is more important than the manganese of the upper Zambezi (E) and includes Quitota where some iron ore has also been produced, Quiaponte, Gungungo, Quiculo and Quichuinhe, 100 km to the NNW. Production peaked with 30,000 t ore in 1956 and ranged between 4 and 15,000 t prior to 1972.

Other resources

Hidrominas is headquartered in arid Huila (SW) since desertic conditions prevail on the Namib coast where clayey sands imbibe some water. The Basement only averages 1-7 m^3h under the Kgalagadi at a depth of 5-7, locally even 30 m. There are also fossile rivers under 2-600 m of sand in the Cunene basin (S).

The Gove and Calueque dams, financed 50 and 100% by South Africa, divert water from this river to Namibia. Lastly, light-colored sandy soils covered the highlands (CW).

Roremina, also located at Lubango, managed in principle twenty odd marble quarries (SW). Its sister, Minaquartzo, exported in 1975 $1.1 M worth of high purity quartz for art and glass from Pocarica and Conda (W). There is some mica at Conda too and at Mondangel (W) (where vermiculite also occurs) and the 200 km long belt includes Cassalengues, Quizambilo, Muxexe, Cambaça, Dui and Fandongo (NW).

Production rarely exceeded 10 tons in the fifties. Similarly, little gold was washed in the Cabinda enclave and at Cuenguè (CW).

Following the XVIIth century sovereigns of Congo, Mavoio (NW) supplied 1000 t copper in 1950 and neighboring Lueca 25 t vanadium in 1956-9. Other occurrences include :

Bauxite : Donde 53-60% Al_2O_3, chromium and nickel Riviera Salgado, Morro Vermelho, copper Brutuè, Iona (SW), lead and zinc Zombo-Mavoio, niobium, rare earths, thorium . Bailundo, and Chilesso carbonatites (CW), sillimannite near iron belt VI (SW), tin and tungsten Cunhanguana (CE).

55 ZAÏRE

Only South Africa rivals Zaïre's variety of minerals (Fig. 36,38-44), which are not processed, though to the same extent. Zaïre also has multiple sources of energy. Although only a fraction of these has been developed, the lower Zaïre could meet alone the requirements of all Africa if networks were interconnected. With a half of the world's cobalt reserves, the country also holds strategic importance, conversely, its predominance as a supplier of industrial diamonds has been reduced by Australia. Zaïre ranks fourth in the world as an exporter of copper which also matches two thirds of its own export revenue. It was insufficient, nevertheless, to cover the debt service, frequently rescheduled by the Paris and London clubs of creditors. Repayment of $ 1.6 B, a third of foreign debt, has been postponed to 1989-95, IMF has provided special drawing rights, the World Bank, IDA and EEC's Sysmin long term loans at 1% which should expand or rehabilitate smelters, harbors and railways.

Zaïre also exports to Rwanda and Burundi their share of hydropower, but it imported 6-8 Mb crude, up to 200,000 t of coke and coal mainly from Zimbabwe. The state's GCM had sales of $1 B and 40,000 men, but mostly declared losses, Petrozaïre, Zaïrefina and ZGO combined had similar sales and reported some pro-

Table 73 Zaïre production and reserves.

Commodity	Production	Reserves	Commodity	Production	Reserves
Petroleum	kb 600	Mb 3p	Oil	Mb 15	Mb 200
Gas		Bcf 40	Coal	kt 300	Mt 80
Methane		Tcf 1	Hydropower	Twh 4-5	Gw 130
Oil shale		Bt 15	Uranium		kt 3
Cement	kt 400		Copper	kt 480	Mt 25
Diamonds gems	Mct 6	Mct 30	Cobalt	kt 10	Mt 1.2
" industrials	Mct 9	Mct 200	Zinc	kt 70	Mt 1.1
Tin	kt 3	kt 32	Cadmium	kt 1	kt 15
Niobium	t 30	kt 180	Silver	t 100	kt 1.5
Tantalum	t 20	t 400	Palladium	t 6	t 150
Tungsten	t 70	kt 1.7	Platinum	t 2	t 50
Gold	t 9	t 150	Rhodium	kg 200	t 3
RE+Th conc.	kg 500		Germanium	t 45	kt 2
Resources			Selenium	t 100	kt 6
Beryl		kt 5	Tellurium	t 100	kt 1
Iron ore 45% Fe		Bt 15	Indium	t 15	t 100
Lithium		kt 180	Rhenium	kg 100	t 2
Manganese ore		kt 5	Lime	kt 160	
Bauxite		kt 65	Sulfur	kt 180	
Phosphate rock		kt 80			

fit with 1,500 employees. The goverment holds all the shares of Okimo, the gold, 80% of Miba, the diamond, 50% of ZGO an oil and only 28% of Sominki, the tin company. Miba employed 5,000 against Sominki's 12,000, but the latter has earned 25% pretax. Based on this model the government was actively seeking investors for other mineral ventures. Meanwhile, diggers washed diamonds and gold, increasing thereby mineral employment to 60,000 people - and illicit exports to a third/half of the total. - A pattern of official exports is shown below in %, there are nevertheless shifts :

	Belgium	France	U.K.	Germany	EEC	U.S.	Japan
Copper	60			5	65	10	20
Cobalt		30		10	10	50	
Silver	9		90		99		
Cadmium	80			10	100		

	Belgium	Germany	France	EEC	U.S.	Japan
Diamonds gems	80			90		10
" industrials	10	50		60		
Tin	30			60		
Tungsten			10	10		90

The reopening of the Benguela railway in Angola will one day reduce bottlenecks in shipment to Matadi, which handled only half of Shaba exports and 10% of its imports. Ten % of exports were routed through Dar es Salaam but the rest were railed to East London and Durban.

GOALS : Rehabilitation of infrastructure and mines, prompt supply of parts, managerial and financial controls, agressive marketing of mineral and energy investment projects.

Energy

ZGO's (Socal) M'bale field produced offshore most of the 30° API oil from 1500-1800 m depth and Zaïrep (Petrofina 54%, Shell 46) set a target of 10,000 bd at Kinkasi, Liawenda-Makekere on land. Lukani (Gulf 50%, Teikoku 32, state 18%) flowed 7,500 bd from 3,700 m depth, Tshinde 6,000 bd, but Libwa, Mwambe and Motoba were smaller. Exxon drilled in the Central Basin (W). Oil shows in the Albertine Rift (NE) and Lake Kivu (E) holds of course, vast reserves of gas (see 'Rwanda'). Nearer to markets, Sozir's refinery (50% Agip) at Muanda has to be rehabilitated to a 20,000 bd capacity. In 1958, 800 tons of bituminous limestones and sands were quarried nearby at Maniangu and Mavuma (W, see also 'Angola') and around Kisangani (CNE) 15,000 km^2 of shales could supply 140 l/t of oil products and 9 kg/t of ammonium sulfate.

Hot springs rise along the Rift and the Virunga volcanoes (E) and near Kamina (CSE) geothermal steam of only 95° has run a 220 kw generator.

Zaïre has been credited with 1.2 B tons of geological coal resources, I at remote Walikale (E), II in the 5-7 m thick seams of the large Lukuga basin, west of Lake Tanganyika which fuelled the Lukaleni (ESE) cement plant. III The Kisulu, Kaluku and Kalube and Luena (SE) deposits are, however, better located. Gécamines extracts here 1.2 Mm^3 in order to produce 220,000 t of raw coal. We might add that lignite shows at Mbanza (W) and in the Kivu Highlands, along with peat at Kibukwe.

Lastly, some uranium is still left in the closed Shinkolobwe mine, which provided Mme Curie's radium, at Shambo and Kalongwe.

Hydropower

Although thermal generation declined to 60 Gwh, hydro, in contrast, increased 40% in ten years to 4.5 Twh. The potential is vast, indeed : officially 13% of the world total, more in practice as the droughts have shown.

Table 74 Hydropower plants in Zaïre.

Lower Zaïre (W)	Mw	Shaba (SE)	River	Mw	East	Mw
Inga I-II	1460	Le Marinel	Lualaba	270	Maniema	20
Zongo, Inkisi	31	Delcommune	"	120	Rusizi	40
Sanga "	25	Cornet F.	Lufira	60	Kiyimbi	20
Kalule, Lubudi SE	7	Bia Falls	"	80	Piana Manga	(35)
					Lubilanji (EC)	16
Maniema plants : Zalya, Ambwe, Belya, Lubiadja, Lubilu, Lutshurukulu.						

More than 90% of installed capacity serves the mineral industry ; even so several plants are closed or require rehabilitation.

A potential of 85 Gw is concentrated between the capital and Matadi. Forty km upstream from the harbor, Inga is the world's only site with a full year capacity of 28 Gw ; it is also attractive since dry beds render diversions and staggered development easy as shown in Fig. 37.

Table 75 The Inga hydropower site.

Development		m^3/s	fall, m	useful Gw	Twhy	Water withdrawal
Inga	A 1-3	2200	45	1	8	Level : Nkokolo
I-II-III	B 4-6	6400	55	3.1	25	Fwamalo canal
Great	C 7	6400	35	4.9	40	Dam
Inga	C 8-16	25300	115	26.3	210	Bundi Valley

The Shongo dam provides 160 Mw for Inga I and 1.1 Gw full year at Inga II. A 1700 km long tenuous 500 kV line supplies additional power to the Shaba mines and hopefully this century, for other industries, in the Moanda free zone. The Maluku steel and other cement projects have thus resurfaced. Although bauxite (35% Al_2O_3) shows at Sumbi and Tshela (W) and east of Tabela, the Alusuisse consortium would import ore for its 150-200,000 ty smelter should infrastructure materialize.

The Copperbelt

Copper shows I in Lower Zaïre (W) at Bamba-Kilenda and Kongo-Diakati, with lead and zinc II Lubi (S) III at Bossobolo near Kisangani (NE), IV Wamikubi (E), V scattered in the Kibara chain VI a small high grade deposit near Kapulo on Lake Mweru (ESE) and VII in the 500 km long copper belt.

The emperor of the Lunda exported copper bars and crosses to the coast in the XVIIth already, followed by M'siri, king of the Bayeke, but it took oxcarts seven months to trek to Ruwe before the railway reached Lubumbashi in 1910. UMHK's production grew from 20,000 to 150,000 and 270,000 t in 1920, 1940 and 1960, and Gécamines shipped 450,000 t in 1980. There are in effect two parallel copper belts which can be illustrated thus :

		West	Center	East	Southeast
Northern	km	25	170	130	130
	name	Kolwezi	Likasi	Lubumbashi	Kinsenda
	from	Dikuluwe	Fungurume	Lukuni	Bw. Mukubwa
	to	Ruwe	Shandwe	Etoile	in Zambia
Southern	km	30	40	50	90
	name	Kalongwe	Tantara	Kipushi	Musoshi
	from	Kengere	Swambo	Kasomba	Luanshya
	to		Shinkolobwe	M'baya	in Zambia

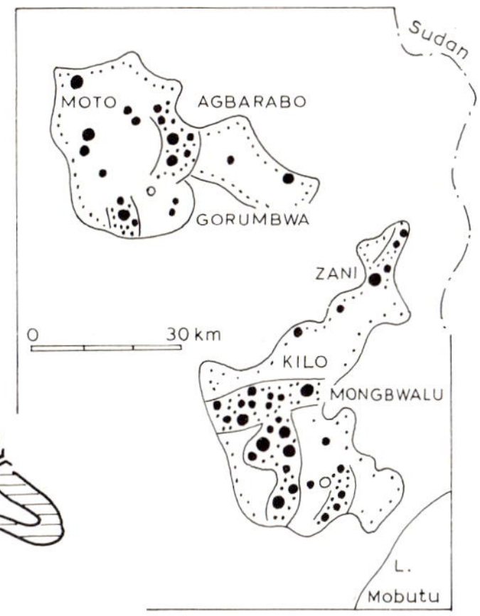

Fig. 39. (Left) Inga. Fig. 41. (Right) Kilo Moto.

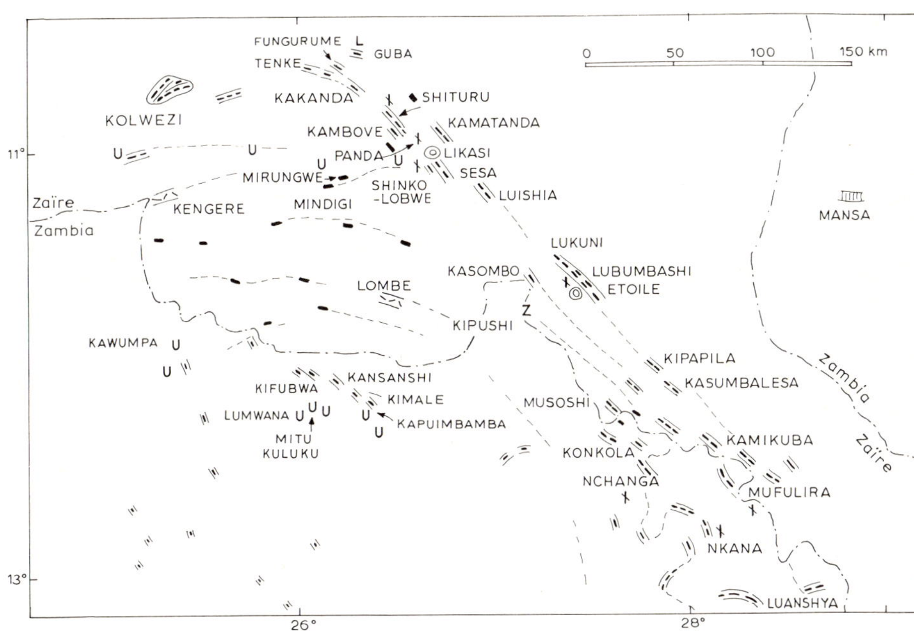

Fig. 40. The Copperbelt. (Legend on page 13.)

Gécamines has taken over the Musoshi and Kinsenda mines where a Japanese group extracted 2 Mty ore grading 2.5% Cu and 200,000 t and 300,000 ty assaying 5%. Their capacity is planned to increase from 100,000 to 300,000 ty of concentrates averaging 40% copper shipped to Japan via South Africa.
Further, a $1.5 B project, SMTF now led by Cogéma, (27%, BRGM 8%, AAC and Charter Consolidated 28%) provided for 130,000 ty of copper and 6,000 ty cobalt from the rich but hard to treat ores of Fungurume and Tenke. Markets and power supply permitting, this scheme might materialize in a scaled down version, like the Dikulushi copper-silver mine, another BRGM project.

Gecamines' copper

GCM employs 800 expatriates and directly provides the livelihood of a quarter of a million people. Two thirds of the production now comes from the opencast mines of Kolwezi in the west : Dikuluwe, Ruwe, Musoshi, Kolwezi, Musonoï and Mashamba, followed by another 20% from underground Kamoto. The remainder is evenly spread between the opencast mines of Likasi in the center : Guba, Kahumbwe, Sesa Kamfundwa and Kamoya and the two other underground mines Kipushi and Kambove which nears the end of its useful life. Conversely, Kov in the west and Kamoya near Kambove in the center are being developed opencast.

A glance at table 75a shows that ores average 4.1% copper and grade 4.4% in opencast Kolwezi where 20 m^3 shovels work previous embankments. Trucks are assisted by cheap electric power. The washery preconcentrates breccia ore and the flotation plants of the west provide three quarters of total copper.

Table 75a Shaba copper-cobalt mines, concentrators and smelter.

Mines	West opencast			Kamoto	Center o.c.	Kambove	Kipushi
Mm^3 excv.	25				2		
Ore Mt	9			3	1	1.6	1.4
Copper kt	400			120	40	50	50
Concentrators	**Mutoshi washery**	**Dima oxides**	**Kolwezi**	**Kam+Dima sulfides**	**Kakanda**	**Kambove**	**Kipushi**
Ore Mt	2.3	0.6	4	6	0.7	1.6	1.4
Conc. kt	50	120	700	550	90	80	150
Copper kt	14	30	150	200	20	40	40
Smelters	**Luilu leaching**		**Kolwezi leaching**		**Shituru cathodes**	**Panda flash**	**Lubumb. blister**
Copper kt	170		2		130	10	160
Co/Zn kt	5 Co		70 Zn		4 Co		

Ores are dressed, according to their oxide, carbonate or sulfide content and recoveries are : concentrators of the west, center and Kipushi 78, 67 and 85% respectively, and 93% in smelting, that is 72% overall. The Lubumbashi smelter is aging, but Gecamines plans to add 1,000 ty of smelting and cathode capacity

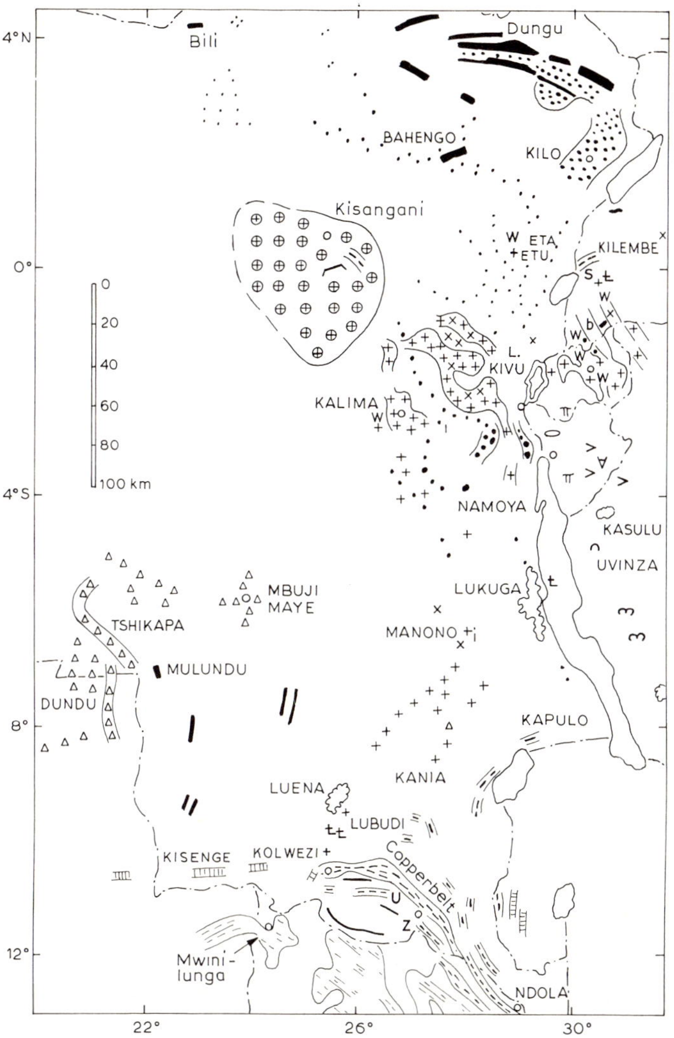

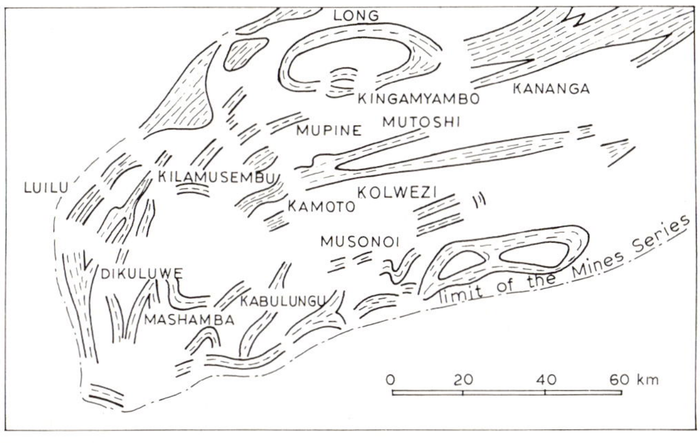

Fig. 42. Zaīre and its neighbors. (Legend on page 13.)

Fig. 43. Kolwezi. (Legend on page 13.)

on top of 40,000 t, 3,000 t cobalt granules and anodes, 200,000 t highest quality wire, 500-1,000 t of copper are already drawn and 500-1,000 t are rolled or extruded.

Other Gécamines products

The Copperbelt is the world's largest cobalt deposit. Grades tend to decrease from west to east. Highgrade occurrences include I Kamoto (0.4%), Musonoï (1%) and Kingamyambo (W), II Fungurume, (0.4%), Kabolela, Kambove and closed Luishia (C), III Luiswishi and Kasombo (S) plus Lukuni and Etoile which are closed. Cobalt and nickel also show at Kalongwe south of Kolwezi, at Kasolo and Mindigi (CS) and further south at Shamitumba. In the west which holds a quarter of the world's cobalt reserves Cu/Co ratios average 10:1, with maxima of 5:1. Depending on the type of ore worked, the leaching plant of Luilu produces 3-8,000 ty of electrolytic cobalt, and the Shituru refinery 1-5,000 t of granules. Up to 3,000 t of cathodes and a maximum of 700 tons in copper-alloys are further refined in Belgium. Apparent refined recovery ranges from 70 to 90%. Production averaged 12,000 ty in the seventies and dipped to 5,000 t in 1982 ; it depends on demand and GSA purchases. While 12,600 t cobalt fetched in 1978 more than copper did, 2,000 t were produced in 1980. Capacity is 20,000 ty.

Kipushi, the continent's principal zinc, cadmium and germanium deposit strongly resembles Tsumeb in Namibia. At Shituru 100-150,000 ty of these sintered concentrates are roasted into 70,000 t zinc. They also produce 100,000 t of sulfuric acid from sphalerite and 50,000 t from imported sulfur for the leach plants. The Kolwezi plant recovers electrolytic cadmium bars from Lubumbashi zinc flue dusts. There is, further, gallite at Kipushi and germanium compounds of this source and the Shituru smelter are further processed by the world's leading producer, MHO in Belgium, into oxide and high purity metal. Oolen also recovers silver from slimes which is further refined at Hoboken. Production declined though to 40-100 ty. Likewise, Hoboken also refines 50-150 kg gold from Ruwe and Kipushi along with palladium and the U.S. Bureau of Mines credits Shaba with a 2 Mt tellurium reserve base. Ruwe and Shinkolobwe have also produced some molybdenum and vanadium.

We might add talc at Kolwezi, fluorite at Kakontwe, barite, crocidolite asbestos and graphite. The principal salt works are at Guba, with other salines at Kalamoto, Mwashya and Nganza. Out of 350-500,000 tons, 250-350,000 t limestone are washed at Kakontwe ; kilns burn more than 100,000 t lime and supply 10-15,000 t of foundry cement for the smelter. Finally, GCM's power needs of 2.5 Twh grow by 2-3% per year as discussed under the heading energy.

Diamonds and gold

George Young discovered Bakwanga in 1916, and in 1946 I. de Magnée identified the kimberlite breccia. Along with its 10 km^2 cover, the world's second diamond deposit already produced more than half a billion carats. Stones are found within a 5,000 km^2 area, and production progressed from 2 to 10 and 18 M carats in 1930, 1959 and the early sixties, and reported output is increasing again from its base of 9 Mct. Indeed, a bucket dragline with its 100,000 m^3Mo capacity provides 1 Mcty of low cost diamonds from the Mbujimayi, to be followed by 20 km of the large Sankuru. Smaller rivers also provide up to a million carats (3-4 ct/m^3), yet hill gravels of Bakwanga, Bozala, Kanshi & still supply 3-4 Mct (4-7 ct/m^3). Simultaneously working of the kimberlite alignment (1-1.5 Mct, 3 ct/m^3) spreads to Tshibua, Tshinyama and the Mudiba.

Miba has modernized with loans from IBRD, its managers, Sibeka, and Britmond an affiliate of the minority shareholder, AAC. 77% of the diamonds are boart , 20% near gem and 3% gem, but the Tshikapa (S) diggers (see MRA pp148-149) recover mostly gems. Miba and licensed buyers purchase a million carats - a fraction - of the total. - The condition of gold is not dissimilar as shown below :

Production	kg	Production	kg	Reserves	t
Buying offices	1-3000	Illicit exports	1-3000	Sominki	
Okimo		Sominki		Twangiza	24
Moto	800	Mobale	300	Mobale	3
Kilo	200	Lugushwa	200	Lugushwa	10
		Gécamines	50	Namoya	10

Hanan and O'Brien discovered gold at Kilo (NE, Fig. 40,18) in 1903 ; by 1920 production reached 3 t, peaked in 1940 with 17 t and declined to 10 t by 1960. Out of this Kilo-Moto contributed 6-7 ty between 1940 and 60. Durba, one of the Moto mines in the northeastern corner covered two thirds of the production. Another underground operation is Adibi at Kilo, and Brazilian Guttierez plans to recover 4.5 ty over 7 years.

Sominki's primary deposits are worked near Kamituga (E), dormant at Namoya and a higher gold price would justify working Twangiza-Kahanda, with assays of 5.5 g/t and possible resources of 65 t in grades of 3 g/t.

Tin group

Production hovered around 15,000 tons between 1940 and 1960 (1945 : 19,000). By 1970 it was down to 6,000, mainly because the placers have been worked out at Manono, where tin was discovered in 1910. Recovery from tailings therefore declined to 250 t of tin, 50 t tantalum bearing slags and 10-20 t of columbite - tantalite (Ta/Nb 3.5:1). However, 160,000 tons of tin and 5 Mt spodumene (6% lithia) are left in rock grading 70 g to 3 kg/t tin, average 350 g/t. Other data indicate 18,000 t proved, 15,000 probable and 76,000 possible in all of Shaba. Yet, since abundant water is too far, the Société Minière de Kania (SE, 80% Coframines) only washes 500 ty of tin with some coltan. Kalimbi is to start too but most of Mitwaba is exhausted and for more details the reader is referred to MRA pp 123-126, 464-9, 473-7, 483-4.

Well managed Sominki, a subsidiary of Cogémines (an Empain company) has embarked on an investment program which will increase production by 20%. The increment will come from the hardrock deposits of Makundju (Moga) Saulia, Mususa (Kaïlo), Tshamaka, Bunza (Kalima) and from the reworking of tailings at Isongo, Tshonka (Lulingu), Bilu (Ona), etc. Separate plans call for mobile plants served by airstrips in the Idiba (Ona) columbite-tantalite deposits.

Table 76 Sominki production and reserves (rv.).

Mine	Tin prod. t	rv. kt	Mine	Tin prod. t	rv. kt	Wolframite	pr. t	rv. t	Coltan	prod. t	rv. t
Tshamaka	270	3	Kaïlo	180	.3	Bishasha	12	800	Ona		130
Saulia ∷	230	2	Moga	400	2.2	Kadubu	10	260	Lulingu	7	
Ona	120	6	Kalima	240	3	Kaïlo	10	700	Ulindi	20	
Lulingu	280	3	Mikonsi	220	3.5	Kalima	8		Nzovu	3	
Kamituga	180	5	Isongo	500	4	Moga	5		Obaye∷		150
N+C+E	1080	19	West	1540	13	Total	45	1660	Total	30	1450

Saulia also produces 500 ty of high europium monazite and Obaye has 1,000 t in reserve. These metals also occur at Kavaronge and in the carbonatites along with apatite. Somikivu (Metallurg of the U.S. and Germany 70%, Sominki 10%, State 20%) has delineated at Lueshe (E) 3 Mt of minute ore containing 75,000 t Nb_2O_5 out of which 50-60% can be recovered through cyclons and flotation. Production is to treble from 500 td. Bingo adds another 2.5-3.5 Mt of reserves assaying 2.5-3.5% Nb_2O_5. Meanwhile the Zaïre tantalum and tungsten reserve bases reach 5 and 2 kt.

Beryl shows in the eastern pegmatite fields, e.g. at Kobokobo ; production peaked with 2,000 t during the Korean War. Tourmaline and amethyst occur in the same areas and amblygonite, a lithium mineral was occasionally produced around Kamituga. Lastly, some bismuth was washed to the south at Ngussa.

Other resources

The principal water problem is quality and the supply of the big cities. Typical yields are 2-20 lbs at Kinshasa at 10-60 m depth, mostly from alluvia, but groundwater of the coast has been polluted by the sea. In the central basin which is covered by Kgalagadi sands evaporation is particularly intense. Ter-

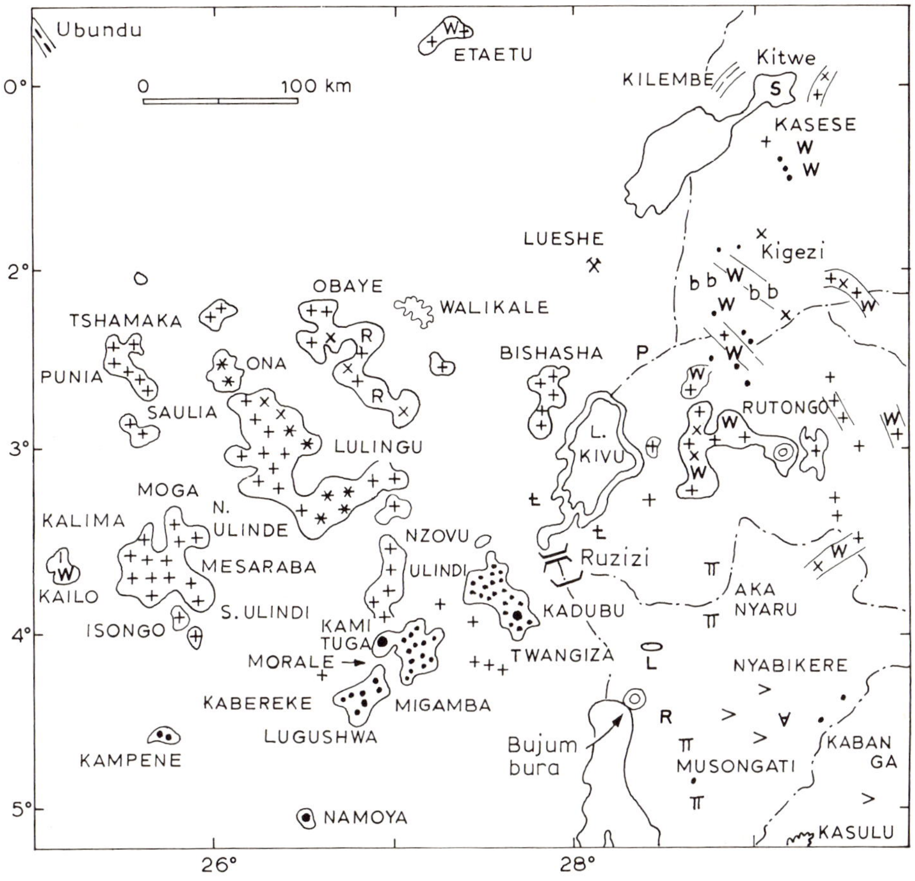

Fig. 44. Zaïre, east Rwanda, Burundi, SW Uganda and NW Tanzania. (Legend on page 13.)

tiary sands provide better aquifers in Kwango (WSW) and the northeastern Kasaï (CE) has good aquifers. The Basement in the east is not as well endowed, but there are free aquifers in the Kundelungu and Roan Series of Shaba (SE) and at Lubumbashi wells are only 5-60 m deep.

Stone etc. is mainly quarried near the cities and waterways and cement plants operate at half of their nominal capacity as shown in table 77.

Table 77 Zaïre cement capacity (1000 t).

Company	Shareholders %	Plant		Capacity	
CINAT	Klöckner 25, state 75	Banana, Kinshasa	W	d	300
CIZA	Egecim 51, state 15	Kinshasa	W	d	420
Cimshaba	Egecim 60, Gécamines 40	Lubudi	SE	w+d	150
Ciments du Lac	Katana (E), d, 35 kt, Kabimba coal, w, 45 kt.				

Iron ore is widespread in the northeastern corner (Fig. 18) I around the Kilo-Moto goldmines (NE) and II 500 km in the Uele (NNE), with grades ranging from 43 to 63% Fe. The iron mountains Duru, Tina, Ibi, Lai, Kai, Bahengo, Asonga, Kuma, Ami, Zani-Kodo and others range in length from 1.5 to 10 km and in width from 0.5 to 1.5 km ; Duru alone would hold 2 Bt of ore ! III Other iron mountains rise south of Bangui (NW). IV Luebo (S) 100 Mt of resources grading 30% Fe. V Another belt extends 110 km from Mutuntu east to Musombwishi (SE), with isolated deposits at Musaka and Kasumbalesa along the frontier and VI between Kambove (Kolwezi) and Kisanga (Likasi, Fig. 38) with 50 Mt , 56% Fe, 9.5% SiO_2 and 0.1% P, UMHK worked the latter and Kasumbalesa for flux.

The closed Kisenge mine (SSE) on the Angola border averaged 400,000 t metallurgical manganese in the fifties. Kamoto and Kasolo are extensions to which should be added Kasekelesa and occurrences as far as Lake Bangweulu (ESE). In the south, N'konko and Lutshasha grade 4.6% Cr and 0.5% Ni, with a 4 Mt concentration of nickel at the latter. Finally, other occurrences include :
Ituri (NE) : garnet, Kivu-Maniema (E) : agate, andalusite (Lueshe), asbestos Mahagi (Lake Mobutu), corundum (Kamabea), garnet (Idiba), potash (volcanoes), travertine and spongialite (diatom like : Lake Kivu), topaz (Mokama), Shaba (SE) anhydrite, malachite, opal, Kasaï (E) agate, chrysoberyl, Lower Zaïre (W) asbestos (Matadi), feldspar (Boma, Busin, Kongo), opal (Kinshasa).

56 ZAMBIA

Although more dependent on copper than Zaïre, Zambia does have hydropower, emeralds and more coal (Fig. 40, 42, 45, 54). With similar tonnages, Zambia employs 60,000 people against Zaïre's 40,000 including three times as many (2,000) expatriates. Workable reserves have, therefore, been halved and production will follow suit in the year 2000.

> GOALS : Strict cost and managerial controls, foreign currency allocations for parts purchases. Substitution of imported oil by coal and hydropower. Opening of new small mines, processing of minerals for the domestic market.

Zambia ranks second among exporters of refined copper, which also covers a third of revenue and 95% of export earnings. More than half of the exports went to the EEC (France, Germany, U.K. and Italy) and a quarter to Japan. The country is also the second producer of cobalt. The Copperbelt and Lusaka, 15% of the area, are well served by roads and railways, but shipment via South African harbors is still safer and cheaper than through Tanzania and Mozambique.

Zimco has a turnover of $2 B and a small loss ; it employs 130,000 people, owns the cement and coal companies and 60% of the shares of ZCCM who break even; another 7% are held by ITM of Mr. A. Sardanis and 27% by ZCI, an AAC subsidiary who have written off this investment in their books. The World Bank has financed with EEC's Sysmin and ADB rehabilitation and modernization of equipment, which should ensure steady output. Additionally private investment in small workings is encouraged.

Table 78 Zambia production and reserves.

Production			Production			Resvs.		Production			Reserves	
Emeralds	kg	300	Hydropower	Twh	12	Gw	5	Copper	kt	50	Mt	18
Amethyst	t	40	Coal	kt	500	Mt	300	Cobalt	kt	3	kt	300
Cement	kt	150	Petroleum	Mb	7			Zinc	kt	20	kt	100
Limestone	kt	500	Uranium			kt	8	Lead	kt	10	kt	50
Lime	kt	200	N in ammonia	kt	30			Silver	t	30	t	750
Stone	kt	200	Phosphate rc.			Mt	50	Cadmium	t	(1)	t	50
Sand	kt	200	Gold	kg	300			Selenium	t	40	kt	7
Talc	kt	1	Tin	kt	20	kt	100	Tellurium	t	2	kt	2
Feldspar	t	300	Nickel resourc.			kt	10	Sulfur	kt	100		
Manganese	kt	10	Iron			Mt	30	Vanadium resources			kt	4.5

The Copperbelt

Following the writeoff of 16 Mt copper in grades of 3.5% at Bwana Mkubwa and of 5 Mt at Chingola (Nchanga), 60 Mt of resources, in situ reserves of 40 Mt, a reserve base of 34 Mt, reserves of 25 Mt and workable ore of 480 Mt, assaying 2.6%, i.e. 12.6 Mt copper are left which sums up the problem along with increasing heat and water.

In addition to the seven areas shown below, there are 7 Mt copper in the dumps, much of which is recoverable and 80,000 ty are already leached.

I Mwinilunga (NW)

II A 500 km long swath extending ESE from Kansanshi to Kasempa, Kalengwa, Karenda (Zn, Ag), Chifumpa beyond Kabwe.

III The Silver King-Antelope, Hippo group west of Lusaka.

IV The Lusaka-Munatti-Chongwe east group (1 Mt assaying 1.5% Cu).

V Mokambo, the end of the Kinsenda belt of Zaïre, is still dormant.

VI The Mufulira-Bwana-Mukubwa copperbelt extending southeast to Mtuga

VII and parallel, the Konkola-Luanshya branch.

Bemba chiefs maintained an active trade of copper and in 1902 Africans led prospectors to the ancient workings of Bwana Mkubwa. 'The big master' opened in 1913 and closed in 1984, since sales covered only half of the costs. Meanwhile, the Copper belt produced 250,000 t in 1940, 650,000 t in 1960, more than 700,000 t in the seventies and 600,000 t are the aim until 2000.

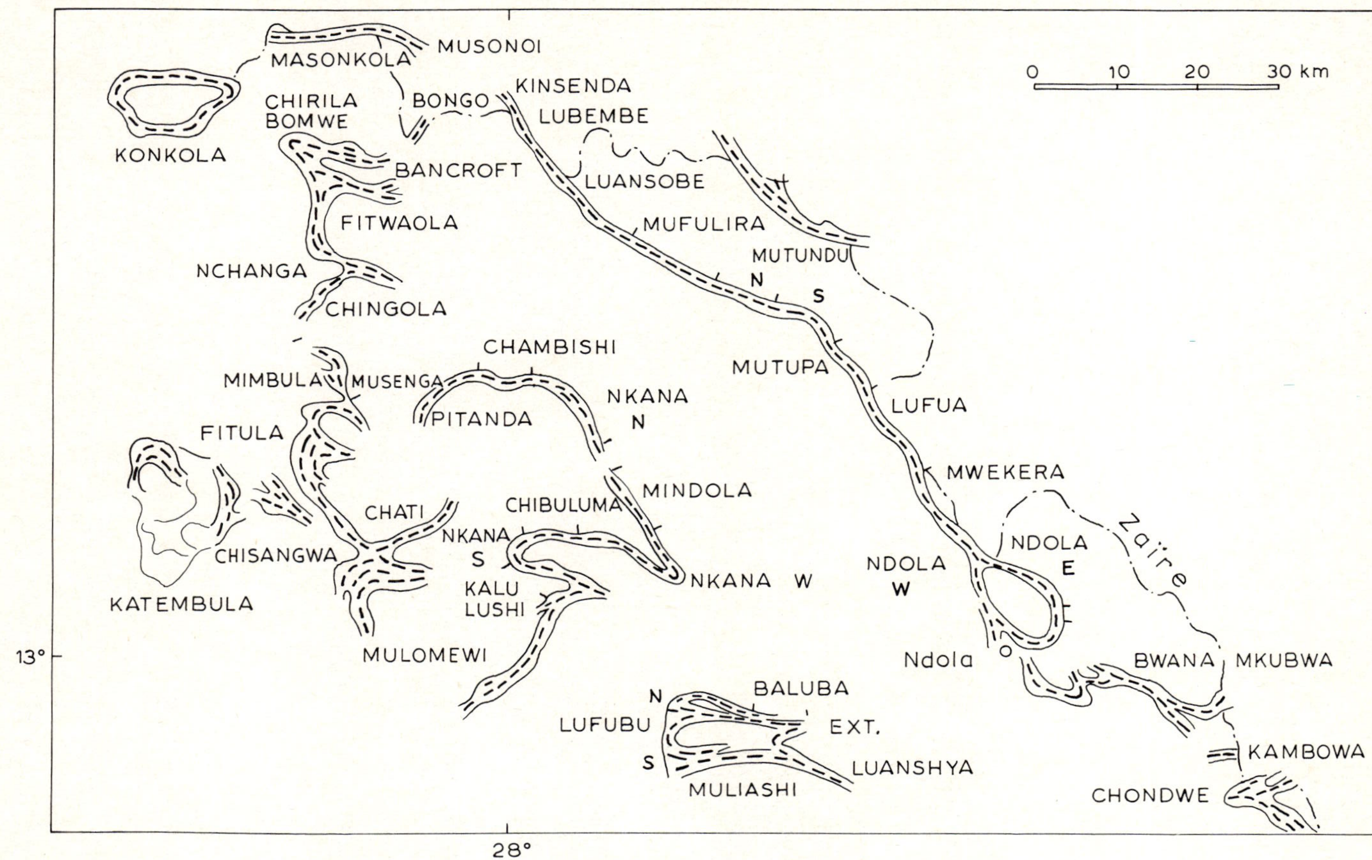

Fig. 45. The Zambian Copperbelt.

Table 79 Zambia copper reserves, mines, leach, plants, smelters, and refineries.

	Kon-kola	o.c. Kan-sanshi	Nchanga o.c. mg	Mufu-lira	Cham-bishi	Chi-buluma	Nka-na	Ba**ł**u-ba	Lu-anshya	Total
Reserves Mt	2	0.1	4.5	2.8	0.7	0.25	2	1.3	1	12.6
Grade %	3.8	2.9	3.7	3	2.8	3.5	2.4	2.5	2.5	2.6
Ore milled Mt	1.7	0.1	6.7 3.2	5	2.4	0.6	4	2.3	3.5	30
Grade %	2.9	5.5	2.8	2	1.4	2.9	1.6	1.7	1.3	2
Concentrates kt	110		750	230	80	50	180	90	170	1550

Smelters/leach	tails Nchanga	leach	Mufulira	Chambishi	Nkana	Luanshya	
Material kt	9500	470	470	roast	630	250	11.3
% copper	1.3	12	33	leach	33	30	
% recovery	57	66	97		94	96	
Copper kt	80	40	150	10	200	70	550
Cathodes kt	Ndola	100	200		160		460
Production kt	Electrorefined cathodes 440						

Nchanga contributes a third of the copper and has 40% of economic reserves, followed by Mufulira. Kansanshi ore is trucked to Nchanga too from the north-west, where the overburden ratio is only 1:4. In practice, twelve mines are managed as six divisions. Each mine has its concentrator but there are only three leach plants for oxides. Their solutions precipitate after electrolytic decoppering. Three smelters produce blister from sulfides and three refineries handle these and leach cathodes. The old Latreca foundry operates at part of its capacity, so does Cablecom (1-2,000 ty out of 8,000). Splendor and Congagec draw and isolate wire and aluminum wire capacity might expand from 1 to 6,000 ty.

Mines and divisions

For each ton of ore 60 m^3 water have to be pumped and some consider Konkola as the world's wettest mine. Its concentrator recovers only 85% and the division employs 5,500 people, Chililabomwe and Kansanshi included. Although smaller than Kolwezi, the Nchanga openpit ranks (with the underground) among the biggest copper mines. The overburden ratio is almost 3:1 and a one ton bucket wheel excavator has a capacity of 20 Bm^3y. Unfortunately, the concentrator recovers only 63% of refractory ores. The division employs 10,000 including Chingola, to which has been added Chambishi. Mufulira, on the other hand, is perhaps the world's largest underground copper mine. As such, the lack of spares maintenance affects it particularly. Along with new shafts and tailings dams should help though. Concentrates move here to converters and anode furnaces. The smelter also treats dump slags and part of the copper is refined at Ndola. Together they employ 9000.

Although reserves have hardly changed at Nkana since 1960, grades declined from 3 to a reported 2.4% and only 1.6% at the millhead. Shafts reach deeper than 1,300 m. There are oxide and sulfide concentrators but the Torco process proved to be too delicate. Nkana has also absorbed Chibuluma (which is extending to the west) including its acid plant. Finally Luanshya (Roan) is being worked out, but Baluba is expanding ; anodes from the reverbatory furnaces are refined at Ndola.

Cobalt, lead, zinc, silver, &

Cobalt reserves generally stand at 3-400,000 t but the full reserve base is 600,000 t, mostly at Nkana, Nchanga and Baluba and at Luanshya cobalt/nickel ratios alternate from 5:1 to 12:1. Cobalt precipitates from the solutions of the leach plants, electrolytic refining and degassing produce the cathodes or granules and the sulfide smelters reject cobalt bearing slag. During oversupply only a third of mined cobalt has been recovered, as shown in table 80.

Table 80 Zambia cobalt reserves, production and plants.

Reserves	Baluba	Nkana	Chibuluma	Nchanga	Konkola		Total
Ore Mt	50	80	7	13	50		200
Grade %	0.16	0.13	0.26	0.66	0.08		0.12
Cobalt kt	80	104	16	90	40		330
Production				Plants	Nkana	Chambishi	
Grade %	0.14	0.1	0.14	Grade %	2.4	2.15	2.2
Concentrates kt	70	80	50	Cobalt kt	1.5	1.5	3

To be sure, Zaïre had to curtail its output, yet Zambia increased capacity to 5,000 t.

While few precious metals and selenium are left in the Mufulira slimes, silver recovery peaked with 40 t in the sixties and gold extraction increased from 100 kg in the fifties to 500 kg in 1975. Kabwe produced in 1984 most sulfur from Nampundwe concentrates (near Lusaka) as shown below (in kt and %) ; but more and more comes from dumps.

Reserves	Ore	Zn%	Pb%	Cu%	S%	Milled	Zinc	Lead	Silver	Copper	Pyrite conc.
Kabwe	1100	22	11			230	25	15	3		15
Nampundwe	11000			0.8	17	200				2	60

Production of zinc and (lead) also had its ups and downs at Kabwe : fourties 20,000 (10,000 t), fifties 50,000 (15,000 t). Both reached peaks between 1969 and 1973 with 75,000 and (35,000), but by 1989 the mine will become uneconomical.

Cadmium output plummeted meanwhile from 130 to 20 t and the recovery of vanadium ceased in 1953 ; on the other hand Kabwe still produces 10-15,000 t clinker.

There is also zinc at Sable 2 and Star (C). Silver accompanies base metals at Kalenda and in groups III and IV and the search for gold is on a) between Changwe and the Luangwa (E), and b) east and west of Lake Bangweulu (N) at Senga

Hill and Mporokose.

Energy

By 1990, Zambia will import 60,000 t of coke from Zimbabwe and 4.5-5.5 Mb of oil for the Ndola refinery via the 1700 km long Tazamo pipeline. Hydropower capacity of 5.5-7.4 Gw and 600-850,000 ty of coal will also be required then depending on substitution at the mines which use more than a third of the oil, half of the coal and three quarters of the hydropower. The Nkana smelter can be converted to coal firing or replaced by an electric furnace. Likewise, vehicles are being electrified. Moreover the fertilizer plant needs 220,000 t coal.

I Between 1957 and 59 Nkana (Mindola) produced 230 t of uranium oxide which is
II also scattered over 230 km from Mwinilunga (NW) east to Solwezi, at Kabompo, Kawanga, Lumwana and Mitukuluku, Kifubwa, Kansanshi, Kimalé, Kapijimpamba and Dumbwa. III Other uranium anomalies are strewn over 10 km north of Lake Kariba (SC) between Chizwabowa & Mwanangalu at Njame, Mutanga, Dibwe, Nyamakanda, Siavonga, etc.

Lastly, a hundred thermal springs rise mainly along the Luangwa (E). While higher temperatures would probably be reached only at 8 km of depth, the Lochinav springs are hotter.

Estimates of the potential range up to 12 Gw. Hydropower sites could therefore be expanded and other sites developed on the Luapula (N) and the Zambezi which loses 90 m altitude from Victoria to Devil's Gorge, then 90 m to Kariba, 100 m to M'Pata and finally 55 m to Cabora Bassa in Moçambique.

Table 81 Zambia hydropower plants, extensions and sites.

Operating	Mw	Twhy	Ex-	Mw	Twhy	Zambezi	Mw	Twhy	Other sites	Mw	Twhy
Kafwe I-II	900	5.2	ten	450	2	M'Pata	500	3	Itezhi-Tezhi	80	0.5
Kariba N	600	5	si	300	0.6	Batoka	800	4.2	Lusiwasi III	40	0.1
Victoria F.	110	0.8	ons	250	1.9	Devil's G.	620	3.5	Luapula	880	5
including Nangweshi.											

Zimbabwe shares the Zambezi and buys a third of Zambian output. Moreover, Zaïre borders the Luapula ; for all of these only the Zambian share is reported above. Further, the installation of micro-hydros (5-100 kwh) with cross flow turbines started at Nkandabwe and Chavuma.

Coal shipments peaked with 940,000 t in 1973 and 74 and have fluctuated between 5 and 600,000 t ever since, that is 40% of nominal capacity. To be sure, plans call for rehabilitating the Maamba colliery near Siankondobo (S). Estimates indicate here 90 Mt in the Kanzize and Izuma basins, 100 Mt at Mulungwa and 80 Mt at Nkandabwe. True, coal measures extend over 250 km along the Zambezi. There is still a small basin at Luana (C) and lowgrade coal straddles the remote Malawi border (NE) at Pangala, Songa, Luwuwushi, Luhoka, Lusangani, Luwumbu and Mwili. Yet, the thickest seam has only 1.20 m.

Other resources

Kundelungu limestones also constitute the best aquifers and best soils. Dolomites supply 800 m^3h per well near Lusaka. Elsewhere in the Copperbelt, flow is only 3-15 m^3h, or along fractures 80 m^3h. As usual, problems arise on the Basement and 53% of wells would not meet U.S. Public Health Service requirements.

Table 82 Zambia groundwater rate of flow (Mm^3), recharge (Bm^3) and storage (Bm^3).

Basin	Flow	Recharge	Storage	Basin	Flow	Recharge	Storage
Luapula NW	830	40	380	Kafwe SC	960	20	250
Luangwa E	1630	30	240	Zambezi S	220	60	1740

The best storage are though the Kgalagadi sands which cover 200,000 km^2 (W).

The ammonia plant at Kafwe has been expanded to a capacity of 100-150,000 t nitrate and sulfate. Phosphates are available on Nkombwa Hill, at Kaluwe and plans call for an output of 200,000 t rock at Petauke (E). The carbonatites also contain rare earths, thorium and niobium, and extraction of aluminum sulfate from clays has been considered. Copperbelt carbonates extend north and south beyond Lusaka to Maseki and Mazabuka. Lime is burnt here in Chief Mapanza's fief as a fertilizer. Yet, most limestone is used as a flux, in the leach plants and copper smelting. Lime burning is to increase therefore at Ndola to 350,000 ty and the coalfired dry process cement plant (state 67%, AAC 23%) has a capacity of 320,000 t. Conversely, the Chilanga plant near Lusaka uses the wet process ; it is rated at 180,000 ty.

Emeralds show at Miku. Nkurulu recover them at Kamakanga (E) and the state's Reserve Mineral Corp. at Fwayafwaya (S). Recorded productions of 20 kg are a fraction of the total. The same holds for tourmaline and for amethyst mined e.g. at Namazombe (N). Market value of the precious stones might, therefore, range up to $150 M, but the 10-30 ty of tinstone washed at Mazabuka, Chaboola and Nuchange (S) are worth less.

Production of manganese at Kapiri Mposhi (C) and around Mansa (N) peaked with 50,000 t in 1958, it leveled out and later plummeted. Yet, processing has been considered. Similarly the U.S.S.R. is to build a 150,000 ton steel furnace at Kafwe The country has been meanwhile credited with 280 Mt of iron ore resources grading 50% Fe.

Table 83 Zambia iron ore resources (Mt) location and grade (% Fe).

Occurrence	km	from	Mt	%	km west of Lusaka		Mt	%
Chisala	240 W	Chingola	50	37	320	Kahara	20	50
Mwinilunga	360 W	Chingola		55	135	Nambala-Sonkwe	20	62
Lufubu	240 W	Kitwe	60	>60	115	Namatombwe	160	48
Chibote	360 E	Ndola	50	50	50	Sanje-Pamba	15	58

Lastly, other occurrences include :
Fluorite : Karibu, glass sands Kapiri Mposhi, glaze clay Kitwe, graphite : Njoka (E), Petauke, gypsum Lochinvar, kyanite Nabwale, mica Mulungushi, salt Lake Mweru, talc SE of Lusaka, PGM and nickel Munali (S).

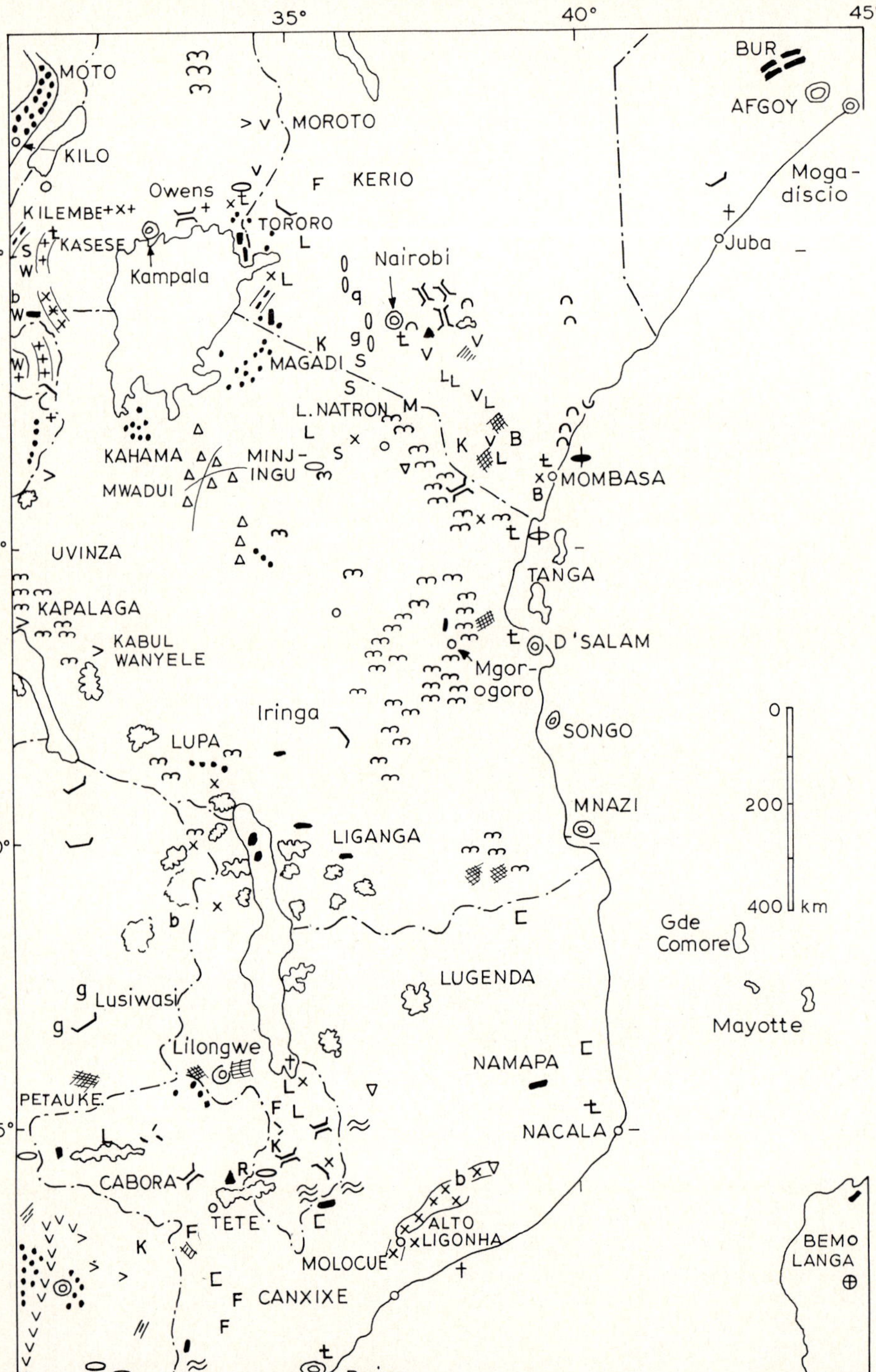

Fig. 46. East Africa. (Legend on page 13.)

6 EAST AFRICA

The least endowed part of the Continent nonetheless has 60 Gw hydropower and a certain oil potential along the coast and in the Rifts, which are filled with water-bearing sediments. Their lush vegetation was the cradle of mankind. Next to these, podzolized soils surround the Highlands ; two other types of red soils can be distinguished : lateritic and non-lateritic, the latter covering acid rocks of the Basement.

East Africa is also the Continent's principal producer of trona and tungsten and well managed countries enjoy prosperity (Fig. 44,46-50).

6 1 BURUNDI

While the U.S. receive most coffe grown by Hutu farmers, led by Tutsi, Belgium supplies most goods. Although petroleum imports from Kenya drain the treasury, the Republika y Uburundi has hydropower and peat.

> GOALS : Fertilizer production, reduction of the fuel imports, exploration for precious and rarer metals. Agressive marketing of the vanadium and nickel resources.

Table 84 Burundi production (present and past) and reserves.

Commodity	Resources		Production				Production		Resources	
Phosphate	Mt	2	Gold	kg	5	Hydropower	Gwh	50	Mw	800
Potash	Mt	3	Tungsten, past	t	3	Peat	kt	15	kt	500
Platinum &	kt	1	Coltan , past	t	3	Lime(stone)	t	300	Mt	4
Cobalt	kt	300	Resources			Kaolin	kt	3	Mt	3
Nickel	Mt	1.2				Tin past	t	50	kt	4
Copper	Mt	1	Chromium	kt	600	RE conc., past	t	70	kt	5
			Vanadium	kt	150					

Water, food and gold

The Ruzizi feeds the water table north of Lake Tanganyika, but quality is not always satisfactory. Apart from the laterite covered peneplains such as Mosso, there is perhaps too much water. Consequently, erosion of good red soils enhanced by steep volcanic slopes is a major problem. Yet, the situation is somewhat more favorable in the valley of the Malagarasy (S).

The phosphates of Matongo-Bandaga (SW) would have to be mined partly underground. A 1/2 km^2 area averages only 11% P_2O_5, but a 33% concentrate could be mixed with lime, magnesia or alkaline rocks or alternatively with nitric acid if the Lake Kivu methane project materializes.

The carbonatite grades 42-45% CaO under 40 m of phosphate sand, rubble and clay and plans call for a 70,000 ty plant which could compete with costly cement imported from Zaïre. There is also travertine in the Ruzizi (W) and limestone in the Malagarasy valley (SE). Next to Matongo potassic rocks of Mvumvu could be processed into glaze, but, unfortunately, the bauxites of Mpinga are subeconomic.

The principal gold belt extends 300 km beyond Muhinga (NE). Gold was also washed on the Nile/Zaïre divide, southeast of and in the Lugogo, north of the capital. Recovery of gold, alloyed with 10% silver, peaked with 100 kg in the fifties.

Metals

Similarly, tin production declined to a few tons at Mogare, Ndora and other occurrences of the big divide (SW) where columbite-tantalite, bismuth and lithium minerals also show.

There is also some tin in the district of Muramvia (W) and between Muhinga and Rwanda (NE). Furthermore, tungsten is occasionally recovered at Busomi and Bubanza and at Murore (NE) where tourmaline is abundant. Conversely, production of bastnäsite ceased in 1978 at Gakara (Karonge, W) because of the irregular distribution of grades. Subsequently, BGR has pointed out concentrations of europium a high priced rare earth.

Magnetites of the 30 m thick and 100 km^2 Mukanda, Buhoro and Nyabikere (EC) complexes range up to 1.7% vanadium and 22% TiO_2 according to the UNDP. Elkem (Norway) has outlined 14 Mt of ore grading 1.1% V_2O_5. However, the reserve base attains only 13,500 t.

More than 300 Mt of laterite average at Musongati (SE), Waga and Nyabikere near Buhinda 1.2% Ni, 0.1% Co, 0.3% Cu and 3.5 g/t PGM and within a zone containing 0.8 to 4% nickel with Ni/Co and Ni/Cu ratios of 7-40 and 5-80, 12 Mt of ore averaging 2% Ni have been delineated. Even a 20-30,000 ty nickel plant has been considered fuelled by peat from Karuyange and Sende. Additionally, the lower laterite zone assays 0.6% Cr_2O_3 with traces of cobalt, underlying hard rocks grade on the Rabare plateau 0.15-0.2% Ni and sulfid samples assay 1% Ni.

Energy

Unless oil is found next to the seepages northwest of Bujumbura and the test 120 km to the south, Burundi might import by 1990 3-500,000 by of costly petroleum products from distant Mombasa (or Dar es Salaam) and 30-60 Gwh of inexpensive electricity from Ruzizi (NW) - or a combination thereof depending on substitution. Five isolated hydropower plants, Mugere (Near Buja'), Mugera (C, near Gitega), Gikonga (Muramvya, CW), and Kayenzi (E, near Muyinga) and Nyemanga (Bururi, SW) have a combined dry season capacity of 10 Mw to which ten mini-hydros will add up to 500 kw. The Rwegura plant (NW) built with the assistance of the EDF, ADB and KfW adds another 18 Mw and 64 highcost Gwhy.

Lastly, the Malagarasy (ESE) has a potential of 28 Mw which might be harnessed in the XXIst century.

At 35% moisture, the calorific value of peat is 25-30% of that of oil and an Irish estimate specifies reserves of 100-200 Mt in the flooded Akanyaru basin (N), but according to Finnish Ekono resources might reach 1.5 Bt. A third of these are in the drier Buyongwe, Kirundo and Ndurumu valleys.

Air drying can reduce moisture content to 30%, pelletizing to 25% and a 3.5 km^2 area could supply 150-300 Mboe of dried peat. Alternatively, Buyongwe could provide 100 Mt of slurry containing 6% peat. Easily accessible reserves may reach 5 Mt out of which Nyamuswaga, east of Ngozi, may have 4 Mt. Since cultivated fields cover more than 10% of the bog only 1 m of height can be drained in order to avoid flooding.

Along the Nile/Zaïre watershed, Kishubi, Kitanga, Nyacijima, Kashiro and Kurunyange are also in densely farmed areas. Here too, more than a meter of waterabsorbent peat has to be left after extraction. Because of wide variations of ash content this peat must be macerated and mixed before briquetting. A 12 ton machine might increase output from the current 3,000 t.

62 RWANDA

With 250 people per km^2, the Hutu republic is overpopulated ; it is also the 'highest' and the Nyakabingo tungsten mine is at an altitude of 2,000 m. To be sure, Somirwa (49% owned by the state) is the leading tungsten, second columbite-tantalite and fourth tin producer in Africa. The company manages three mines and employs 8,000 people ; it also buys concentrates at a moderate price from 12,000 part or full time tributors, some of whom run medium sized workings. Although Somirwa is not profitable, mainly because of the overpriced franc, the austere government favors foreign investment : the 3,000 ty tin smelter at Kururama, near Kigali, and a cement plant are its first achievements in processing.

> GOALS : Reduction of oil imports, development of hydropower and methane resources and of a building materials industry. Establishment of a mineral map. Recovery of semi-precious stones and gold.

With a quarter of the total, tin and tungsten rank second and third among the exports - between coffee and pyrethrum. The principal trading partner is Belgium, followed by the U.S. and U.K. for exports. In contrast, petroleum products trucked through Kenya rank second among imports. Output and reserves can be summarized as follows.

Table 85 Rwanda reserves, resources and production.

	Production	Resv.		Production	Resv.		Production	Resources
Tin	kt 1.5	20	Beryl	t 50	kt 8	Hydropower	Gwh 150	Mw 600
Tungsten	kt 0.3	12	Gold	kg 40	t 10	Peat	kt 1	Bt 1
Coltan	t 50	700	Lime	kt 10		Methane		Tcf 1
Lithium conc.	t 30	kt 2	Travertine		Mt 7	Cement cap.	kt 50	

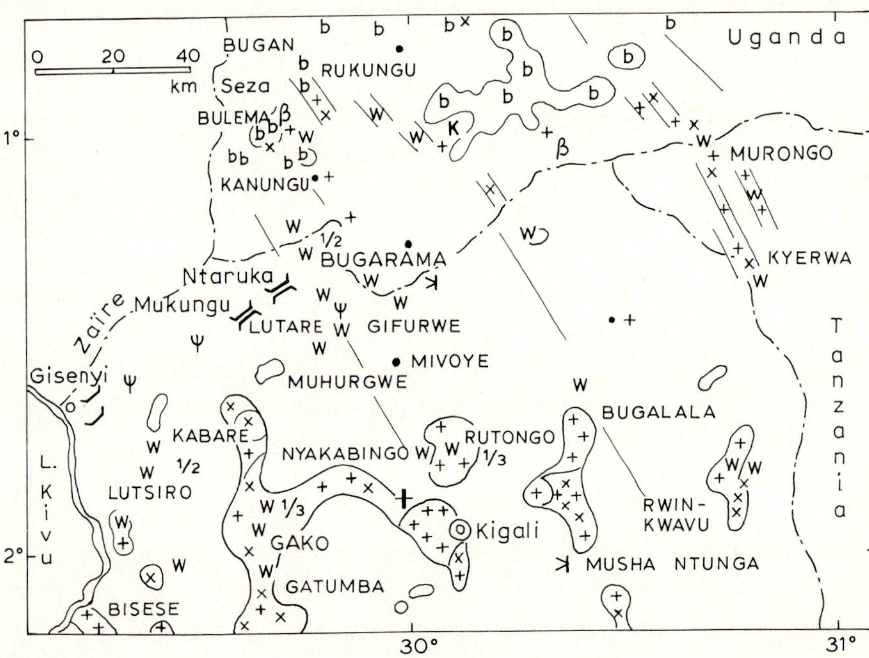

Fig. 47. Rwanda, SW Uganda and NW Tanzania. (Legend on page 13.)

Metals and pegmatites

Tin production ranges from 1 to 2,000 ty and is increasing again after a dip to 100 t in 1962. Tungsten is back to its level of the fifties, tantalite has not risen yet to its wartime heights and beryl and lithium concentrates have declined from their 500 and 3,000 ton peaks of the sixties.

Deposits follow six NNW-erly trends on both sides of the Gatumba massif (CN) :

I Scattered on the Zaïre-Nile divide (W) : Tsharwa, Nyankuba, Kabaya, Lutsiro 4,000 t, tin, tungsten 1:1 , Bisesero 500 t tin.

II The Gatumba Lutare (W) and Muhurgwe 'finger' (NW).

III The Gatumba-Tshubi-Kigali 'aureole' (N) including Gako where 7,000 t were produced (W 1/3).

IV The Kigali-Nyakabingo-Uganda axis (NE) with Bugarama 2,000 t, Sn/W 1:1, Gifurwe 3,000 t, W 1/3 and Kidwa island (W).

V Mbuye-Rutongo (CE) : 35,000 t W 1/3, N'tunga 700 t (mainly tin)

VI The Bugambira (500 t mainly tin)-Rwinkwavu (1000 t) axis and

VII Karagwe.

The principal ferberite deposits are at Bugarama and Gifurwe (N) and tungsten ratios are indicated above. True, the Rutongo vein is the largest remaining tin and tungsten deposit, followed perhaps by Rwinkwavu. These three are the only industrial mines, but there were all in all some fifty workings.
Most columbite - tantalite has been recovered from area III and M'buye (V), most beryl from the tailings of Gatumba, Bijejo and Nyarigamba nearby and at Kibuye, but the Rongi lithium (amblygonite) deposit has closed. While bismuth only showed at Buranga, tourmaline, mica, feldspar and kaolin are of course, widespread. There are also garnet pegmatites and subeconomic graphite schists are often host rocks of the deposits .

The gold trends (N) are wedged between the tin alignments. The principal deposit, Kalenda Hill, near Myove contains 2.5 tons. Furthermore, gold is disseminated on the Nyongwe (SE, 750 km). Reported recovery has grown to 20-40 kgy. Finally, copper shows at Kibuye (SW) and there is some iron ore on the Uganda border.

Energy

By the time 400,000 barrels of petroleum products reach Kigali from Nairobi, they cost three to four times the posted price. The Dar es Salaam-Isaka route is surely less expensive.

The Ruzizi loses 685 m of altitude between Lakes Kivu and Tanganyika and the Mururu (SW) hydropower plant generates in Zaïre 150 Gwh from 28 Mw capacity. Downstream the Panzi, Kitimbo and Kamaniola sites have capacities of 36, 36 and 270 Mw and the 40 Mw of Ruzizi II, 20 km from Bukavu can satisfy the demand of EGL. A competitor, KBO's Rusomo site on the Kagera (SE) could provide 4-600 Gwh from 80-130 Mw. Nonetheless, this power would be more expensive and 30-40,000 people's homes would be flooded. At any rate, Rwanda will use 150 Gwh by 1990 to which Ntaruka (NNW) and Mukongwa downstream contribute 20 and 48 Gwh from 11 plus 12 Mw. Gihira (NW) generates 8 Gwh, to which should be added some 50 mini-hydros and the Akanyaru Basin peat deposit which extends from Burundi into southern Rwanda, perhaps on a larger scale. One of its bogs, Busoro (CS) may contain 2 Mt peat (dry basis) with a 5-20% ash content - it already supplies 1,000 ty. Other bogs are Nyanza near Gitarama (CW), Gishoma (SW) and along the Rugezi (NW).

There is undoubtedly geothermal steam in the Virunga volcanoes (NW) and monazite is widespread in the tin fields. Likewise, granites of Gitarama (CW) and rocks of Kibirira near Gatumba (C) are highly radioactive. Finally, uranium also shows at Shore, Buranga, and Karago.

Gas, water and cement

Methane and carbon dioxide are absorbed by the waters of Lake Kivu, the prettiest in Africa, below 70 m of depth and from 180 m to its maximum depth of 490 m 2.4 Tcf of methane accumulate. Out of these, 1.8 Tcf are recoverable below 270 m of depth in Rwanda and Zaïre which share the lake. Moreover, biological and/or volcanic phenomena partly or fully regenerate this unique resource. Its exploitation might even enhance the formation of gas and the risks of a turnover are not great.

Each m^3 of deep water holds 1.4 Nm^3 gas consisting of 73.5% CO_2 and 25% methane. Although proportions are variable, combined gas content and the layering of the lake is not. Life has only been observed above 60 m ; fossiles undergo anaerobic fermentation, their water is warmer, therefore saltier and heavier : it does not rise and keeps the methane absorbed. It can be released easily though by a pipe and degassed as atmospheric pressure decreases. A small plant at Cape Rubona (W) extracted gas this way and sold it to the Gisenyi brewery, and a new plant supplies 1.5 MNm^3y containing 70% methane to a 700 kw generator. A 20-40 times larger methanol plant might even compete with imported gasoline, but the production of ammonia and urea requires larger markets than Rwanda.

There is phosphorus and nitrogen in the waters of Lake Kivu and some lavas of the Virunga chain (NW) assay up to 8% K_2O. Salt brines also rise along the Rifts.

The needs of fast-growing Kigali, Gitarama and Gisenyi are met by watertable wells. But the juvenile relief and intensive erosion hinder the development of adequate aquifers. Soil is washed away for the same reasons. Since valleys are narrow little water seeps into their gravels. Conversely, the moors of the north and south absorb, store and release water, but there is little surface water at Kajesera.

Cement ranks fourth among imports and the Mashyuza plant fed by Ruzizi travertines could be expanded to 100,000 ty. Since pumiceous tuffs are abundant light pozzolanic cements can be produced as well as lime. The plant might ultimately use domestic peat or methane along with some hydropower.

6 3 U G A N D A

In the past, minerals provided employment for 8,000 and covered half of government revenue. Copper alone contributed a quarter of the budget, but electricity is only a minor export item now, far behind coffee, which is mostly shipped to the U.K.

On the other hand petroleum from Kenya ranks high among imports. The reader may be therefore interested in outputs of peak years :

Table 86 Past production and current reserves of Uganda.

Past production			Reserves			Past production			Reserves	
Cement	kt	300	Iron ore rc.	Mt	100	Hydropower	Gwh	650	Gw	12
Asbestos	t	50	Vermiculite	kt	120	Phosphate	kt	200	Mt	10
Lime	kt	30	Zircon	kt	500	Fertilizer	kt	5		
Salt	kt	10	Diatomite	kt	100	Niobium ox.	t	500	kt	300
Gold	kg	700	Soda carb.	Mt	18	Tantalite	t	10	t	300
Silver	kg	100	Soda chlor.	Mt	2	Tin	t	150	kt	2
Bismuth	t	50	Soda sulf.	Mt	2	Beryl	t	40	kt	15
Mica	t	15	Potash sulf.	Mt	.7	Tungsten	t	100	kt	5
Lithium ore	t	300				Copper	kt	17	kt	120
						Cobalt	kt	1	kt	8

Current production comprises 50 kt cement, 5 kt lime, 5 kt salt and 50 tons tin and tungsten.

Energy

Fuel imports of more than 1.5-2 M barrels are mostly trucked from the coast rather than shipped by cheaper rail. The hydropower potential of the Nile attains 2 Gw of which only 150 Mw or 650 Gwh are now used. 40% of these are exported to Kenya ; even so, the remainder matches a quarter of fuel imports. Apart from an isolated 1 Mw plant at Kabale (SW), the outflow of the Nile from Lake Victoria at Owen falls provides all this capacity which could be uprated to 210 Mw and 980 Gwh based on 630 m^3sec firm available flow. The Nile loses 100 m altitude down to Lake Kyoga and another 410 m downstream to Lake Mobutu Sese Seko (or Albert). The following potentials could be developed should additional export sales materialize.

Table 87 Nile power sites in Uganda.

L. Victoria to Kyoga			Kyoga to L. Albert		
Site	Mw	Gwhy	Site	Mw	Gwhy
Bujagali	180	910	Kamdini	230	530
Busowoko	160	840	Ayago	540	2900
Kalagala	110	590	Kabalega	520	3300

Oil seeps along the Albertine Rift near Lake Mobutu and showed in 100 and 1200 m deep drillholes. There are lignite seams at Kisegi on Lake Edward, and steam also rises from 1500 m depth near Toro. Furthermore, the Albertine Rift may have a geothermal potential of 450 Mw, mostly at Katwe, but also to the NNE At Buranga and at Kibiro (NW)

South-eastern and eastern Uganda

The beaches and islands of Lake Victoria are covered by glass sands (98-99.9% silica !), e.g. Kome Island at Entebbe and Kyaggwe. While potters mostly worked at Buku and Mukono, Ugandans manufactured brick and tiles e.g. at Nansana, Kisabi and Bugungu. In 1970, 170,000 t of limestone were mined at Tororo and Ucemo's dry kilns rated 150,000 ty. The company also produced asbestos cement. TCC has rehabilitated this plant with assistance of the IDB. An additional 5 Mt could be quarried on Cave Hill, better limestone is available on Bukiribu Hill of the Butiriku carbonatite, 40 km north, and Turkish experts were to build a plant at Budude.

The Busumbu Mining Co. has shipped 6000 t phosphate to Magadi (Kenya) assaying 25-34 P_2O_5 at Bukusu and tailings can be reworked for P, Nb and Ba. Three valleys of Sukulu also include 130 Mt grading 13% P_2O_5 plus 70 Mt of additional resources. Ticaf produced here 22,000 t of single superphosphate in 1969 and projects for a new 50-80,000 ty PNK plant might materialize. Resources of these two carbonatites total 200 and 50 Mt, but there is also apatite under Tororo town, at Bugisu, Bukiribu, Napak and Toror (NE). Other carbonatite resources include: barite North Valley of Bukusu 3.7%, beryllium Napak, Sukulu, copper Nambala Hill at Bukusu 180 Mt ore per 30 m depth assaying 0.06-0.1%, fluorite Bukusu, Tororo, iron ore Sukulu 30 Mt 60% Fe, 0.3% P_2O_5, molybdenum Napak, niobium Sukulu 130 Mt assaying 0.2% Nb_2O_5, Nb/Ta 1500:1, potash Toror 15% K_2O, rare earths Lolekek, strontium Central Vent of Bukusu 1% SrO, 13% K_2O, zinc, lead traces at Napak, zircon 200 Mt of Sukulu soil grade 0.25%, Zr/Hf 100:1, vanadium samples of Surumbusa Hill at Bukusu assay 0.1-0.3% V_2O_5 and 20% TiO_2, vermiculite Namkara (Bukusu) soils : 500,000 t to 15 m depth, top quality 300 kg/m^3

yet after screening the coarse weighs only 60 kg/m^3 and fines 110 kg/m^3.

In 1944 some low grade mica was exported from Karamoja (NE), gold has been panned at Amonikakine and Tira Busia and the marbles might be used for terazzo. Further, geochemical traces dot the border (E):
Morongole Range and south of Rom : Cr, Ni, Co, Cu, Zn ; Lochomo : Cu, Cr, Rom : Zn, Ni - beryl and graphite, Kaabong : Zn, Loybro and Bobong (50 km north of Moroto) Zn and 0.1-1% copper over 500 m. As usual, talc accompanies the Moroto chromites and the ores grade 3-5g/t of platinum.

Other areas

The oil fired Hima cement plant (CW) rated 300,000 t, but the two rehabilitated kilns operate below capacity. In the east bay of Lake Katwe carbonates and sulfates precipitate before trona and salt from waters containing 120 to 460 g/l solutes in the wet and dry seasons. Unfortunately, there are mud layers and the salt produced costs twice imports.

Buhweju (W) was the principal gold field. Similarly, lead and zinc production ceased at Kitaka and Kampono in 1959. More important, Kilembe (W) closed in 1977. 6.5 Mt averaged 1.7-2% copper, 0.17% cobalt with traces of nickel to which another 6 Mt of resources can be added. There are more oxides in the north of the mine, where the thickness of the orebody doubles to 12 m and tailings are said to assay 1.3% of cobalt. Both the mine and Jinja smelter (E) could be rehabilitated, funds and markets permitting.

Hundred fifty beryl workings were scattered in Kigezi (SW). At Mutaka recovery was only 14% underground, but handpickers recovered 25-85%. Lithium minerals and columbite-tantalite were sorted from the same ores at Rwentali, Lyasa, etc. A third of the tin concentrates came from Mwirasandu, followed by Bulunga and Naniankoko. Nyamuliro, Kirwa and Kilembe were the principal sources of tungsten and bismuth was mainly recovered at Kayonza, Kitawulira and Rwanzu. On the border 30 Mt hematite grade 78% iron, Muko ridge (Kobule) will be evaluated and other occurrences include :
Southwest: beryl Bugangari, Rwabuchacha with coltan, Ishasha, Bulema and Nyabukweri with feldspar, Kasumu, fluorite Shingo, many pegmatites ; garnet and kyanite bearing rocks, tungsten Bahati, Ruhizha, Mubolere, uranium pegmatites; west brines Kibero hot springs, with bromine and some iodine, Lake Bunyampaka, graphite Kilembe, gypsum Mobuku nearby, Lake Katwe, limestone Muhokya, Dura, tungsten Mituana; northwest diatomite Panyango under 4.5 m cover, Atar, Alui, 10-80% kyanite Erusi and Azi hills; south beryl, columbite, uranium traces Nampeyo Hill.

6 4 K E N Y A

Farming, mainly coffee, good management and services created prosperity. In the same spirit, Kenya draws maximum benefit from its limited mineral potential : it imports oil and exports refined products, Kenya ranks also first in Africa as an exporter of soda ash and cement and second for fluorite.

GOALS : Further processing for the domestic market. Oil and groundwater exploration, in the northeast. Development of small workings.

Government actively encourages investment and promotes mineral projects ; it controls more than half of the shares of the larger companies, but leaves management to the minority partner, mostly from the U.K., where many gems go. Asia takes a third to a half of soda, and the U.S.S.R. imports half of the fluorspar, followed by Japan, the Netherlands, and the U.S.

While inland neighbors are also customers for soda and petroleum, cement works import less expensive, Swazi coal.

Table 88 Kenya production, capacity and resources.

Building materials prod.			Production			Production			Resources	
Cement	Mt	1.3	Aquamarine	kg	50	Hydropower	Twh	3	Gw	13
Limestone	Mt	0.5	Garnet	kg	250	Petroleum	Mb	20		
Coral	Mt	1.5	Ruby	kg	400	Soda ash	kt	250	Mt	55
Kunkur	kt	120	Sapphire	kg	150	Salt	kt	60		
Lime	kt	30	Tourmaline	kg	40	Fluorite	kt	80	Mt	15
Vermiculite	kt	2	Gold	kg	5	Barite	kt	1-7		
Volcanic ash	kt	2	Silver	kg	0.5	Iron ore	kt	15	Mt	15
Gypsum	kt	30	Diatomite	kt	1.5	Rolled steel[c]	kt	100		
Shale	kt	300	Feldspar	kt	0.7	Manganese			kt	200
Magnesite	kt	2	Wollastonite	kt	1	Lead+	kt	3	kt	220
Talc+	kt	1	Carbon diox.	kg	3	Zinc			kt	20
Kaolin	kt	1.5				Niobium			kt	460

Building materials, metals & barite

Cement production grew fivefold in 20 years. At Mombasa (SE) the Bamburi Cement Co. uses the dry process, corals, shale and gypsum from Roka (E) and Mida for a capacity of 2.8 Mt. The Blue Circle group and the state are also the key shareholders of the wet process Athi River plant (S) of the East African Portland Cement Co. which uses Sultan Hamud limestone, volcanic ash from the Rift and gypsum from Tana. Gypsum is even transported to the coast from the Tula Valley nearby. Most marbles are not in the populated areas, at Kajado near Turoka they even serve for lime burning and the obsidians of Magadi and Naivasha expand like perlite. The Iwatania-Lonata Vermiculite Co. exports from the largest deposit at Wasini near Lodonit (W). In contrast, the open pit Kenze mine (S) supplies crushed vermiculite for the local market and other occurrences include :

Asbestos Kinyiki, W. Suk, Baragoi, Taita, carbonatites Koru, Songhor, green clay Kisui, Lodwar, gypsum Wajir el-Wak, Mandeca, Gougoni (evaporated), Homa kaolin Karatina, Machakos, Voi, Eburru, limestone Mombasa, Koru, Turoka, Kajiado, West Pokot, magnesite Kinyiki, Magongo, Merti (N) mica, ruby : Matthew's Range, Embu, sheet Tsavo, Taita, Kierra, pozzolana Highlands, pumice Naivasha, Longonot, Menengai, Rift Valley, talc Machakos, W. Suk, Ortuna, Soka, vermiculite Sultan Hamud, Machakos, Taita, Kinyiki, Kipiponi.

Gold shows around Lake Victoria (SW) and some can be leached from the McAlder dumps ; this mine was also the principal producer of silver (1-2 ty), and base metals, more of which show in the carbonatites : Vitengeni Ba, Pb, Cu, Hg, Kéwaba Mn, Simba Mn ; Kinangoni originally 3 Mt assaying 7.6% Pb and 0.6% Zn, with some Ba and Ag ; Changombe Pb, Zn, the Mwachi vein, Zn, Pb (near Mombasa) ; Mrima Nb, Ba, Mn, Pb, Zn, Ag, Fe and Lunga-Lunga Ba, Pb (on the border). Chloride Metals Ltd had a capacity of 30,000 t copper concentrate at Vitengeni. Occasionally, some barite has also been exported, but most of the niobium is hard to extract at Mrima where 600,000 t ore also assays 20-30% and 15 Mt grade 7-36% Fe. Moreover Bamburi produces magnetite at Ikuta and Taïta (SSE) and 17 Mt of pyrite rate at Bukuru (SW) 30% Fe. Lastly in the northwest I 60,000 tons assay 49% Cr_2O_3 (Cr/Fe 3) with traces of nickel at Telot, and II 100,000 t grade 54% Cr_2O_3 (Cr/Fe 3) and 1% nickel at Sekeri and other occurrences include :

Barite Sabaki, Didimtu, beryl Baragoi, Sebit, chrome Kangura, Debel, Dudati, Raboli (N), cobalt tr. Macalder, copper Eldoret-Barkugal (N) , gold : Kibigori, Kisii, Turkwell (W), Suam, iron McAlder, Uyoma (W), lead Athi Valley (Katulani), Kaimosi, molybdenum tr. Kavirondo Bay (SW), W Maragoli, Maralala, Taita, Mrima (SE), vanadiferous ilmenite Formosa Bay, Malindi.

Finally, the Kenya Steel Co's (Marubeni, Chandari & Sha) cold rolling mill at Mombasa produces sheet steel.

Industrial minerals

Salt is evaporated on Fundi Isa on the coast but an ICI affiliate is the principal producer at Magadi (S) in the Rift. In the northern gulf of the lake of 3 x 17 km and at Amagad the principal trona bed is 20 m thick ; it thins out though to 7 m towards the center (MRA pp 145, 585-6). Production of soda ash containing 97.4% Na_2CO_3, sodium bicarbonate and sodium fluoride (bath salt and washing soda bases) increased thus : 1917 50,000 t, 1941 100,000 t, 1961 150,000 - current capacity 300,000 t. Buckets dredge the trona, and the slurry is piped from the crushers and electromagnetic separators to the shore where it is centrifugated and calcinated before cxport via Kilindini. In some parts of Magadi there is 20% villaumite (NaF) and fluorspar is widespread in Nyanza (SW), the gold fields (SW), the carbonatites and at Rata. Most of the production of the Kenya Fluorspar Co. in the Keiro valley (WSW) is of acid grade, but metallurgical grade fluorite is separated by gravity. Grades are high $\geqslant$ 50% CaF_2 and reserves reach : 6 Mt proved, 4 Mt probable and 6 Mt inferred at Elgeyo-Marakwet and Baringo along with Kamnaon, Chof and Kimwarer.

Rubys and chrysoberyl are recovered at Chania (Tika), sapphires at Kinyiki, opal at Thomson's Fall, and melanite, the black garnet in the western carbonatites.

The best known diatomites of the Rift grade 84% silica at Kariandus. Output, used in local soaps peaked in 1958 with 5,000 t ; even filtrates and insecticides are exported, and cerussite is used as a filler; carbon dioxide is also tapped on the Rift at Kerita (E) and Esegeri (W) under higher pressure. Near Tsavo (S) corundum accompanies kyanite which has been converted to mullite at Taveta on an industrial scale. Likewise, graphite production peaked with 1000 t in 1958 and ITDC designed a new flowsheet for the high carbon Nharura material. To conclude, other occurrences comprise :

Alum Rift fumaroles, magnesium alum Voi, apatite (phosphate) a third of Ruri, Homa and Rangwa carbonatites (SW), also Mrima (SE), Ikutha, carbon dioxide Kedong (S), corundum, diamonds Kakamega goldfield, Ndarugu, Kinyiki, Taïta, Loldaika, Baragoi, graphite Tsavo-Mtitc Andei, Machakos, Voi, Namanga, Loperot, Merti, Kyanite Kevas, Loosoito, Sultan Hamud (SE), nepheline syenite, Homa, ceramic quartz Kinyiki, soda Elmenteita Nakuru, sulfur Naivasha, L. Turkana, wollastonite Epiya, Chapeyu, Homa Bay, Kajiado.

Water and energy

In the Highlands (S,SW) the walls of the Rift facilitate the tapping of groundwater and cashcrops grow on volcanic soils. In the west the Basement yields as little water as in Uganda, but runoff is abundant. Unfortunately, the sedimentary areas of the east and northeast have none of these and Somali (NE) and Turkana (N) nomads cover 20 km to watering points.

Until oil will be found Kenya will import 23-30 Mb crude. It already exports $>$ 10 Mb of products refined at Mombasa mainly to the land-locked interior and as aviationfuel. Albeit demand for oil will increase longterm, there is a respite. In order to compensate, coal imports might grow to 300,000 ty. Although Uganda would wish to increase its power exports from the current 250 Gwh, much of the 2.5-3 Twh demand projected for 1990 can be met from domestic hydropower, based on 750 Mw installed, 570 Mw peak and 450 Mw average capacity mainly on the

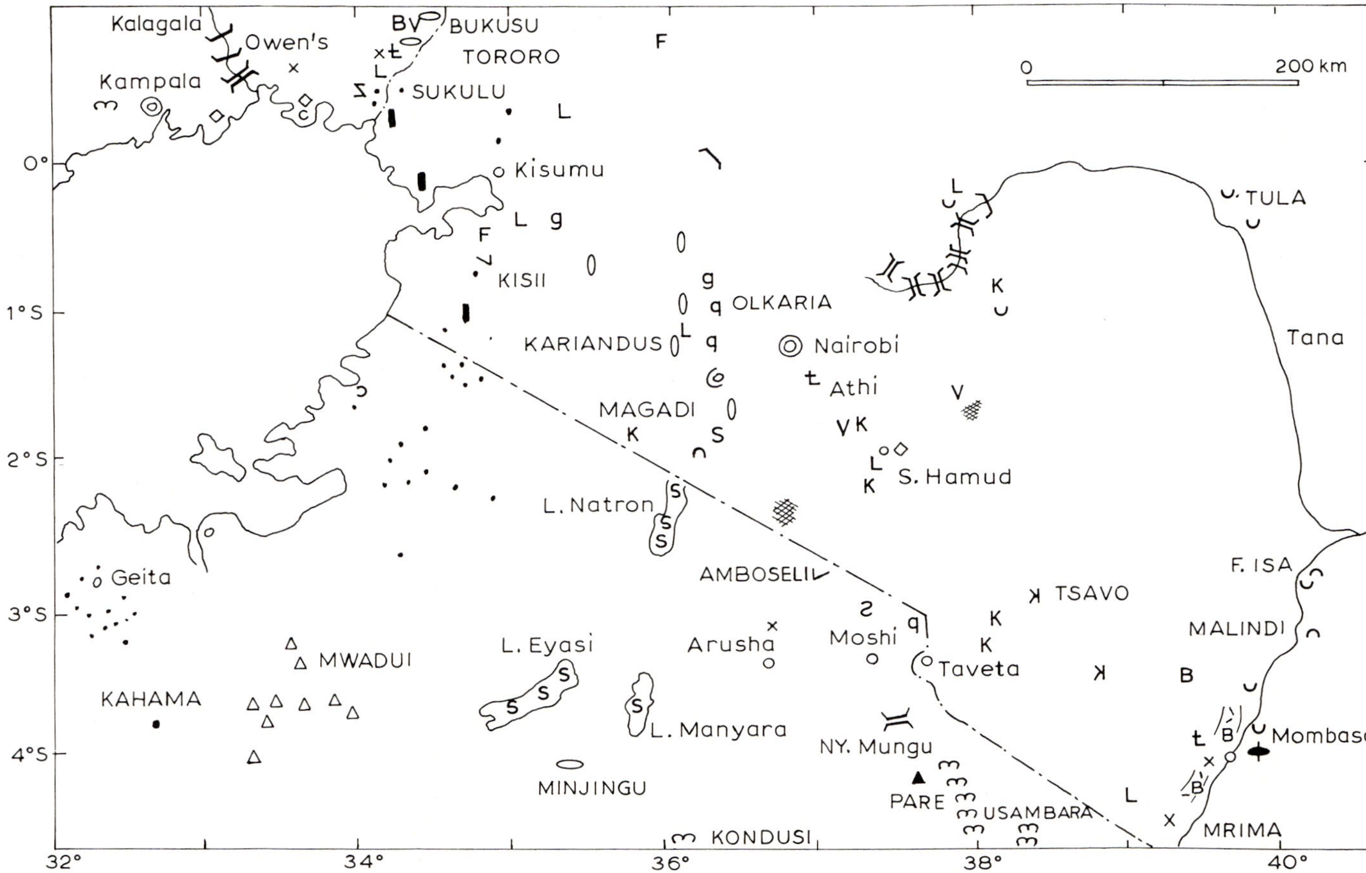

Fig. 48. SE Uganda, southern Kenya and northern Tanzania. (Legend on page 13.)

Tana (CS) at Tana 14 Mw, Wanji 7, Masinga 40, Kamburu 90, Gitaru 145 and Kindaruma 45. To these should be added 140 Mw at Kiambere, the Mutanga and Grand Falls sites downstream, and hopefully before 1990 the 120 Mw Kerie scheme on the Turkwell (W).

Out of a potential of 170-500 Mw, 30 Mw are used in the 1000 km^2 Olkaria (SC) geothermal field. Additional but larger sites are being developed south of Mt. Eburru and JICA and the UNDP survey the northern slope where steam dries pyrethrum and low pressure vapor is condensed for the watering of cattle. They also study Lake Bogoria where fumaroles and springs boil at the intersection of a fault with the Rift. Other southern sites include the Menengoi crater, the southern shore of Magadi, picturesque Lake Naivasha, etc.

6 5 T A N Z A N I A

When the gasfields and new hydropower plants come on stream, new hope will rise for the centrally managed economy of the United Republic of Tanzania, whose GDP declined in recent years. Even though the country has vast low grade and remote coal and iron resources diamond reserves declined.

> GOALS : Efficient investment incentives, further exploration for oil, groundwater and precious minerals. Organization of gem and mica recovery, development of the fertilizer industry.

Up to 5000 are said to make their living from minerals, Stamico assits in principle small workings and the state's TPDC is a silent partner of Agip, Amoco and Elf.

Diamonds are flown to the U.K. and Belgium, but they rank only eighth among exports, behind coffee, cotton and sisal.

Water resources are inadequate on the Masaï steppe (N) and the surrounding area, mostly covered by the Basement. In the north, the Kilimanjaro, Arusha and Ngorongoro, and the Iringa (CS) and Mbeya (SW) highlands retain more water - and soil. Yet only 12% of the country is cultivated and 8% of irrigable lands are watered.

Table 89 Tanzania production, projects^{p}, past^{+}, and reserves.

Production	kg	Production				Production		Reserves	
Aquamarine	500	Petroleum	Mb	5	Gas			Tcf	4
Amethyst	50	Salt	kt	30	Hydropower	Mw	330	Gw	21
Ruby+sapphire	10-20	Bentonite	t	80	Coal	kt	10	Mt	30
Rhodonite	5-10	N in ammonia	kt	5	Phosphate	kt	20	Mt	5
Zoïsite	10	Cement	kt	500	Soda ash	kt	30^{p}	Mt	150
Scapolite	10	Sheet mica^{+}	t	10	Diamonds gem	kct	200	Mct	3
Chrysoprase	10	Vermiculite^{+}	t	20	" " industrial	kct	140	Mct	2
Resources		Lime	kt	4	Limestone	Mt	1	Mt	50
Soapstone	kt 150	Gypsum	kt	12	Kaolin	kt	2	Bt	2
Magnesite	kt 150	Corundum	kg	10	Glass sand	kt	1	Mt	1
Ilmenite	Mt 3.6	Zircon	kg	10	Gold	kg	200	t	40
Nickel	kt 200	Beryl^{+}	kg	70	Silver	kg	100	t	35
Niobium	Mt 1	Tin	t	10	Garnet	kg	10	kt	30
Rare earth cc.	kt 300				Iron ore	kg	2	Mt	300

Energy

By 1990 Tanzania might consume > 1 Twh hydropower and 5-800,000 by of petroleum refined near Dar es Salaam.

Proved and probable offshore gas reserves of AGIP-Amoco's Mnazi Bay, Vermilion Bay (SE) and TPDC's Songo Songo (ESE) reach 4,200, 600, and 900 Bcf respectively, but the Kimbiji (E) wells onshore are not commercial. If fully proved, Mnazi Bay gas might be compressed and/or liquefied and compete with petroleum transported to the distant southeast. Furthermore, a 6", 220 km long pipeline connecting Songo Songo with Dar es Salaam industry might bring additional benefits. Meanwhile exploration continues along the coast a) along the limit of the sedimentary basin (CSE) and b) in the Lake Rukwa Rift (S) where a) Songwe-Kiwira coal heats 5500 cal/g. Although low in phosphorus and sulphur, it contains 25-30% ash. Underground production of the Ilima colliery might increase here to 30,000 ty should the Liganga pellet plant materialize (q.v.). China has promoted another underground project at Ivogo Ridge, but mining opencast 500,000 ty at Yukuyu, on Kabulo Ridge, near Kiwira, might prove to be less expensive. Other Rukwa fields include :
Mkomolo-Muze and - Namwele. Except for Kasulu (NW) all other fields are in the remote southeast.

Table 90 Tanzania coal reserves, Mt, measured reserves in brackets.

Rukwa	Mt	Basin	Mt	Basin	Mt	Basin	Mt	Basin	Mt
Ufipa	20-60	Songwe (30)	610	Liweta	30	Lumecha		Mhukuru	20
Galula	50	Keţewaka (190)	680	Ngaka	150	Mbamba (30)	110	Njuga	130

The Mtera hydropower plant and the Iringa dam (CS) come on stream in 1990. Other sites are also mainly on the Ruaha, on the Rufiji nearby and on the Ruhindji.

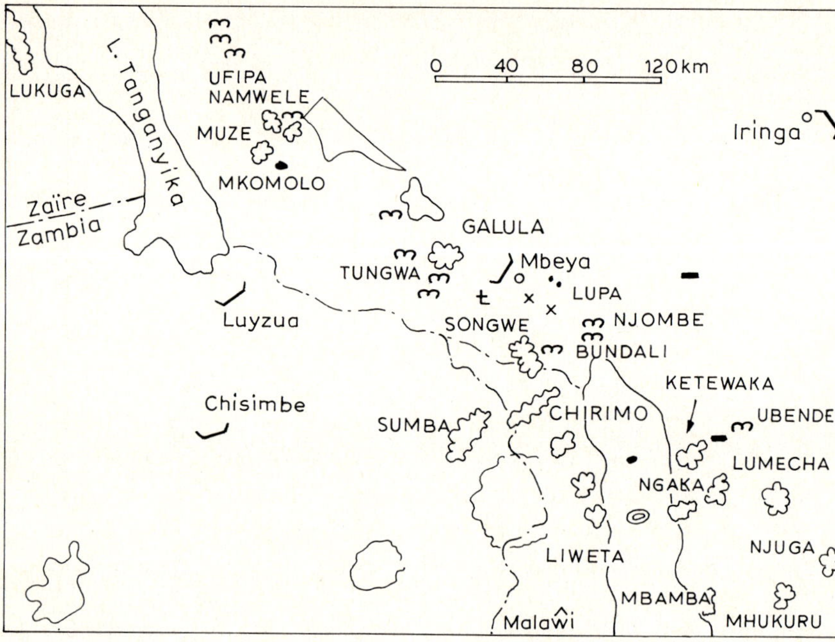

Fig. 49. Lakes Tanganyika and Malawî. (Legend on page 13.)

Table 91 Hydropower sites in Tanzania.

Developed	Mw	Developed	Mw	Site	Mw	Site	Mw	Site	Mw
Kidatu	200	Ny. Mungu NE	8	Ruhindji	680	Nkive	160	Malagarasi	100
Mtera	80	Kikuletwa	1	Ikondo	340	Taveta	150	Mara	90
Halem	21	Site		Kihansi	290	Lukose	130	Ntembe	60
Pangani NE	17			Kishanda (NW)	180	Wami	120	Songwe	50
		Stiegler's G.	2100						

Although Stiegler's Gorge on the Rufiji would control floods south of Dar', it appears to be oversized. The two Kihansi sites on the Kilombero tributary and on the Luwegu might be, therefore, easier to carry out this century.

Food, building and industrial materials

The overburden ratio is 1:6 at Minjingu (N) and with 10 Mt phosphate grading up to 20% to 60 m of depth Stamico has a capacity of 170,000 ty of rock. Following grinding, cyclones and filtering 24.6% P_2O_5 concentrates are trucked to Arusha and then shipped to the Tanga fertilizer plant which originally rated 25,000 ty P_2O_5. TFC also started small scale production of ammonium hydroxide. Another project of 1300 td ammonia and 1700 td urea is based at Kilwa (CSE) on 60 Mcfd of Songo Songo gas.

Salt extraction at Uvinza (W) should expand to 100,000 ty in order to serve Zaïre, Zambia and domestic markets of 150,000 ty to which Nyanza contributed

20,000 t. Soda salts also accumulate in the Tanzanian Rift - 150 Mt at Lake Natron and 500,000 t at Lake Balangida, both in the north. A million ton plant would require unfortunately costly rail links and the fractional crystallization process. Unido has recommended therefore a 60,000 t works.

Finally, 250 cfd helium ascends at Manyeghi, 130 cfd at Maji ya Moto ('hot water') with other smaller vents scattered in the Rift.

The state's Saraji Corporation received advice from Danish F.L. Schmitt and Swedish Cementa at Tanga (NE) and Wazo (E) which produced 150,000 t each through the dry process. True, Wazo rated 500,000 t originally, similarly, Mbeya (SW) operated at fifth of its 250,000 t capacity. Meanwhile limestone reserves reached 20 Mt at Wazo and 6-10 Mt each at Lugoba and Tarawanda. At Pindiro (SE) each meter of depth represents 50,000 t of high grade gypsum. Furthermore, at neighboring Mbaru and Mkomore beds extend to several hundred meters of depth.

Plans call for increasing kaolin recovery to 10,000 tons at Pugu Hill (E) out of 2 Bt resources and magnesite output at Chambongo to 25,000 ty. 130,000 tons of this material have been inferred at Gelai (NC) and 150,000 t soapstone around Kilombo, which was worked prior to 1940 at Hedaru. Meerschaum was discovered on the shore of Lake Amboseli (NE) in 1953. Because of the lack of spares extraction at Sinya stopped and the Arusha factory, which can produce 300 pipes a day imports sepiolite from Somalia. Another firm has exfoliated vermiculite from Mikese, Kwekivu, Lufusi, etc. and other occurrences include : Asbestos Ikorongo, Mbembe, bentonite Gelai (NC), diatomite Makutapora (Rift), Kagera (NW, 76% SiO_2), fluorite the Sengeri (Mbeya SW) and Ngualla carbonatites, glass sand Bukoba, Lake Victoria (N) with 2 Bt kaolin Pugu Hill (E), graphite Nachingwea (Uluguru Mts) Idete, Daluni gypsum Mkomazi, Msagali, Itezi, kaolin Matamba, Malangadi, kyanite Chankulu, Uguruwa, magnesite Haneti, Itisio, Merelani, ochre Monduli, phosphate Chali, Chamoto, apatite Mbeya, Zizi, potash springs of western (and central) Rift, pumice between Tukuyu and Mbeya (SW), salt Changwahela, Ivuna, Kitame, Kunduchi, Lindi, Mtwara (SE), Saadani, sulfur Wingayongo springs, Kilimanjaro crater, volcanic ash to the east.

Gems and mica

While there are also kimberlites in the Ruhuhu valley (SE), the four principal alignments I Usongo-Udutie-Mwadui, II Kisumba-Ilole, III Mabuki, IV Iramba are south of Lake Victoria, Germans discovered the first diamond near Mabuki in 1913 and Dr Williamson located Mwadui in 1940. Shabani was also an important diamond producer and 10-20,000 carats are recovered from the Alamasi outcrop. Germans also started the extraction of garnet in 1907 at Namaputa, but recorded production has declined from 300 t in the fifties to 50 t. At some distance in the Pare Mountains, geologists have estimated in the cover of Lolubokoi, Buiko and Kichaa, respectively, 25,000, 800 and 100 tons of garnet with 10,000 tons in harder rock at Sangarama. In addition, at Nyarumba (C), 10 million tons of overburden grade 15% garnet and the underlying hard rock assays 30-40%.

Although recent data on gem sales are missing, workers recovered rubies and sapphires from the Umba River in the Lushoto district.

'Anyosite' (ruby and corundum) of Longido fetched good prices but low grade corundum reserves are larger at Mleha and Ilende ; production peaked here with 30 t. 'Tanzanite' resources are also adequate, but German Zoïsite stocks limit sales of the state cutting laboratory at Moshi (N). There are, further, gems in the mica fields where production started in 1897 and peaked with 100 ty during the war and in 1963-5. Mica remains in the dumps but few workings are left at Morogoro (CW), in the Pare Mts. (NE) and Karagwe (NW)and in belts I-VII :

I Mahenge-Lushoto (E),600 km, including Ruaha, the Uluguru Mts. (tourmaline, garnet), Mikese, Kwanandu-Munga (tourmaline, amethyst, sapphire).

II Iringa-Kwekivu (CE), 300 km, including Rubeho (piezo quartz, kyanite), Ukaguru, Ngaru (vermiculite), Kwekivu (piezo quartz).

III Lushoto-Moshi (NE) 250 km : Usambara Mts. (beryl, aquamarine), Pare, Lelatema.

IV East of the lakes (SW) : Njombe, Ukende, Bundali, Tungwa, Ufipa.

V Mokanjiro (beryl)

VI Hombolo (tourmaline), Maganzere, Zoïssa (sunstone), Itiso (chrysoprase, green opal), Kondoa.

VII Mbulu, etc. (tourmaline, garnet, corundum). Lastly, there is : chrysoprase and amethyst at Kilosa, emeralds at Manyara (N), iceland spar and amazonite at Luguruni, piezo electric quartz at Idibo and zircon at Mashowa.

Metals

Production still reached 3-4 ty during the war in the Geita-Musoma (N), Iramba (NC) and Lupa-Mpanda (SW) goldfields. Mpanda also produced 70 t silver and there still is activity on the Buck Reef (Geita), at Lupa, Nzega, and on dumps. Further, Dar Tadin al Umma is to contribute a third of the 2ty Kahama (N) production to the state out of 3.6 Mt ore grading 10.6 g/t gold and 10 g/t silver.

Liganga (SW) holds 45 Mt iron ore assaying 50% Fe, 1.3% SiO_2 and 0.2% V plus resources of several hundred million tons. Extraction of 1.5 Mt, beneficiation for 500,000 t pellets and infrastructure would cost $2 B at current prices and even a small steel mill has been discussed. There are also 70 Mt at Manyoro grading 30% Fe and up to 11% Mn, but Hundusi (E) only holds 8 Mt. Aluminum shales have also been sampled here, near Mombo (Usambara Mts.) and at Amani (NE). To continue, Kabanga (NW) might have 20 Mt ore 'reserves' and 60 Mt resources assaying 0.7-2.3% nickel and 0.2-0.3% Cu. With 5 Mt containing 0.7% Ni Kabulwanyele (W) is smaller ; so is the 12.5 m thick Kapalaga ore body (W), with samples of 0.3% Ni and 0.1% Cu.

The principal lead-copper mine was Mukwamba and the largest tin mine was Kyerwa. Output peaked with 400 t in 1938 and 300 t of tungsten were also mined in the northeastern part of Karagwe (NW) at Chamuyana, Ihunjare and Karugu. Finally, the Mbeya carbonatite (SW) contains 800,000 t niobium and other occurrences include :

Base metals, gold, silver Sikito-Kapapa (W), chrome Nkenza, cobalt Chenzema-Mtenga Hills (E), Ukinga, Kaberinga, Mongalisa, copper Kigoma, Kigugwe, Kidete, nickel Twamba (0.9% Ni samples) ; with chrome Mwahanza (E), Kabanga, disseminated pentlandite and massive pyrrhotine, Kemuyarika, Ngasamo (N), Zanzeri, niobium Oldonyo Dilli, Galappo, Hanang, Ngualla, Musensi, Wigu Hill (E), with rare earths, pyrite, pyrrhotine Samena Hill (NC), tin Ilama, Kaitambuzi, Katera, Mutongo, titanium beaches of Bagamoyo, Kunduchi and Mbwemmi (NE) ; Milembula.

66 MALAWI

The principal exports of prosperous Malawi are tobacco and tea ; petroleum ranks second among imports.

> GOALS : Development of hydropower and coal. Further exploration for oil and precious minerals. Promotion of existing resources and improvement of communications through Moçambique.

Table 92 Malawi resources and production.

Resources			Resources			Resrcs		Production		
Ilmenite	Mt	1.2	Phosphate	Mt	1.3	Twhy	5.3	Hydropower	Gwh	500
RE oxide	kt	330	Bauxite	Mt	60	Mt	800	Coal Lime	t	300
Thorium ox.	kt	33	Iron	kt	200	kt	300	Kyanite Cement	kt	80
Niobium	kt	200	Manganese	kt	300	Mt	330	Limestone	kt	120
Fluorite	kt	2	Garnet	kt	4	kt	90	Vermiculite		

Energy, water and food

Unfortunately, three small basins with oil potential are under 500 m of water between Manda and Chilumba (N). Further, petroleum and coal imports of 1 Mby and 50-80,000 ty through (and from) Moçambique have suffered interruptions. The Tete field extends though to Sumbu-Nkombedzi (SW). Further east, at Chiromo the high-ash seams are even subanthracitic. According to MDC in situ reserves and resources of the Rumphi-Livingstonia (N) fields reach 100 and 800 Mt: the 2 m thick Khondowe seam, grading 15-18% ash lies underground, Kaziwizi holds 0.5 Mt, Nkana is up to 2 km wide, and Kapembe Hills and Chombe (Mt. Waller) coal contain 30% and 12-40% ash.

Some 140 Mw installed mainly at Nkula on the Middle Shire (S) generate most of the 500 Gwh capacity, Nkula can be built up in stages to more than 220 Mw and Tedzani downstream has 56 Mw power. The 120 Mw Kapachire project is also dependent on the variable flow of Lake Malawi from which the smaller Zoa Falls on the Ruo are isolated. There are also numerous sites, such as the Bua (C) and the South Rukuru (N, 100 Mw), but less than half of the country's 5.3 Twh potential will be used this century.

There is not enough water in the dry season in the Rift and in the southwest, where salty water is sometimes found above sweet water. Also in the south, shallow wells may yield 1-5 m^3h, but the quality is rarely good. Up, in the highlands deeper wells are often required for cattle. The rule of 1 to 10 also holds : cultivated/total area, irrigable/cultivated, irrigated/irrigable...

Phosphate rock reserves attain 3 plus 1.5 Mt averaging 24% and 32% P_2O_5 and Tundulu Hill includes to 30 m depth 190, 110 and 30,000 tons of hard phosphate in grades of 20, 10 and 5%. However, the resources of Chingale, the Mlinde soils and the guano islands of Lake Malawi are even smaller.

Other resources

Coowned by MDC and ADMC, the Changalumi cement plant (S) produces 90,000 tons of clinker per year. Yet this material is barely suitable for lime burning. Nor are the magnesian limestones of Matope and the Middle Shire. Domestic production, therefore, covers only half of the requirements. Nonetheless, 300 Mt of limestone have been delineated in the Machinga district.

BPB has worked the Mwanza vermiculites, but only 5% exfoliated at Kapirikamodzi (S) out of the 90,000 tons available to a depth of 50 m. In 1952 and 54 General Refractories extracted 3,000 tons kyanite from rock grading 15% at Kapiridimba, near Ncheu ; between 1942 and 46 1,000 tons of corundum were exported from Tambari (S) and the blue corundum of Chimwadzulu can be cut. In the Mauze complex 12 km^2 grade 30% fusible nepheline and 150 km^2 of Zomba assay 50%.

In 1960 geologists estimated 60 Mt bauxite on Mt. Mlanje, and in the southwestern segment 5 km^2 of Lichenya total 30 Mt on a dry basis with grades of 44% Al_2O_3, 15% SiO_2 and 14% Fe_2O_3. A scheme for an aluminum industry based on cheap Cabora Bassa power has therefore also been mooted.

Around Mzimba and Nyika, Africans have smelted iron for ages ; there are also 50,000 tons ore at Dzonze, 80,000 at Mindale and 500,000 on Chilwa Island (S) along with 300,000 t manganese 2,000 t fluorite and less than 100,000 t low grade niobium. There is probably more uraniferous pyrochlore at Ilomba (NW) and most of these contain more zirconium than niobium.

The Tengani titanium sands grade 0.05-3.5% TiO_2 ; they also contain 4,000 tons

of abrasive garnet. On the eastern shores of the Lake, there are moreover, a million tons grading 20-46% ilmenite and 650,000 t of ilmenite and rutile cover Monkey Bay (SE) along with 300,000 tons monazite. In addition to fluorite and bastnäsite there are 30,000 tons monazite at Kangankhunde in grades averaging 4.5% with more strontium. Resources are even larger. The extent of areas containing 6.5% monazite is, however, smaller and veins carrying 50% barite are subeconomic. To conclude, smaller shows include :
Barite Nathace, Bangala, ceramic clay and graphite Dedza (Lobi) cement clay and diatomite Lake Malombe (S), copper and nickel Lisandwa, Cholo, Blantyre area, Likabula river, Mpemba Hill, fluorite Nkalonje, Songwe, Nsengwa carbonatites, 2 m thick layers on Chilwa, gold Lisungwa, Dwangwa, graphite Lilongwe and Dowa 10%, Wamkurumadzi, gypsum Mponela, - marls : dambos (C), Chirimoare, kaolin Rivi Rivi, Chinteche, Kumbanchenga, kyanite Chingowa, Chakulambela (Neno), lead Dowa, magnesite Middle Shire, mica Tanzani, Cholo, Senzani, Neno, Kasho Hill, Mt. Hora, Lombadzi, &, nepheline Nyika, Nathace, pumice & uranium traces Karonge, soapstone Ntonda.

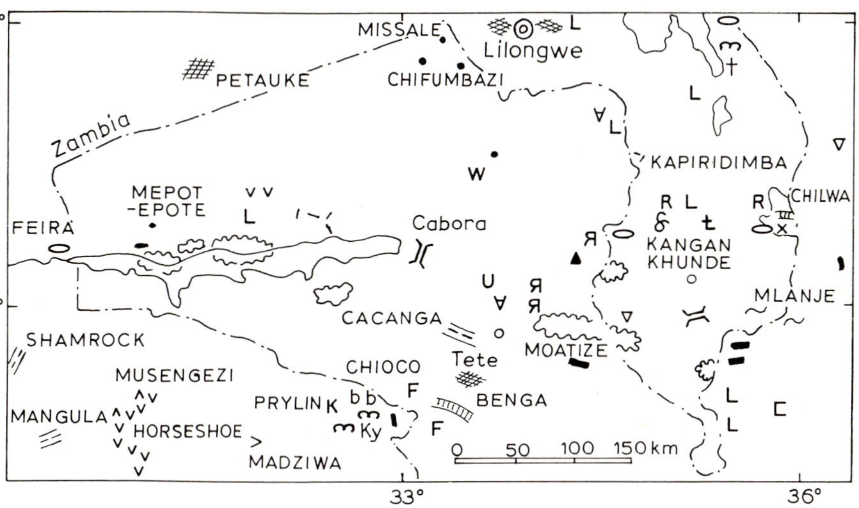

Fig. 50. Tete, south Malaŵi and NE Zimbabwe. (Legend on page 13.)

67 MOÇAMBIQUE

Gas shows on the southern coast, but hydropower, coal and most minerals are concentrated in the Tete 'beak' (W) which juts toward Zambia. Merchandise transits with difficulty from here, Zimbabwe and the Transvaal to Beira and Maputo.

GOALS : Rehabilitation of roads, mines and small workings (Table 93) after the civil war, hydrocarbon exploration, updating of the mineral maps. Marketing of hydropower. Systematic drilling for water.

Out of 5,000 miners, half work in the state's collieries including East German and other Comecon experts. Others are seasonal workers, but even more are migrants in South Africa. The Socialist government encourages investment and its mineral code, drafted by Arthur D. Little, provides incentives through generous write-offs.

Table 93 Moçambique production, reserves and capacity[c].

Past[+] output			Production			Production		Reserves		
Gold	kg	500	Cement	kt	300	kt	500	Coal	Mt	400
Silver	kg	90	Petroleum[+]	Mb	3	Gw	2[c]	Hydropower	Gw	12
Copper	kt	2	Coltan[+]	t	70			Gas resc.	Tcf	2.4
Chromite	kt	10	Beryl gem	t	2	t	20	Beryl conc.	kt	6
Graphite	kt	2	Garnet	t	10	t	30	Salt	Mt	5
Asbestos	kt	3	Mica scrap	t	2-900			Fluorite	kt	900
Corundum	t	5	Kaolin	t	300	re		Soda carb.	kt	150
Tin	t	200	Bentonite	kt	1	so		Ilmenite	Mt	12
Feldspar	t	900	Steel rolled	kt	8	ur		Rutile	Mt	2
						ce		Iron	Mt	50
						rvb		Vanadium	kt	8

Energy

Three oil seepages have been observed at Inhambane (S) and the Cretaceous fields of Pande, Buzi and Temane might contain more than 2 Tcf EUR gas (Fig.72). Interest has also focused on the northern Rovuma basin, but the Zimbabwe pipeline starts at Beira and the refinery has a 17,000 bd capacity. Finally a 120 kty ammonia export project is based on Pande Island. Coal measures crop out between Caringe and Chicoa and over 80 km east of Tete in the Morongodzi, Revugo and Moatize basins where the thickness of seams ranges up to 7 m with an aggregate of 12 m. The Rio Mepotepote and Ilomba fields straddle the Zimbabwe (CW) and Malaŵi borders ; there are other basins between Lake Malaŵi and the Lugenda (NW) and the Chiasa-Pangura field holds 150 Mt coal. The governor of Rios de Sena discovered coal here in 1836, and extraction started in 1932. Output of Carbomoc's four collieries has bottomed out with 25,000 tons out of a rated capacity of 2.5 Mt which could grow though to 5 Mt should export projects to Brazil materialize. Similarly, if railways were upgraded, the cement plant at Dondo (Beira) could also convert to coal.

The Cabora Bassa site still has three times the potential of Aswan and the hydropower plant is a smaller version of Inga with its tenuous 220/533 kV 1400 km powerline to South Africa.

Hot springs are most abundant along the Shire and Malaŵi (NW) border, and along a line connecting Pebane harbor with the Limpopo, with concentrations between the Beira railway and the Save (SE).

Table 94 Moçambique hydropower sites and projects[p].

Zambezi	Mw	Site	River	Mw	Sabié	Mw
Cabora	2250	Mavuzi	Mavuzi	65	Corumana[p]	14
Lupata	0.6	Chicamba	Rebua	50	Lucheringa	0.7
Boroma	0.4	Massingi[p]	river	40		

Water and food

The Basement covers two thirds of the country and retains some water only in the zone of weathering. Since sandstones and shales alternate in the Karroo, the Zambezi valley does better. A continuous free aquifer also spreads in the Cretaceous and post-Cretaceous sediments of east-central and southern Moçambique. South of the Save alternating diatomaceous and clayey sediments and the Tertiary limestones of Morrumbane yield abundant water. Unfortunately streams vanish in the dry season and the quality of surface water deteriorates. Even at a depth of 30-50 m groundwater is frequently saline and in the coastal belt of favorable gravels salinity extends beneath 150 m. Unfortunately, soils are inadequate in the south, slightly leached soils spread into western Moçambique and red-brown soils cover the north.

Population has doubled during the last twenty years, but salt recovery has declined to a third. The major producer was then the Nacala district 60 kt, followed by Zambezia 30 kt, Maputo 20 and the far north 15 kt. These salts averaged 0.1-2.7% impurities and 1.6-5.2% other salts, but the brines of Chibuto (S) contain 10% of these. Further, 5 Mt of salt and 150,000 t of sodium carbonate might be recovered from 350 Mm^3 of subsurface water in Gaza province. The Nova Mambone flats (S) add another 7,500 ty capacity.

At Vilanculos and the Save River, birds have covered caves with a meter of guano representing a volume of 20,000 m^3. With depth, the phosphate content increases from 8.5 to 10.6% P_2O_5, concomitantly nitrogen grades augment from 5.7-2%, but potash content remains fairly constant at 0.8%.

Building and industrial materials

Limestones and marbles are widely scattered in the country, above all near Maputo, at Mossurize (W) near Maravia, Tete, in Niasa (C) and around Naguema (N). Rather as an exception, three cement plants operate at capacity and even export to Malaŵi and Tanzania : Nacala semidry (NE) 40 kt, Dondo wet process (E, 300) 170 kt and Matola (S,600) 160 kt. Figures in brackets indicate clinker capacity. Vermiculite is reported to occur at Tundumula and Matema (Tete). Production of short chrysotile asbestos oscillated between 200 and 300 tons in the fifties at

Serra Mangota (W). While low grade fibers of Tete are of the same type, the tremolite of Muhinga (Manica) does not fiberize. In the same district on the Chimolo road and in the Muza Valley there is magnesite and talc. Near the Goba railway (S) diatomite assaying 87% silica is also abundant, the Imamputo plant has produced 8,000 ty bentonite and perlite straddles the border (S). Itotone produced 2,000 t graphite in 1961 and Evate 300 tons in 1951. Angonia (Tete) is a larger deposit than Otaco-Ancane, Metochéria and Nacoto. At Mt. Snuta in the Moriangane Range (W) extraction of refractory bauxite started in 1935. By 1960 5,000 ty, assaying 59-64% Al_2O_3, 3-14% SiO_2 and 2% Fe_2O_3, were shipped to Zimbabwe and South Africa where they were processed into aluminum sulfate. Unfortunately, the three m thick bauxites of Mt. Muto are of lower grade : 41-53% Al_2O_3.

In 1955 9 t of corundum were exported from Revue (SW). Reportedly 1,000 ty might be recovered 50 km south of Chimolo. Other occurrences include Zobuè and Conchoeira and Cachissene in Niasa, garnet at Baruè, Ancuabe and Zobuè (SW), kyanite Zobuè, Muaguide (N).

The principal fluorite quartz veins cut mostly Precambrian gneisses in Tete and Sofala (C) Reserves containing more than 35% CaF_2 include at a) Chioco-Djanguire : measured 71,000 t, probable 95,000 t and possible 530,000 t, b) Canxixe (Mt. Geramo) 160,000 t and Macossa (Mt. Djalira) 43,000 t. Other fluorspar veins are at Changara (Mt. Domba) ; in contrast Mt. Muambe, also in Sofala, and Xiluvo (CS) are associated with carbonatites.

Metals

A few pegmatites of Alto Ligonha and Ribaue (N) have been major sources of tantalum and beryl since World War II. Marropino, the largest, was followed by Muíane, Boa Esperança, Macoteia and between the Namirrue and Metuisse : Nanro, Nacuissupa, Marrapane, Nihira, Mirrucui, Nahia , Piteia, Injela ; Nahora : Namicaia, Namitaca, Alata, Mocachaia, Macula, Muacuamano , Nahavarra.

Recent and (1962) production levels compare thus : tantalum and niobium minerals 70 (110), beryl 30 (450), lithium minerals - (200), scrap mica 200-900 (10), bismuthine 5 (15), feldspar 900 (-), kaolin 300 (60)t.
Boa Esperança and Muíane were the principal bismuth producers and several minerals contain some uranium, thorium and rare earths. Tourmaline is also abundant but topaz is rare. Mt. Zungi and the Curumana Mts. yielded agate and more gemstones occur on the Upper Luiro (NW) too.
Some tin also shows in the pegmatites, but most of the 200 t exported prior to 1952 came from Mt Doeroi (W). Africans washed gold at Manica (W) for centuries, but Missale and Chifumbazi are not as ancient. By 1914 production reportedly reached its high of 500 kg and silver peaked with 90 kg even earlier.

Only the Isiaca Hill mine is left near Edmundiavi (W) although as much copper came from Cacanga (Tete, NW). - Production plummeted to some 200 tons... Moreover , 10,000 t of refractory chromite have been shipped from Nhanthreze (Tete) in 1946.

Similarly the titaniferous iron ores are subeconomic, but 2 Mt of black sands grade 78% ilmenite, 2% rutile, 13% zircon and 5% monazite (Th, RE) at Moebane and beachsands of Pebane contain 70-85% heavy minerals. Finally, other occurrences include :

Copper M'Panda-Unkua, Conua (N), Massamba, Chichie, iron ore Anamapa (N), Tete (W) and the southern shows, manganese Chundize, Benga (Tete), Nhantreze, Mt. Munuca (N), Alto Chimezi, nickel Achiza (Tete), Tungsten Machingoa, Catoa (Tete) vanadium Angonia.

Finally, at Maputo a mill rerolls billets imported from Zimbabwe.

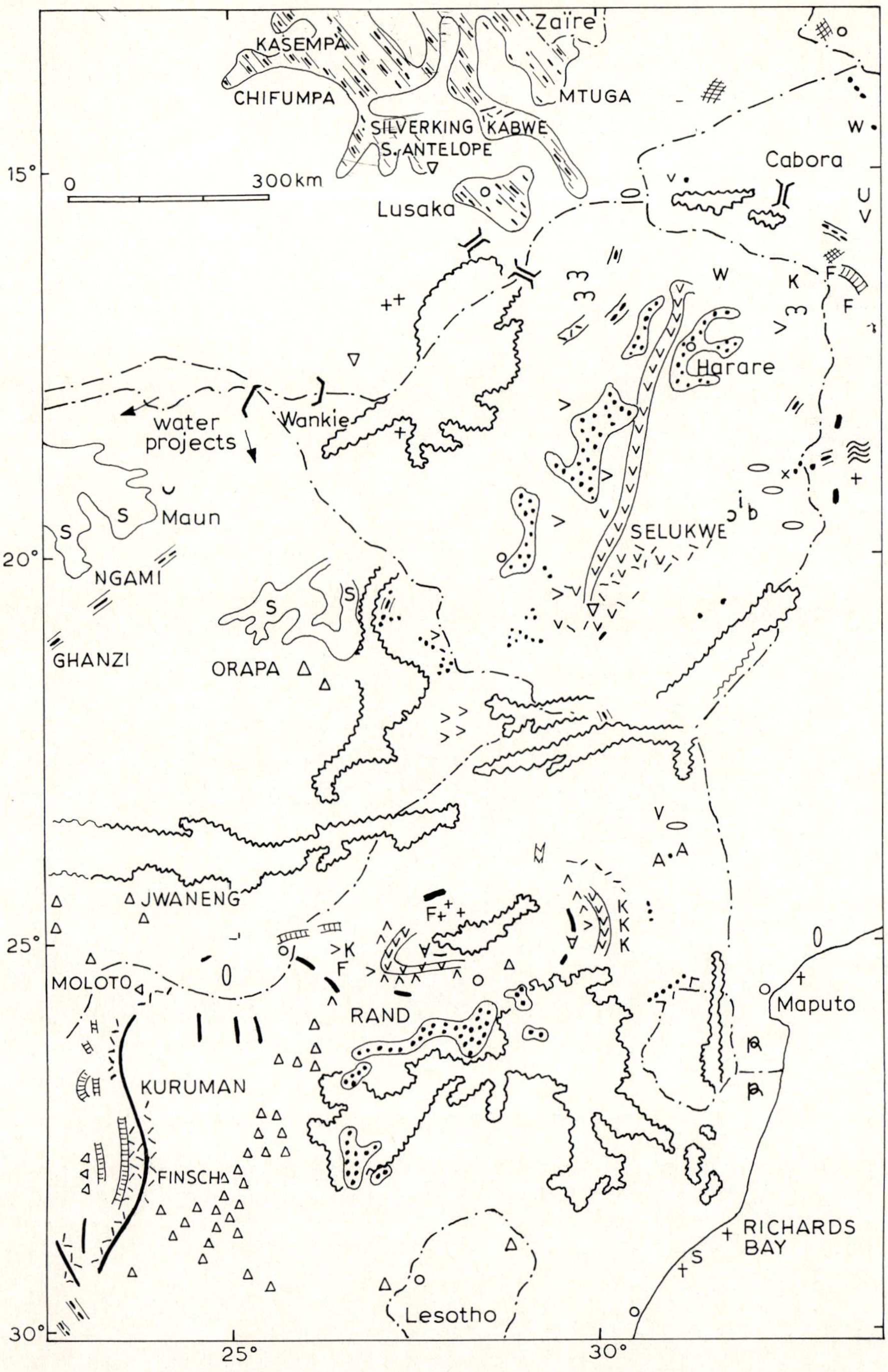

Fig. 51. High Africa from Zambia to Lesotho. (Legend on page 13.)

7 SOUTHERN AFRICA

This subregion (Fig. 50-59,65) ranks first for gem diamonds and second for coal asbestos, chromite, nickel, PGM and gold most of which are exported via South Africa, its principal trading partner. SADCC countries have also benefited from the high level of industrial activity in the Republic, but combat, of course, its racial system.

71 ZIMBABWE

Chromite, ferrochrome and spinnable asbestos are of strategic importance. With incipient industrialization the pattern recalls South Africa - on a small scale (Fig. 50-55). There are differences though : less concentration and 200 small workings on the flanks of the Great Dyke, less (but growing) involvement of the state which provides employment for 70,000 people. Zisco's (steel) turnover ranked third, Rio Tinto (gold, chromite, nickel, emeralds), fourth, followed further on by AAC's Wankie, Zimalloy and Bindura (40% + JCI 9%) with coal, chrome and nickel. These groups employed respectively 3,500, 2,300 and 6,000 people. To these should be added Lonrho (Corsyn, gold), Turnall (asbestos), Union Carbide (Zima's co, chrome). Lastly, the state's IDC owns 50% of Zimasco 40% of Wankie, 90% of Kamativi Tin, 47% of Quarrying Enterprises and its Mineral Marketing Corp. sells all products except for gold.

> GOALS : Control of soaring costs. More substitution of imported oil by coal and hydropower. Agressive marketing of mineral products. More drilling for water in the arid west. Further organization of small workings and buildup of products for domestic use. Incentives for marginal mines, managerial and mining education.

Minerals contribute 40% of export and fuels require a third of import revenue. Gold is shipped to the U.K. The major customers for ferrochromium (and nickel) are also the EEC (Germany, France), and the U.S.

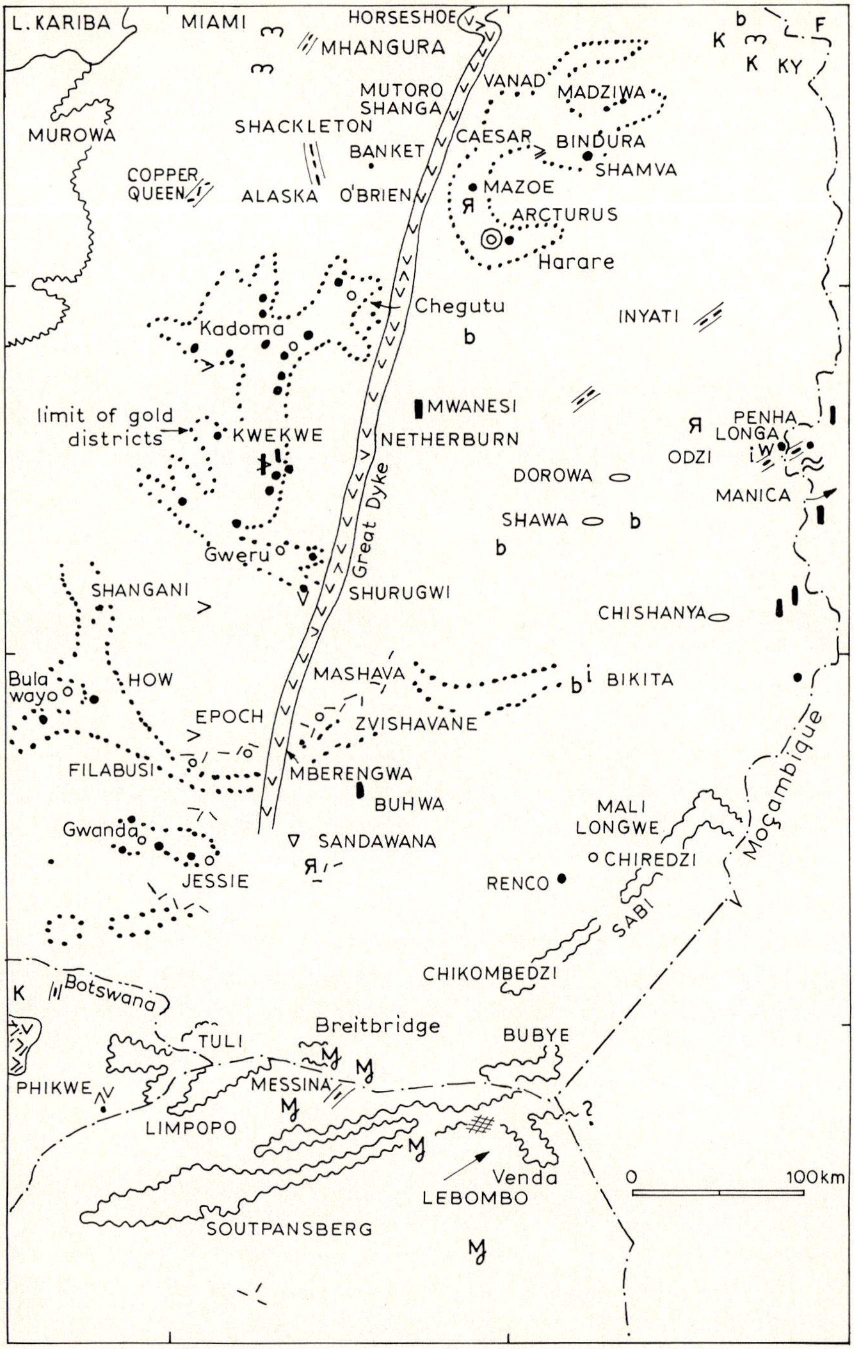

Fig. 52. Zimbabwe and its neighbors. (Legend on page 13.)

Table 94a Zimbabwe production and reserves.

Production			Production			Production			Reservs.	
Corundum	kt	9	Cement	kt	400	Gold	t	15	t	600
Quartz	kt	600	Limestone	Mt	1.3	Silver	t	30		
Magnesite	kt	60	Fireclay	kt	12	Asbestos	kt	150	Mt	5
Bentonite	kt	80	Ferrochrome	kt	200	Chromite	kt	500	Mt	500
Talc	t	300	Ferromanganese	kt	2	Hydropower	Twh	8	Gw	4-5
Graphite	kt	10	Iron pigment	kt	1	Coal	Mt	3	Mt	250
Pyrites	kt	60	Coke	kt	200	Iron ore	Mt	1	Mt	100
Sulfur bypr.	kt	5	Steel	kt	500	Copper	kt	20	kt	200
Barite	t	800	P_2O_5 in fertilizer	kt	20	Phosphate P_2O_5	kt	40	Mt	2
Sillimanite	kt	2	N in ammonia	kt	70	Fluorspar	t	500	kt	200
Emeralds	kg	10	Kaolin	kt	3	Lithium	t	140	kt	25
Amethyst	t	2-5	Feldspar	t	800	Cesium	t	10	kt	25
Beryl	t	0.3-1	Tin	kt	1.5	Tungsten	t	30	kt	1
Aquamarine	kg	-50	Beryl conc.	t	50	Mica	kt	1^+	kt	50
Chrysoberyl	kg	- 2	Ta-Nb conc.	t	40	Nickel	kt	10	kt	220
Garnet	kg	20-150	Antimony	t	200	Cobalt	t	100	kt	8
Topaz	kg	2-8	Bauxite	kt	7	PGM	kg	150	kt	1
Tourmaline	kg	5-500								

Precious metals

Africans have mined gold for centuries and in 1880 King Lobengula granted Cecil Rhodes his concessions. Excluding peaks of 20 t in 1913 and 1940, output averaged 16 ty between 1910 and 1950 ; it is increasing again after a low of 11t in 1980. The Chegutu district (C, Hartley) contributed more than 40% of the output, Gweru (CS) 17, Bulawayo (SW) 21, Harare (N) 17%. Out of 4-5,000 occurrences, 500 were worked, 30 report 95% of the output, and the six biggest mines half of total production.

Table 95 Principal Zimbabwe gold producers (100 kg), grade (g/t), employment x 100.

Mine	100kg	g/t	employ.	remarks	Mine	100kg	g/t	employ.	remarks
Renco R	18	8	9	good pot.	Freda	3	2	4	low grade
Dalny F	18	8	20	pressure & heat	Patchway R	2	6	4	end life
How	7	5	5	big orebody	Roma	2	3	2	develpt req.
Independence	7	3	7	tonnage	Brompton	2	7	4	development
Shamva	7	4	9	large, price	Old Nic	2	10	3	steady
Blanket B	6	4	5	good potential	G. Valley	2	2	4	good reserv.
Muriel C	6	6	6	reclamation	Lennox	2	6	2	new reef
Venice F	5	4	9	successful	Sabi	2	2	2	irreg. prod.
Arcturus C	5	9	8	good reserves	Tiger Reaf C	2	3	3	low grade
Vubachikwe	5	6	6	old mine	Bell Riviera	1		2	irreg.,roast
Mazoe	4	4	6	scattered pay	Cam R	1			leaching
Athens	3	6	5	promising	Jessie	1	11	1	high grade

With 2.5 ty Falcon mines (F in table 95) has increased the milling rate at Venice to 60,000 t. In contrast some Dalny reserves became marginal. At Corsyn (C) production of 1.5 ty is stationary, but costs are not : Muriel has the lowest costs with Shamva and the highest grades with Arcturus.

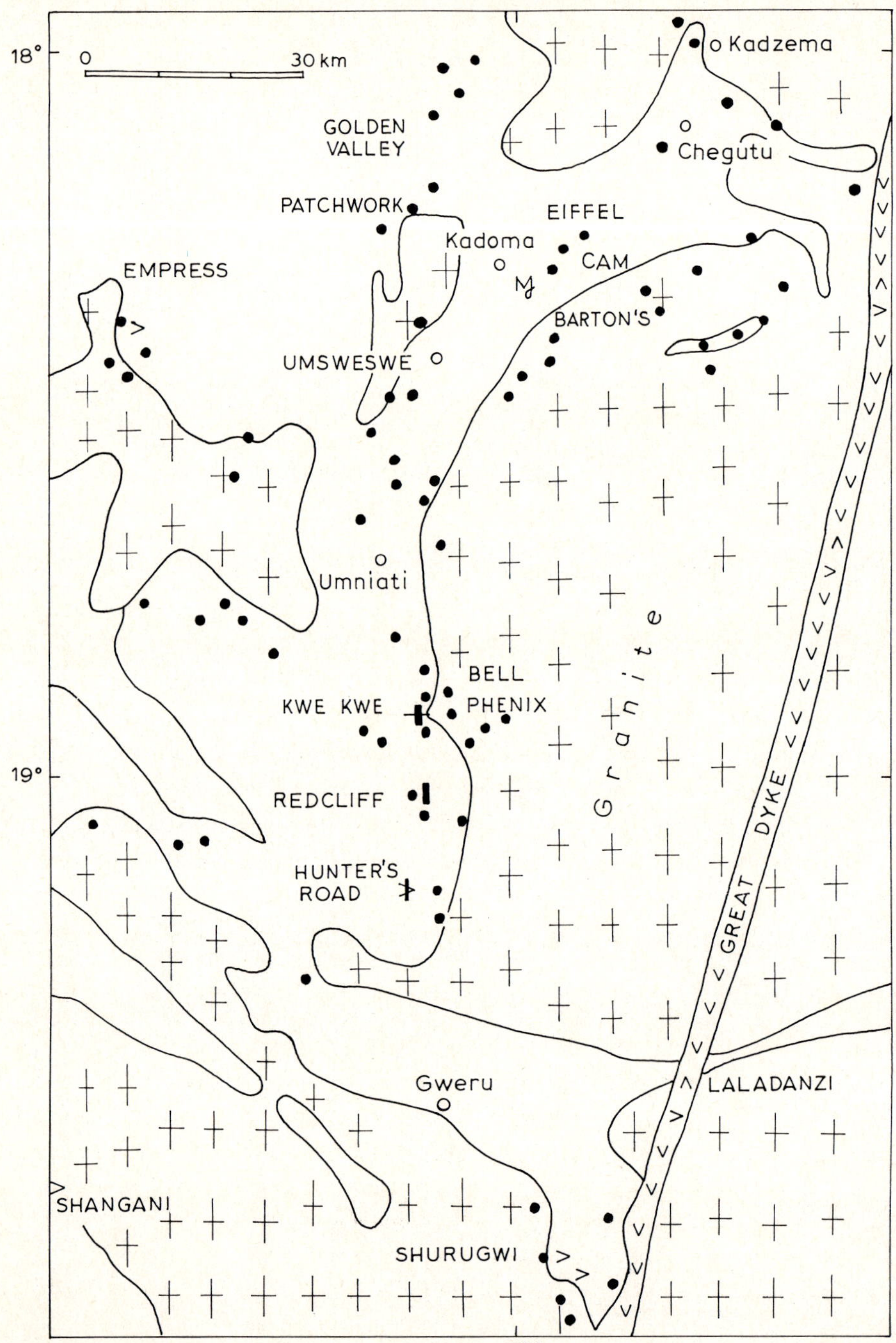

Fig. 53. Central Zimbabwe. (Legend on page 13.)

Corsyn also produces 1.5-2 ty of silver and 3,000 t copper at Inyati (and Muriel) ; its later developments include Legion, Tigre Reef and Mashona Kop. RTZ (R in the table) and Corsyn employ 3,000 and 3,000 and had a 20% return on equity. Indeed Renco (SE) operates at full capacity and also supplies 300 t copper. Further, exploration continues to the east and at Erin. Finally, Falconbridge (B) reopened the Golden Kopje, near Chinhoyi with a capacity of 6,000 tMo.

Recovery of silver, a byproduct of base metal and gold mining, rose sharply and perhaps peaked with 30 t. Should palladium prices improve RTZ might develop Zinca (CNE) with assays of 3-5 g/t and a capacity of 2 Mty or 6-10 ty PGM with accessory nickel, copper and gold. Union Carbide has a smaller prospect, Mimosa (S), the Empress nickel dumps and the massive nickel ore of Shangani assays 1.9 g/t Pd, 1.6 Pt, 1.1 Ru, 0.64 Rh and 0.22 Os + Ir. Indeed, pyroxenites of the Great Dyke invariably contain PGM ; its Chegutu and Wedza complexes average 4g/t in 2 Bt of ore...

Building materials

Asbestos deposits are located along a 150 km long arc (S) which stretches from Filabusi and Norma to Zvishavane and thence to Mashava, where Gath's and King contributed 40% of the total tonnage. Among 20 mines let us mention Boss, Rex, Pangani and Vanguard of Pangani Asbestos Ltd. and Essexvale's Kudu at Esigodini. Smaller occurrences are aligned further south between Geta and Gwanda, and Ethel in the far north was the only producer on the Great Dyke.

Kerr pegged his claim at Zvishavane in 1915. Output reached 70,000 t in the fourties and doubled in the fifties with an average of 260,000 ty between 1975 and 1981. Eight to ten silky fiber bodies are spread over 10 km here. The flowsheet comprises drying, screening, the opening of fibers and grading. Each ton of ore required 12 kw and 5 t of air for mining, drying, and beneficiation. Yet only 15% of the mill feed is recovered as fiber. At Temeraire, therefore, even the waste dumps are retreated.

The dry process cement works are fired by coal, but the 60,000 ty Bulawayo plant of United Portland used gas, Cleveland limestone and clay from Korbut (Gwanda).- It was originally rated 400,000 t. The Blue Circle group controlled 65% of the 290,000 t Ha re plant where a semi-dry kiln also operates and Colleen Bawn near Gwanda (S,50% Premier Portland Cement) has a capacity of 230,000 t. Lime is burned at Snow White (Harare), Shamva, Redcliff and Lalapanzi. High grade flux limestone comes from the Early Worm and metallurgical calcite from Sambawisa (W). Dolomitic limestone is quarried at Rushinga, Springbok, in Chinhoyi (N) and travertine at Moosgwe. In a different vein, Zoe hematite is shipped to South Africa as high priced iron oxide pigment and granulated slag is a byproduct of coking.

G + W Industries produces talcs at Kwekwe. There are 25,000 t of reserves to the south at Tritan along with barite, but Hawkshead (E) is smaller. Talc and magnesite are also byproducts of asbestos mining, but magnesite also shows at Barton's near Kadoma (NC) and vermiculite at Mutoko (NE) where some black 'granite' is quarried. Because of rising labor costs Mitmar closed Turning Point (NNW) which rated 150 t flake. Smaller mica producers of the Hurungwe field were Pat and Simluck, Gil Gil was further north and Coronet at Karoi (NW).

Energy

By 1990 Zimbabwe might import 7 Mb of oil. It has though 30 hot springs, thorium in the carbonatites and uranium shows, e.g. at Chirundu, Molly and Cripmore. 800 kg of uranium concentrates have even been recovered from 300 t ore but only the Sabi in the southeast and the Zambezi in the north have well balanced flows over the year. The first hydropower development, Kariba, now provides 70% of electricity needs - with a quarter, 2 Twhy, imported from the north. Both countries will also need before 1990 the 2 x 300 Mw extension and the other sites would have to be developed jointly during the last decade (see Zambia chapter).

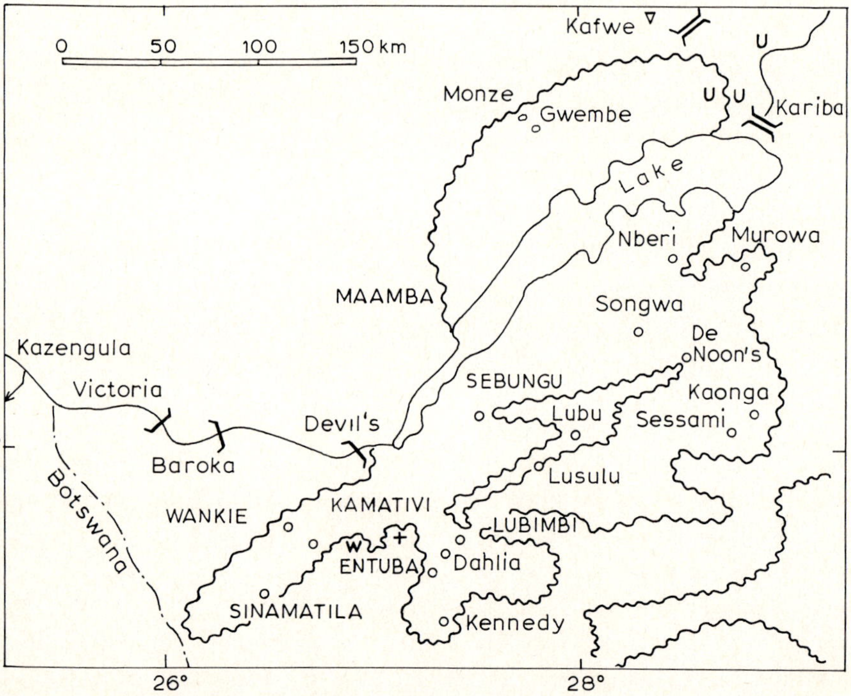

Fig. 54. The mid-Zambezi coal basin. (Legend on page 13.)

Meanwhile, coal from the Hwange II project will have to provide 800 Mw extra power before 1990. Both Hwange I and II burn previously discarded high ash coal. Initially, coal output declined from 4 Mt in 1960 to 3 Mt in 1980, but it should double, and might reach 15 Mt by the end of the century. Out of 2.2 Bt reserves and 27 Bt resources in the Sabi and Zambezi valleys (Table 96) WCC simultaneously produces steam and coking coal. About 100,000 tons of coke are used by ZISCO and ZMSC and 200,000 t are normally exported to South Africa and Zaïre. Although production plummeted recently, more will become available for export as coal output grows. Lastly, coking also supplies 30 Ml tars, ammonia and 1.5 Bcfy gas which are not fully used.

Table 96 Zimbabwe coal reserves and resources, Mt.

Zambezi	resv.	resc.	Zambezi	resv.	resc.	Sabi	resv.	resc.
Hwange conc.	600	390	Hankano	160	20920	Mkushwe	150	160
Hwange West	410	?	Kennedy		1200	Malilongwe	-	260
Entuba	160	?	Lusulu	250	2750	Bendizi	10	?
Sinamatila	-	170	Sessami-Kaonga	-	1000	Tuli : Massabi	-	30
Marowa	10	10	Sengwa	400		Limpopo : Bubye	30	30
			Lubu-Sebunga	-	80			

Resources of the Lubimbi inlier (N) predominate now in the 40,000 km^2 mid-Zambezi coal triangle (see also 'Zambia'). Intensive exploration may change though this pattern. The same holds for the Tuli, Kwasini and Sabi valleys of the Limpopo basin (SE,S). Tuli coal (SSW) is of good quality, the lowest 14 m thick seam at Mkushwe is even subanthracitic. However, most Limpopo coal has 35% ash.

Chrome, nickel, cobalt

With a reserve base of 750 Mt and resources of 11 Bt chromite, Zimbabwe ranks second in the world. Exports increased steadily from 1908 to 1930-50 when they leveled at 300,000 t. From new peaks of 700,000 t output declined to 4-500,000 t with the switch to low chromium high carbon steels, which did not justify anymore the price of high grade ores. Concomitantly, Zimbabwe's world share dipped from 20 to 5%. Out of a maximum of 11 thin seams, in the Great Dyke, rarely more than 3 were worked at one mine. The upper seams are generally of chemical grade, but metallurgical grades with 2.8 Cr/Fe are achieved up to seam 3; similar characteristics prevail in the smaller Mberengwa and Mashava (S) deposits. Nonetheless, Zimasco's Shurugwi and neighboring Mtoroshanga, Ngesi, Lalapanzi and Darwendale represent 70% of capacity since they can be mined opencast.

Table 97 Zimbabwe chromite reserves, economic RIE, marginal RIM in Mt and % Cr_2O_3.

	R1E Mt	R1E %	R1M Mt	R1M %		R1E Mt	R1E %	R1M Mt	R1M %		R1E Mt	R1E %	R1M Mt	R1M %
Shurugwi	11	47	76	met.	Rutala	11	46	273	46	Vanad A	4	51	9	51
Bee	19	47	172	48	Noro	9	51	16	51	Cesar A	3	51	19	51
Sutton, R. Camp A	15	50	61	50	Cambrai	9	47	8	47	Divide	3	51	17	51
York, Windsor	14	51	20	51	Glennap	6	50	10	50	Umsweswe	2	46	36	46
Netherburn A	14	41	12	47	Impinge	6	50	12	50					

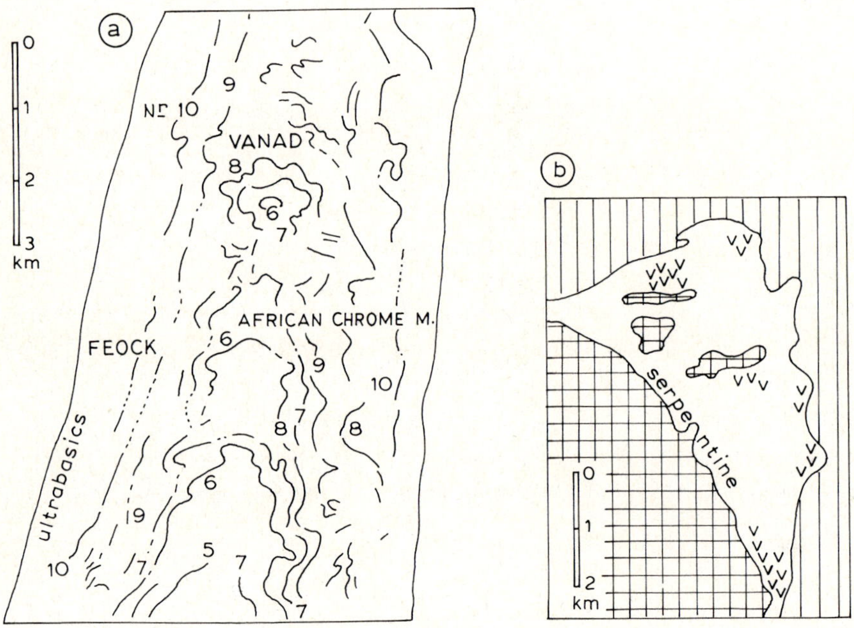

Fig. 55. (a) Chromitite seams in the Great Dyke north of Harare. (b) Chromite pods at Shurugwi.

Zimasco is also the largest ferrochromium producer with a combined high and low carbon capacity of 230,000 t at Kwekwe which might expand to 350,000 t. Zimalloy's (A in table 97) Netherburn expanded, but most other mines were temporarily shut. Yet the furnaces had capacities of 52,000 ty high carbon, 32,000 t low carbon and 25,000 ferrochromium silicon at Gweru and buildup to a total of 160,000 t has been considered. Finally, RTZ produces chrome tanning salts at Eiffel Flats.

Most nickel came from small deposits, which were expensive to mine. Larger, low grade resources exist in the Chegutu and Wedza complexes of the Great Dyke (5 Mt, grade 0.25%), at Hunter's Road near Kwekwe (100-150 kt, 0.75%) and in the laterites of Musengezi.

Table 98 Zimbabwe nickel reserves (1000 t) and % Ni.

Mine	kt	%	Mine	kt	%	Mine	kt	%
Trojan AAC	900	0.75	Epoch AAC	20	0.75	Empress RTZ	250	130
Madziwa AAC	100	1.08	Shangani JCI	100	0.61	Perseverance, Noel, Damba		

Total capacity was probably in excess of 25,000 t metal and the Empress, Bindura (N) and Kadoma refineries also produced 1-300 t cobalt.

Other metals

Iron ore is scattered at Kwekwe, Mamba, Kadoma, Nyuni, Mberengwa and in the carbonatites, but Ripple Creek and Buchwa supplied the Redcliff steel mill at Kwekwe. Sintering capacity exceeded 800,000 ty, exports will hopefully increase again, since local demand will only reach a quarter of a million tons. Similarly rod and wiredrawing at Lancashire Steel's plant was not profitable. In contrast, export of pyrite continues from Glendale (NC, Mazoe) and AAC produces some high carbon ferromanganese at Gweru. Indeed, the Dan and Ripple Creek deposits 16 km from Redcliff, assay 29% Mn and 25% Fe. Manganese also occurs at Black Mamba and Yank, near Kadoma and some bauxite has been extracted north of Penhalonga, Moçambique. A new foundry at Gweru is to produce, furthermore, brass, steel, aluminum and copper castings.

Copper production declined from its 1975 peak of 50,000 t (not withstanding the Alaska refinery) and might be summed up thus : a) MTD's Miriam and Norah at Mhangura (N) 10, and 5 kt, plus 150 kg gold, 130 + 25 kt reserves (assays : 1.3 and 1.1%), b) Shackleton 5 kt, 40 kt reserves with Angwa, plus c) Corsyn and Renco. d) The nickel mines also produced copper and the Great Dyke contains $>$15 Mt (0.25%...). To these might be added 6 kt and 5 kt zinc and lead resources, but MTD has shelved the Copper Queen (NW) project because of high power costs.

Production of cesium concentrates at Bikita (E) peaked with 280 t in 1976 ; vast reserves of rubidium are also available. Lithium minerals peaked with 60,000 t at Bikita in the sixties, but the Enterprise belt (N) and Mutoko (NE) are smaller. In 1953, 1700 t of beryl concentrates were recovered, but output dipped. There remain, nonetheless, four workings each at Hurungwe and Karoï and one at Mutare (E) and Darwin.

Kamativi (W), the principal tin mine also operates a custom smelter, tantalite- columbite is recovered here, at Benson and at Bikita, and tungsten production peaked with 500 t concentrates in 1952. Finally, other occurrences include :
Antimony Sebakwe, Koodovale (SE), arsenic Kwekwe roasting plant, beryl Gutu, Buhera, Sabi, copper Gwai (W), Umkondo (E), Shamrocke (N), lithium Glen Cove, molybdenum Seigneury, Chegutu, tantalite-columbite Makanga, Minister, Buhera, tin Darwin, Gwaai, tungsten Nelson, Wallingford, Sangwe, Gwenda, Chizedzi, zinc-lead Elbus, Shanyati.

Other resources

As usual, the populated, Basement covered areas do not have continuous aquifers, but normal years rainfall and run-off provide enough water. There is some artesian water in the Karroo of the Zambezi and Limpopo ; even better, compact Kgalagadi sands yield 70 m^3h. Similarly, black soils appear in this farming country and fertilizer demand is already more than 500,000 ty. Yet, only 160,000 t of phosphate are produced at Dorowa (E) and grades of the other carbonatites, Shawa and Chisanya, are lower. Further, Sable Chemical Industries produces ammonium nitrate at Kwekwe and RFC Ltd mixed fertilizers at Windmill.

Some barite is produced at Kwekwe River and Shamva, but Chiroswa's Dodge Mine is on a care basis. At Tinde (W) a 3 m thick fluorite vein extends over 3 km and in the sixties 400 ty of diatomite reached markets. Crystal Mine (Hurungwe) and claims northeast of Miami provided a few tons of highgrade quartz and Samabhula (Gweru) Industrial Sands has a capacity of 1,500 tMo flux and flotation and 1,000 tMo of glass sands. Ferroalloy flux is also produced at Duzi (C) and R Broadside's (Gweru), Patronage provides two silica grades. Nonetheless, the largest reserves are at Lome Kop (Kadoma). At O'Brien's Prospect (NE), the world's largest corundum supplier, 60% of the ore is abrasive. Chikomo is also near Mazoe (E) and Baobab is near Mberengwe (S). But the Harare area and Umtali deposits are not mined.

Recovery of diamonds ceased at Somabhula (S) in the fifties but Riozim cuts at Zeus 6-8 kg emeralds valued only $0.6-1.2 M. The small peppered Sandawana stones (S) are famous for their color and quality and other occurrences include: Gems, agate, amethyst, topaz, tourmaline : pegmatites (see 'lithium' and Nyamandhlovu (SW) ; aquamarine, heliodor Karoi ; emeralds Mberengwa (S), Athens, Kanya Hlaza, Maharani ; Nyanda (E) : Chikwanda, Mayfield, Popoteke, Renders, Twin Star, Novello with chrysoberyl ; Filabusi (S) : Coen's Luck, Flame-Lily, Mustard ; calcite : Mutoko, clay Hwange (W), Firdon (pipes), Barry, Bemas, Korbut (SW), Tipton (Lalapanzi, refractory), feldspar Mistress, fireclay Hwange (W), graphite - King, Lynx, Zororo, Urungwe, kaolin Tom (Kadoma), quartz Redcliff, Kwekwe, Lonekop, Jopempi (S), refractory sandstone Hwange, serpentinite Rusape (E), silica Gweru, Hangwe (W), sillimanite Gweru.

72 BOTSWANA

In the sixties, Bechuanaland protectorate supplied half a million dollars of commodities (MRA, p 164). Independent Botswana produces a growing half a billion dollars of diamonds (and nickel, Fig. 51, 55-57, 65). To this example of good management and mineral development should be added abundant coal resources which are second only to South Africa. Mmamabula is 160 km from its Ellisras railhead. Moreover, 10 Mty exports might justify one day 800 km of rails linking Kgaswe with Gobabis and Walvis Bay in Namibia across the desert which covers 90% of the country.

GOALS : Development of water resources, of the soda scheme and small workings. Further exploration of the Kgalagadi.

The population lives along the eastern strip, served by the Zimbabwe-Cape railway. Another 1.5 million Tswana live in the South African homeland, Bophuthatswana. Diamonds of Debswana , De Beers' joint venture with the state, are sorted at Gaborone, the capital, and sold each fifths week to a CSO company, which ships them to South Africa for distribution and export.

AAC, the sister finance house, owns most of the colliery and a quarter of BLC. The Bamangwato control 15% of the debt ridden company ; with another quarter Amax refines the nickel-copper pellets. Matte has also been treated in Norway and Zimbabwe.

Table 99 Botswana production and reserves.

Resources			Production		Resrvs.		Production	Reserv.
Soda	Mt 400	Nickel	kt	18	kt 450	Diamonds gems	Mct 4.5	50
Salt	Bt 2	Copper	kt	20	kt 420	" " industrials	Mct 7	170
Talc	kt 40	Cobalt	t	220	kt 2.5	Coal	kt 400	Bt 3.5
Iron	Mt 10	Kyanite+	kt	10		Gemstones	t 1-20	
Manganese	kt 100	Sand	kt	150		Gold resources	kg 10-20	t 6
Uranium	t 400	Stone	kt	200		Silver	kg 540	

Diamonds

I. Gems show in the thin gravels of Middlepits (SW), 20 km north of the Molopo.

II. Associated heavy minerals spread in the Kgalagadi, over 200 km west and 100 km east of Jwaneng and De Beers found diamonds, between Tshabong, the Mabuasebuhe (S), Tshane (SW) and Kokong (SC) and Falconbridge's Gobe concession.

III. The pipes around Jwaneng (SE) were difficult to locate ; DK7 is being eva-

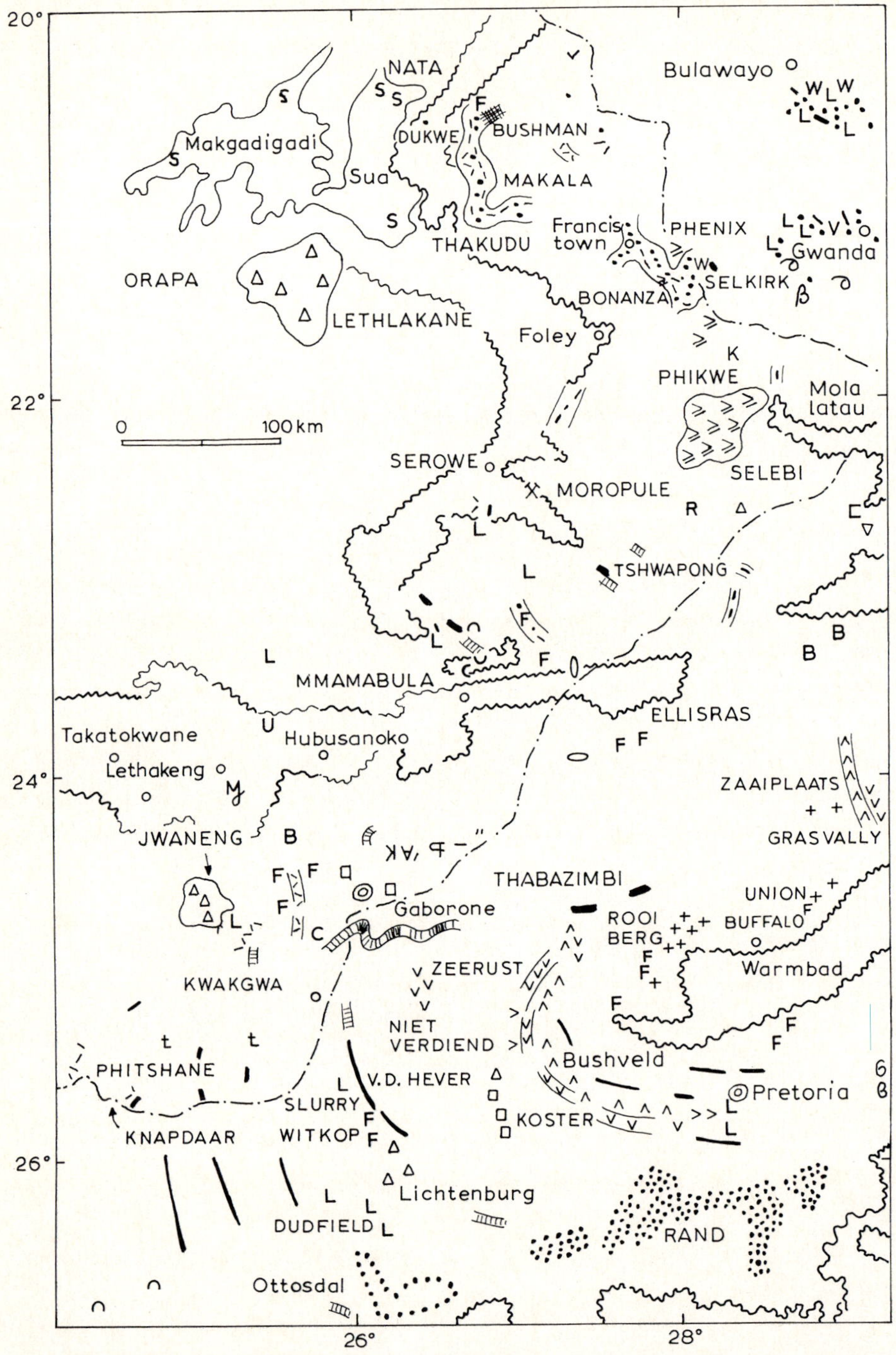

Fig. 56. Eastern Botswana, western Transvaal and southwest Zimbabwe. (Legend on page 13.)

luated and at DK2, the kimberlite is at a depth of 27 m. Jwaneng itself is covered by 4-5m of sand and 45 meters of cretes and marls. The three lobes of this 0.7 km^2 pipe were intruded through Basement and platform rocks. 400,000 tMo of kimberlite are extracted along benches of 1,000 x 80 m whose ultimate numbers will reach 10 at 200 m depth. The material is crushed to 25 mm and sorted into three grades 25 ct/t, 25-50 ct/t and >50 ct/t in the red beds.

Income%	pipe	ore Mt	Million ct	ct/100t	% gem.
65-70	Jwaneng	5	7.5	150	35-40
20-25	Orapa	8	4.5	60	15
10	Lethlakane		0.7	$0.5/m^3$	75

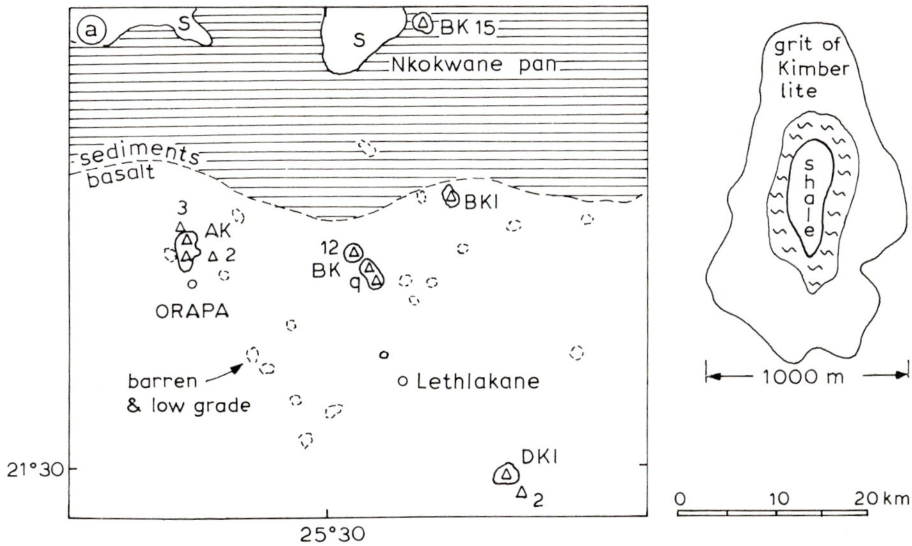

Fig. 57. (a) Pipe of district of Orapa.

IV. West of the Venetia pipe in the northern Transvaal, a few diamonds were found in 1959 at Foley (E), exploration expanded along the old drainage of the Motloutse over 300 km as far as Mopipi and alluvia may even have dispersed under Lake Makgadikgadi. Orapa was discovered in 1966 and the open pit inaugurated in 1972 with proved reserves of 85 M carats to a depth of 37 m plus vast inferred reserves at lower levels.

The upper 85 m of the one km^2 pipe, an altered and re-worked, medium grained kimberlite, contain 2.5 ct/m^3, the fine grained, hard rock is poorer.

All pipes pierced the Karroo and some rise as small mounds of up to 10 m from the desert ; others are covered by calcrete.

Out of 30 smaller pipes Bk 1, 9, 12 and 15 cover 5, 11, 3 and 2 ha and average 20 ct/100 t. Lethlakane Dk 1 covers twelve hectars and contains 2.2 Mct better quality stones to a depth of 37 m. Nearby, the 4 ha of DK2 grade 0.4ct/m^3 to 30 m depth.

Coal

The nickel mines now take 150,000 tons. If washed, coal could also substitute for imports, since a million tons and 300 Gwh will be required by 1990. Alternatively, 180 Gwh would have to be imported from South Africa in addition to 1.5 Mb petroleum, unless oil is found in the west.

Faulting has subdivided the vast Karroo basin with its four good seams.

I. At Dukwe (E) 9, 1.5 and 1 m thick seams lie at 150 m depth, but there is only one 2 m thick seam to the south.

II. West of Foley (E) 2.5 m thick seams are covered by 15-20 m beds. There are 1 Bt here and AAC estimated another billion tons over an 80 km long spread south of Dukwe.

Table 100 East Botswana coals.

Field	thick m	Ash % mined	washed	Moist.%	Volatiles	Heat Mj/kg
Dukwe	2	21	19	6	23%	22
Foley	1	24	11-15	6	31	22
Morupule	6-12	15-23	10-12	5	21-24	22-25
Mmamabule	3-5	15-19	15-	5-6	24-26	22-25
Kwenen W	2-4	15-23		5-8	23-35	22-23

West of Morupule and Mmamabula coal has been intersected over 250 km. In the eastern Kweneng district >1 Bt have been indicated at 50-250 m depth. Further west 3-5 Bt lie indicated or semiproved under thicker Kgalagadi beds. Although seams are thinner at Takatokwane (24°E, 24°S), they have been intersected at Molalatan near Namibia. Total, partly workable, resources therefore reach 30 Bt, resources 17 Bt, out of which the two coalfields hold 10 Bt - 3.5 Bt are recoverable.

III. BP's Mmamabula covers 300 km^2 in the extension of the South African Waterberg field.

Table 101 The five blocs of the Mmamabula coalfield.

Bloc	area km^2	grade	ash	thickness m	cover m	resources Bt
Tuli	60	medium	high	10-12		
East	80					1.5
Central Lower	50	2.5Mj/kg	15-22	2.4	30-60	
Central Upper	32	lower	20	5.4	18m higher	0.4
Central Thin upper	15	lower		1.1	28m above	
South	100			4	10-40	1

IV. At the Morupule colliery, the cover is 15-35 m thick and weathering reaches down to 30-45 m. Resources of the Main Seam attain 5 Bt, half of which is proved or semiproved, including 1.5 Bt minable reserves. Another 1 Bt are in unfaulted blocs. The whole Greater Morupule field may thus comprise 7 Bt. There is also shallower coal to the north and the Main Seam might contain at greater depth another 2 Bt in Morupule West. All seams except for part of N°2 can be upgraded into coal containing only 10% ash with a yield of 35 and 22% for N°1 and 2.

Table 102 The Morupule-Serowe coalfields.

Morupule	thick m	ash%	Serowe	minable	resvs.Mt	thickn.m	depth m	volats%
Main	8-9	20	Morupule	Main	370	7.4	150	24
DirtyN°2	6	30	Lotsane		100	1.8	100	33
Marker, Upper	2	25	Serowe	Bright	190	1.5	50	33

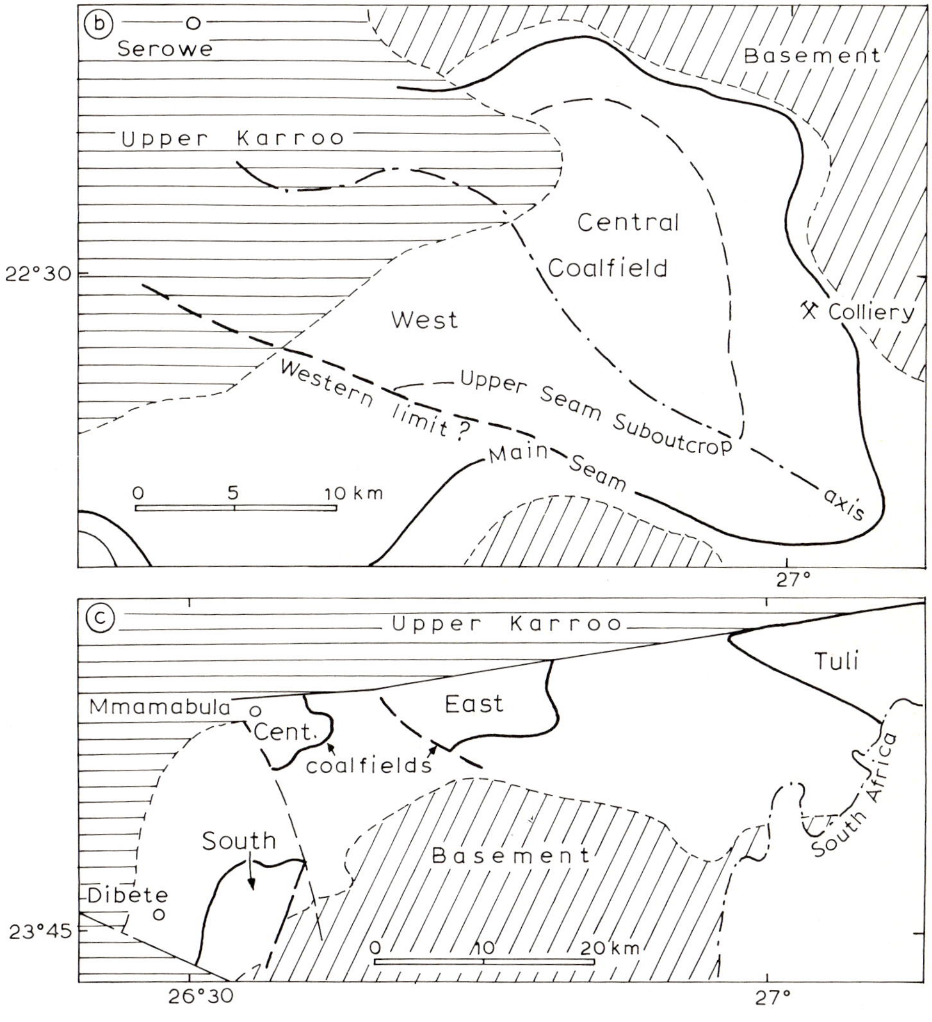

Fig. 57. (b) Morupule. (c) Mmamabula.

V. Further west, at Serowe, a Shell-state venture, the main seam averages 17% ash, 1-2% S and 6,100 cal/g and three underground mines could produce at Kgaswe 7.5 Mt of washed coal, out of 15 M raw, with an ash content of 12%. Lastly, let us add that Amax, EdF, CFP and Ruhrkohle also explore(d) for coal in Botswana.

Metals

Selebi was located in 1963 and production started at Pikwe in 1972. About 200,000 tons of ore mined underground at the two locations provide a 40/40% nickel, copper and 0.6% cobalt matte.

Table 103 Original Botswana copper and nickel reserves in kt and grades,%.

Deposit	type	Reserves kt Cu	Ni	grade % Cu	Ni	Deposit	type	Resourc.kt Cu	Ni	grade % Cu	Ni
Phikwe	prov.	250	320	1.14	1.45	Phenix	indic.	40	100	0.8	2.1
Co 0.1-0.01%	indic.	100	100	1.09	1.13	Selkirk	indic.	30	30	0.8	0.9
Selebi	prov.	160	70	1.56	0.70	Takwane	indic.	4	7	0.6	1.2
Co 0.03%	indic.	30	20	1.28	0.88	Thakadu		50		2.5	
Selebi N	indic.	20	16	0.97	0.86	Makala	indic.	100		2.1	
Parallel trend : Phokose, Mogolodi						Logolo	indic.	15		1.8	
Dikolati	prob.	10		1		Bushman	prob.	40		2	
Lentswe	indic.	10		1		Total		850	670	1.3	1.1

indic(ated), prob(able), prov(ed) resources

The Tati (NE) schist belt consists of 2 trends :

I. Nickel-copper Phenix, Selkirk, Tekwane : a 20 km long north-south belt.
II. Gold a) Monarch-Francistown-White Elephant, a 50 km long arc ;
b) Bonanza-New Zealand 40 km ; c) Rainbow 20 km on the border.

Gauta is the latest of four small mines out of 60 former workings. Reportedly, 2.5 t gold were left at Map Nora, 3 t at Bonanza and 1.25 t at Golden Eagle along with some tungsten. The Monarch may still hold some primary ore ; lead shows at Lady Mary and the belt has supplied a ton of silver.

Further north-northwest, 60 km of the Matsitama belt comprise six deposits and a dozen prospects. The ancients smelted copper at Thakadu and Bushman, where 2 Mt of oxides assayed 2% Cu, along with some gold and silver. Zinc accompanies at Makala (the 'ashes').

Finally, occurrences include :

a) in this belt : Logolo, Somerset, Dihudi, Nakalakwana and Phudulooga ;
b) in the southeast lead at Taupone near Kanye and the 3 km long Ditshukutswane vein with copper, lead, fluorite and some zinc. Uranium : Mokobaesi;
c) the 300 km long Ghanzi trend extends from Lake Ngami to Shinamba Hill along a line connecting Namibian mines with the Copperbelt. Between Ngwako pan and Kwegbe Hill copper is disseminated in 1 to 20 m thick shales.

Prior to 1972 130,000 t manganese were extracted at Kgwakgwe (SSE) and 200,000 t are left. Ootse and Lobatse (SE) supplied 100,000 t and at Ramotsoa, nearby, assays average 35-40% MnO_2 over a kilometer. The shows scattered over 45 km at Lerala (E) resurface with grades of 42% MnO_2 at Chadibe along with 30 Mt ore containing 30-40% iron. Naka-la-phala east of Shoshong (E) has 0.5 Mt, to the west 4 m thick beds extend over 20 km at Legatadimane and to the SSE at Matlhabanelo, the 10 m thick formation grades 30-55% Fe over 1.5 km^2.

Other resources

Assuming that only 1% of 450 mm rain reaches groundwater, the annual recharge is 3.5 Bm^3. To these should be added perhaps 12 Bm^3 in the Okavango swamps (N). Yet tapping the inflow of the Zambezi at Kazengula (NNE) seems to be remote, even though Botswana would receive a third of the flow of this $2 B project with the remainder pumped to South Africa over a distance of 1035 km. To be sure, inhabited areas abut on the Limpopo (E) and the new cities are fed by water table wells. But their needs are growing. It also rains more in the east. Unfortunately, only faults and weathering contacts of the Basement yield 0.4-2m^3h in 28% of the tests. Water is more abundant in middle Ecca (Karroo) sandstones under the thick column of the central Kgalagadi. It contains nonetheless 1% solids in the south-central desert and 0.7% at 300 m depth under the Boritse pan. In contrast, there are 'only' 470 ppm solids in the quartzite aquifer of the Sekhuma-pan on the Ghanzi range. The government promotes, therefore, the distribution of hand pumps and small diesel pumps at watering points every 30 km along the cattle trekking routes, where there is also shallow sandy loam. Only in the southeastern strips can soils sustain some farming with the help of fertilizer and lime. Consequently, only 0.1% of the land surface is cultivated.

The ancients collected salt at the spit of the Nata into the Makgadikgadi (NE) and among its many pans 900 km^2 of the northwestern Sua contain 130 Bm^3 brines. BP plans to collect through a wellfield of ditches, some pumping and evaporation ponds 300,000 ty soda ash mostly for the South African market.

Some 20,000 t of asbestos were produced between 1950 and 66 at Moshaneng (SE). At Khakhea pan long but silicified fiber shows over 3 km and crocidolite extends to Knapdaar. Northwest of Kanye (SE) kaolin patches show over 5 km, and barite lenses crop out over 1.5 km of the Lephephe road (SE) (the best measures 2.5 x 1.8 m). Halfway Kop (Tati, E) was the main kyanite producer and WNW of Foley (E) one meter thick bands average 25-30% gypsum under 1 to 150 m overburden. Lastly, other occurrences include :

Asbestos Keng pan, Oikhe pan, Naka-la-phala, Christmaskop (E) with talc, brick-clay Gaborone, Serowe, Dinokwe (E) corundum NE of the Lotsane, diatomite Sedibeng, Metloje pan (Phitshane), feldspar Prospect mine (E), Bodiakudhu (SE), fire clay Mmamambula, Moropo mudstones, fluorite Taupse River (E), Ditshukutswana (SE), garnet 10 km N of Zanzibar farm (E), graphite, fluorite near Bushman mine, iron ore Mahibitswane, Matsiloje (E), Sebilong, along the Molopo (S), Knapdar (S), Tshidilamolomo, Phitshane-Mogobewa Kgomo, limestone and marble Moshaneng, nepheline syenite Semarule Hill (SE), sandstone Ootse, soapstone hillocks, Powmatsie, stone Gaborone, Kanye, Mahatlane, Lobatse, talc Mabelane, vermiculite Mochudi (SE), wonderstone Dipotsane.

7 3 N A M I B I A

The value of minerals produced increased 20 fold in 20 years time and has peaked. The country ranks sixth among non-fuel producers in Africa and gem diamonds and uranium are its key contributions to world trade. Minerals also cover in normal years two thirds of exports, yet their share of government revenue, needed for better social services, declined from a half to a third. Ovambos, the majority, live along the northern border where more than 400 mm rain fall. The mines only employ 20,000, but labor is short at Oranjemund and Swakopmund near the southern and central coast where most of the resources are (Fig. 58, 59, 65).

Table 104 Namibian production, including past^{+} peaks, and reserves.

Commodity	Production	Reserves	Commodity	Production	Reserves
Copper	kt 50	kt 650	Diamonds gems	Mct 1	20
Lead	kt 40	kt 350	" " industrials	kct 50	
Zinc	kt 30	kt 300	Uranium	kt 3-4	120
Silver	t 100	kt 1	Gas		Tcf 2
Gold	kg 20		Coal resources		Bt 3
Germanium	t 45	kt 4	Hydropower	Mw 260	
Cadmium	t 70	kt 1	Salt	kt 200	
Sulfur	kt 90		Limestone	kt 2	
Arsenic	kt 2		Lime	kt 1	
Vanadium^{+}	t 800	kt 4.5	Quartz	t 150	
Manganese^{+}		Mt 1.5	Mica	t 100	
Scandium capacity	t 5		Tin	kt 1	0-60
Fluorspar^{+}	kt 25	Mt 10	Lithium minerals	kt 1	500
Kyanite^{+}	kt 4		Cesium^{+} minerals	t 300	kt 10
Graphite	t 800		Beryl^{+}	t 600	
Trona	kt 2		Coltan^{+}	t 10	
Guano	kt 1		Tungsten^{+}	t 150	kt 2

The régime encourages foreign investment in minerals. At the outset companies are heavily taxed yet in practice investments can be offset against income. Half of the exports reach the EEC, the other half is shipped to South Africa by rail, whence Namibia imports fuel, equipment and power. Further, De Beers control diamonds and AAC coal, Yskor tin and 51% of Rosh Pinah with 49% to Zincor; GfSA holds 43% of Tsumeb against BP's 14% and the 36% of Newmont companies, the managers. The state's ENOK has 50.3% of the voting rights and 3.5% of shares of

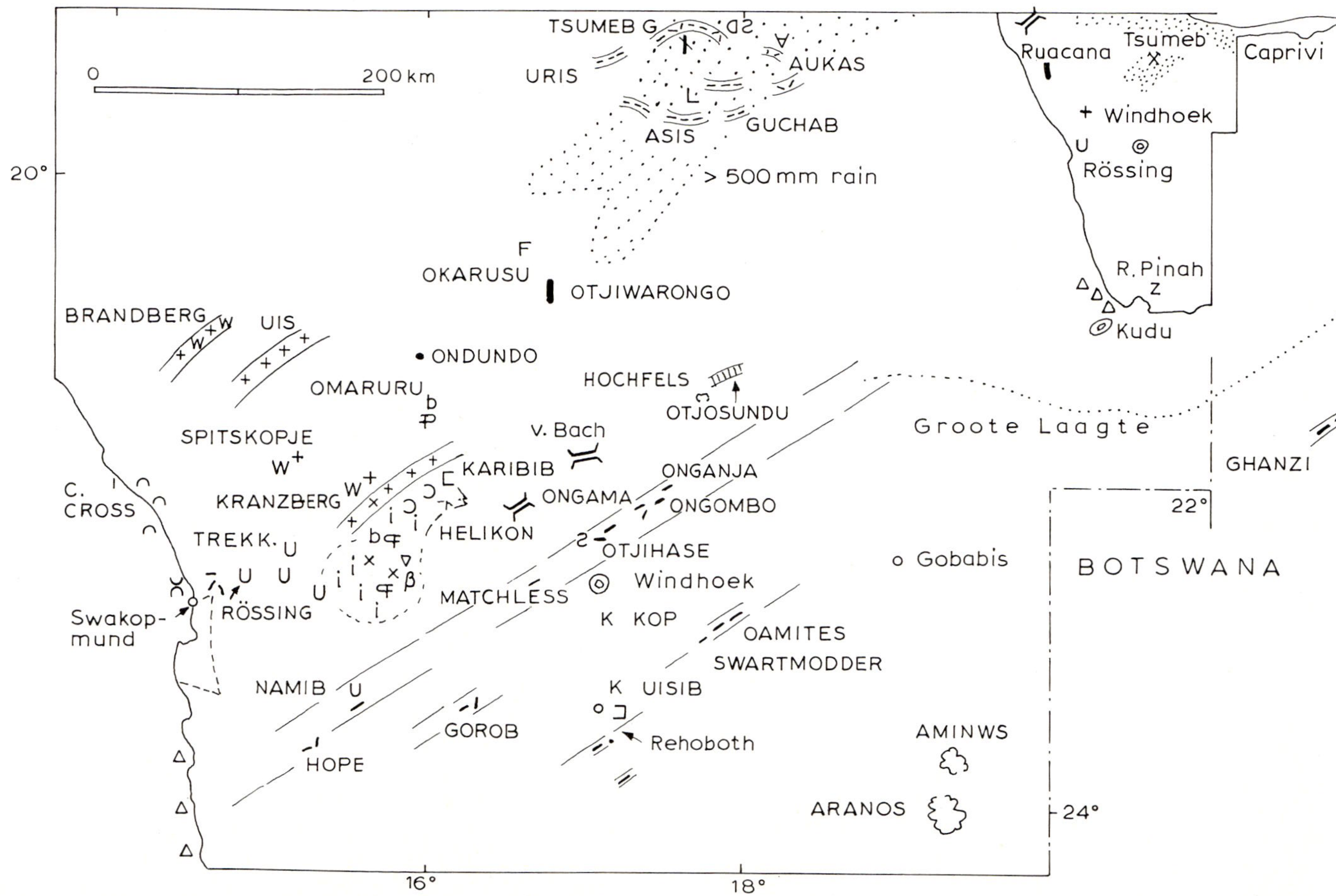

Fig. 58. North-central Namibia. (Legend on page 13.)

the uranium company, RTZ, the managers 26% with 47% of the equity (combined with Rio Algom's 10%); Gencor Minatome and Total hold minorities.

GOALS : Investment incentives for the future. Equal opportunity for all. Aid and advice to small workings for employment. Further efforts to insure water supply and distribution.

Meanwhile, a copper refinery, zinc smelter, diamond sorting and cutting and exploration in the south and center will have to await political decisions.

Diamonds

In 1908 a worker from Kimberley picked up the first gem in the Namib or diamond desert of Pomona, east of Lüderitz Bay, in German West Africa. Production reached 1.6 Mct in 1910, 700,000 ct in 1925, 1 Mct in 1951 and oscillated inbetween around 200,000 ct according to demand for quality stones. Highs of 1.8 - 2 Mct in 1970, 77 and 78 were therefore halved. Kimberlites are found 250-350 km inland, ENE of Lüderitz and east of Oranjemund and the terraces of the Orange River are also mineralized. While diamonds occur along the whole coast, the Calypso sucks here 2-3,000 ct/Mo for Ocean Diamond Mining in the Eiland Diamante concession which surrounds 12 Penguin Islands, claimed by South Africa. Diamonds, also show at Conception Bay, in the Bogenfels and the Chameïs Reef, 450, 210 and 160 km north of the Orange, but the Lüderitz diamonds average only 0.2-0.5 ct with many fragments. While CDM reassesses the old German workings, most activity is concentrated over the southern 100 km of the shelf. At the Orange the concession extends 200 km into the sea, but few easily minable reserves have been developed and the face now reaches 250 m seawards from the high-water mark in areas CDM leases from its sister, Marine Diamond Corporation. Behind a 30 m high seawall pumps raise 900 m^3h of water, bucket wheel excavators remove 750 m^3h overburden and moving belt loaders a total of 17 Mty of material (a half of past years). Bulldozers, hydraulic excavators, frontend loaders and 35 t haul-trucks move 7 Mt of cemented gravel to the crushing plants. The bedrock, 20 m below the sea, is finally brushed. After milling, heavy media and Xray separation, 1-1.2 carats are recovered per ten tons. Namibian diamonds are highgrade, currently 95% and 20 years ago 80% were gems of >1 carat. With the startup of new mines elsewhere, Namibia's importance declines and CDM now covers only 8% of De Beers' net profit. The company pays royalty (22.5%), export tax (15%), but can deduct them from the 60% profit tax ; it employs 4-5000 people directly and is involved in real estate, distribution, insurance, and education.

Energy

Out of 700 Mw capacity 240 have been installed at Ruacana (N) on water diverted from the Cunene in Angola. There are smaller hydropower plants at Mariahöhe, but the 120 and 45 Mw generators use coal and diesel oil from South Africa, which can supply another 200 Mw. Texaco located Kudu, 120 km west of Oranjemund. Now controlled by Swakor, a subsidiary of ENOK, the field could supply up to 1 Bcfd of gas from 4400 m depth. Exploration also continues in Walvis Bay and Btosha Bay (N), but the Etosha (N) finds are subcommercial. Coal seams also underlie this pan. More important, AAC discovered coal on the north bank of the Nossob River (E) near the Aminuis Reserve. CDM then covered the other bank, the Aranos area with prospecting grants and estimated 600 Mt steamcoal. Resources of 3 Bt are under up to 300 m overburden and an underground lake. B & O Minerals, an associate of AAC, also located coal near Hochfels, southeast of Otjiwarongo and exploration continues as far as Okahanja and Herero land.

Some 190,000 tons of uranium resources are scattered in the granites and certain Karroo layers on the coast and under the cover of the Namib desert. Yet the four principal deposits are Rössing(100 kt), 60 km from Swakopmund, Tubas to the east and Goldfield's Trekkopje, where AAC holds the mining rights. Gencor's low grade calcrete at Langer Heinrich exhibits peaks of 10% U_3O_8, but generally irregular mineralization. At Rössing radioactivity has been detected since 1928 and RTZ has delineated the deposit in 1966-73. Ore averages only 0.03% U_3O_8, half of which is uraninite. Complex niobium-tantalum and rare-earth minerals accompany. More than 15 Mt of alaskite can be ground to <6 mesh, and leached, following ion exchange, solvent extraction and ammonia precipitation yellow cake is calcined.

Out of 3,000 workers, 1,600 are Africans and capacity is 5,000 t U_3O_8. The major customer was the British Electricity Board, but contracts still run with Japan (16 kt over 1984-90), France (11 kt 1981-90) and Taiwan (4 kt over 1983-98). RUL has also repaid its creditors, but the tax holiday is over ; corporation tax is 42% and the company appears to make $ 15-20 M profit per year !

Base metals and silver

A hundred occurrences are scattered in the I Otavi Highlands (N), II the Windhoek-Damara trend (C), III Namib Siding (W), IV low grade lavas of Gohoseb, Awahab and Rehoboth (SC) and V the South, where 200,000 t ore of the Rosh Pinah (Fig. 65) openpit provide 80,000 t concentrate smelted in South Africa into 30,000 t zinc, 15,000 t lead with 30 t silver. Grades are 17% Zn, 8.5% Pb and 300 g/t Ag, and recovery is 93%. The Haib river copper workings are further east, near the South African Aggeneys deposits.

Prior to 1850 Africans smelted copper in the north, in 1910 a German company started operations and TCL was set up in 1948 following expropriation. TCL owns the Kombat mines and manages Asis, mostly controlled by TECO (in which GfSA has a 25% share) ; it also owns Matchless (C) and 70% of the Otjihase Joint Venture nearby, with the remainder accruing to JCL. Production and reserves can be summed up as shown in table 105.

Table 105 Namibian metal reserves (rsv), production (prd) and grades (grd).

Reserves	Tsumeb		Kombat		Asis		Matchless		Otjihase	
	rsv	grd	rsv	grd	rsv	grd	rsv	grd	rsv	grd
Copper kt %	170	3.2	30	1.7	20	4	10	2.2	200	2.3
Lead kt %	180	3.5	40	2	10	2				
Silver kt g/t	400	90	30	15	20	30			80	9
Sulfur kt %							290	13	160	19
Production	prd	grd	prd	grd	prd	grd	prd	grd	prd	grd
Copper kt %	13	3.2	10	4	1	5	2	2	14	1.7
Lead kt %	18	4	5	1.8	0.2	1				
Silver t g/t	50	100	10	30	1	30	0.6	6	3	4

According to the U.S. Bureau of Mines the country's reserve base attains 1000 kt copper, 500 kt lead, 400 kt zinc, 1.2 kt cadmium and 900 t silver.

Tsumeb also acts as a custom smelter and total production has reached 55,000 t blister. Conversely lead output declined with depth from a peak of 70,000 t in the fifties, to less than 40,000 t. Similarly, zinc grades dropped to 1.2% and metal in concentrate from a 20,000 t peak to 5,000 t. Cadmium plummeted more steeply from 1,000 t to 50-100 t. Tsumeb was also a key source of germanium ; there are also 1,500 t in the dumps accompanied by some gallium.

However, Kiln Products, a GfSA affiliate, does not report vanadium anymore since the closure of Berg Aukas in 1978. Finally, among smaller occurrences let us quote Alt Bobos, Gross Otavi, Groot Fontein (N) and Ongombo and Onganja (Cu, C). Klein Aub (83% owned by Gencor) provides 200 km south of Windhoek 5,000 t of copper and 10 t silver and Namib (W) still supplies 3,000 t lead, zinc and silver concentrates. In contrast, little copper and silver are left at Oamites and Swartmodder (C) and at Lorelei (S).

Food and water

Light brown steppe soils cover the useful part of the country, yet before the year 2000 its 500 Mm^3 water resources will be exhausted. Since ranching will remain the key activity up to 160,000 hectares may need water then. An undulating band of 30 to 110 cm of sand stores some water in the east, handpumps have been placed every 30 km on the main cattle routes and riverbeds, deltas and gravels of the interior basins are the principal source of often low quality water. Aquifers of the Karroo are even more salty and those of the Basement are notoriously fragmented. Out of a total of 150 Mm^3y limestones and dolomites and their karsts yield most groundwater, some of it stored over 25,000 years. Even though water is drawn from the Cunene and Cuvelai in Angola (N), the Orange (S) and the Kuiseb and Swakop rivers, desalination plants will be required again. Moreover, plans call for the Eastern Water Carrier to channel water from the Okavango (NE) to the Karstfeld watershed near Tsumeb and to move it through 700 km of channels, pipes and the Omatako, Swakoppoort and von Bach dams to Windhoek. Even its storage in the Otavi dolomites and tapping the Zambezi floods in the Caprivi strip (ENE) through a reversal of the Lake Liambezi spillway towards the Okavango has been considered in competition with the Kazengula project of Botswana and South Africa.

Salt recovery from the pans scattered between Swakopmund and Cape Cross has trebled in twenty years. Two thirds are exported to South Africa and the South West Salt Co. produces half of the total, followed by Koes Salt, Snow Salt, also at Swakopmund, Salt and Chemicals (AAC and Rand) at Walvis Bay and the rock salt supplier Salz Gassow. Soda evaporates on the edges of the Kgalagadi and in 1975 Swaco recovered 2000 t trona from the Otjivalunda Pans (NW). Some 1000 ty of guano also accumulated in the sixties on platforms erected on the coast and on South African islets, but cave guano contains little phosphorus.

Other resources

Estimates of fluorspar reserves at Okorusu (NC) range from 8 Mt grading 50% CaF_2 and 7-15% apatite (phosphate) to 10 Mt assaying 35% CaF_2. Production reached here 3,000 t in 1954 and 25,000 ty at Omburu (W) between 1973 and 75. The Eisenberg occasionally supplied flux to Tsumeb and sillimanite sales from Kyanitkop (C) and Uis (W) peaked in 1957. Smaller occurrences scatter around Warmbad and Keetmanshoop (S) and graphite output rose to 800 t in 1955 , 150 km to the southwest. The Bethanie workings were, however, closed in 1952. Similarly, the Otjosundu manganese mine could not compete with South African ores.

Swarms crop out over 150 x 200 km. Nonetheless, Yskor has to mine and crush 700,000 t of pegmatite with a stripping ratio of 3.6:1 in order to recover 1000 t tin in concentrate at Uis (W) out of 100,000 tons of resources. According to the price of tin, reserves may reach therefore from nil up to 60,000 tons. Indeed grades average only 1 to 1.5 kg/t with 2-3% columbite-tantalite. To the west, Brandberg produced up to 230 t tin and 70-130 t tungsten out of reserves of 6.6 Mt averaging only 1.5 kg Sn/t and 800 g W/t and to the east, Krantzberg supplied 150-200 t tungsten in concentrate. Production of 3-5000 ty wollastonite has been considered in the Karibib field where Helikon produced petalite and Rubikon the other lithium minerals and beryl. Productions peaked in 1954 and pollucite in 1959. Indeed, large reserves of cesium and rubidium await markets. The pegmatites of Aurora are in the same area, but Tantalite Valley and Dassiefontein are in the Warmbad field (S). Lastly, lapidaries have recovered from all these sources 100 kg to 5 ty gemstones :

Aquamarine, heliodore, morganite : Spitzkopje, Rössing, asbestos Swakopmund, Keetmanshoop , barite Khomas (C), Otavi, bismuth Erongo, diatomite North of Orange, garnet pegmatites and Swakopmund, gold Andundu (7 g/t) with antimony

Rehoboth, Iceland spar Gibeon, Rehoboth, Swakopmund, iron ore Otjiwarongo, Otavi, Ongaba-Owihenda (NW), marble Damara and Otavi highlands, molybdeny Natas (W), rutile Giftkuppe.

74 SWAZILAND

The Liqoqo rules the lovely country of the Ngwane in the name of the King (Fig. 64). Sugar-cane and wood for pulp, the major export products, grow with the rains. Enough water is thus available for farming except for the driest seasons.

GOALS : Expansion of coal extraction and hydropower capacity. Exploration for precious minerals and the development of small mines.

Albeit 4,000 people find their livelihood in the mineral industry, much will depend on future developments at Nidis assisted industrial ventures : the Tibyo Taka Ngwane fund owns 51% of the shares of the brickworks and the Roberts Construction Co. of South Africa 17.5%. TTN has also acquired 48% of the colliery from AAC and from Turnall most of the asbestos company which has contributed 10% of exports revenue.

Table 106 Swazi reserves and production.

Iron ore reserves	Mt	%Fe	Resources			Reserves			Production		
Gege (SW)	55	40	Barite	Mt	1	Hydropower	Mw	700	Hydropower	Gwh	10
Maloma (SE)	75	30	Fluorspar	kt	2	Coal	Bt	1	Anthracite	kt	150
Forbes R. (NW)	6	38	Andalusite	kt	30	Kaolin	Mt	1	Asbestos	kt	30
Nickel rsc.	kt	150	Pyrophyllite	kt	200	Silica	kt	250	Stone	$1000m^3$	100
			Corundum	kt	500	Calcite	kt	10	Tin	t	1-5

Energy

Key data of energy production are :

	1970	1975	1980	Peak	Year
Coal kt	140	130	180	180	1980
Thermal Gwh	40	50	160	180	1981-2
Hydro "	100	110	130	130	1978

The first hydropower plant, at Edelwini on the Usutu generates 10 Mw and after its 1974 low of 117,000 t anthracite, capacity has grown to 200,000 t. While 20-30,000 t are used locally, Kenya and Moçambique are good markets and exports move through Richards Bay and Maputo. - Coal measures extend over 140 km. The eastern region and the upper coal zone contain bright, ash rich anthracite. Three minable seams are 1-1.5 m thick, but the main seam averages 7 m. In the south, along the Stegi-Manzini road coal measures assay and heat :

	Fixed C %	Ash %	Moisture %	cal/g
Upper Zone	71	21	1	6,700
Nsalitshe	60	16	8	5,300
St. Phillips	65	15	5	5,800
Nhloya	54	13	11	5,000

Geological reserves reach 5 Bt, out of which 1 Bt are accessible and 200 Mt easily worked. At Mhlume (NE) Shell has considered a 800,000 t colliery based on 100 Mt anthracite and Japanese groups show interest in Lubhuku (S) where 190 Mt of reserves extend to 600 m depth.

Other resources

In 1929 H.H.D. Castle found asbestos at Havelock in the high ridges of the northwest and the mine exported since 1939 35-40,000 ty of fiber through a ropeway to South Africa. Output declined with demand, so did reserves. 3,000 t of soapstone and 60-800 ty of ceramic grade diaspore have also been exported respectively to South Africa and Germany from Concession 50 of the Mankaiana district, where there is also andalusite. The clinker grinding plant of Matsapa (W) has a capacity of 70,000 ty and the Langa works near Mpaka can ship 65 million bricks per year to South Africa. Formerly, a few hundred tons of barite were exported each year from Londosi (W).

In the seventies De Beers located diamonds in the Ehlone (N) near the Dohol-wayo kimberlite ; industrial diamonds were found in the south too and Trans-Hex held a licence near Manzini. In 1882 Mbandzeni, King of the Swazi, granted a concession to exploit the Forbes gold Reef. There are two other major occurrences in the northwest : Piggs Peak and Horo. In 1910 a production of 400 kg was reported, in 1950 50 kg, in 1963 - 65 180 kg and in 1975-6 : 8 kg. In 1908 tin concentrates from Ezulwini, Sinceni and the Mbabane River peaked already with 600 t, declined to 200 t in 1930, 40 t in 1950 and to a total of 80 t in the sixties and 50 t in the seventies. Iron ore formations extend 10 km from Clary's Camp to Ngwenya or Bomvu Ridge, cross into South Africa over 20 km and reenter Swaziland further NNE at Iron Hill. They average 49% Fe, 22% SiO_2, 0.07% S and 0.03% P. AAC provided 50 Mt of ore assaying 62% Fe on Bomvu and extracted 18 Mt between 1964 and 1978 which were shipped to Japan. Furthermore, deposits of the southwest grade only 31% Fe at Gege and 40% in the Hlatikulu and Mankaiana districts, but they contain 8% manganese. Lastly, other occurrences include :

Calcite : Nsalitshe, columbite : Pigg's Peak, corundum : Hlatikulu d., graphite: idem, fluorspar, opal : Conc. 31 Hlatikulu d., Hluti, Mlosheni, kaolin : Mankaiana d, Mahlangatsha, Ngane, lead and silver : Ngwasuma (Hlatikulu), magnesite : South of Kaapmuiden, molybdenum : Komati, nickel-copper : Forbes Peak 0.11% Ni

Usushwana, perlite : Conc. 49.74 Stegi, Conc. 11 Hlatikulu, silica : Pigg's Peak, Insuzu Series Mankaiana d., talc : Havelock area, tantalum, rare earth concentrates : Kelly mine, Mbabane River.

75 LESOTHO

Some 50,000 Basotho worked on the Rand, where they gained fame since they specialize in such difficult tasks as shaft sinking. Skills are thus available and the government encourages foreign investment.

GOALS : Development of water and hydropower resources, detailed exploration for semi-precious stones.

In 1959 workers recovered already 9,000 carats of diamonds. There are also kimberlite pipes at Ngopetvou, Tsepine Berea, Molibi Bere and Thabantvonyana(W) but the principal producer, Letseng-la-terai (NE) has been closed by De Beers since grades declined to 3 ct/100 t. A fifth were gems and more than 12% were larger than 10 ct. Production reached 30,000 ct in 1969 and 40-70,000 between 1977 and 1982 ; 5-10,000 carats have been recorded in other years. Manual work continues meanwhile nearby on the cooperative pits of Liqhobong and Lemphave in the Maluti Mts. and reportedly resources would still reach 4 Mct.

Agates of the Drakenberg also find ready markets. Lesotho also imports tiger's eye from South Africa for polishing and reexport. Each year 20-30,000 tons of stone are quarried, some limestone at Mohale's Hoek, the red and Molteno sandstone for building, bedded Cave Sandstone for paving and dolerite for the surfacing of roads. Finally ochre has been recovered in the Sephongong and salpeter shows in the Mosikolelo cave.

Water and energy

It is difficult to live off the 'roof' of Africa (Fig. 51, 64) : most of the woods have been felled and erosion is progressing in the highlands where water is generally abundant, but the populated areas of the Northwest use shallow wells of the Caledon plain.

Uranium shows in kimberlites and thin seams of coal bend along the border over 150 km from Qacha's Nek towards the south and then beyond Maseru to the northwest. At Butha Buthe and Mohale's Hoek their calorific value is 6,100 g/cal. Similarly, the high-ash peat of Mothae near Mokotlong (E) is sub-economic.

Consequently, 700,000 b of petroleum, 100,000 t of coal and 200 Gwh of electricity will be imported from South Africa by 1990, unless some of the 450 Mw or 2 Twhy hydropower is harnessed providing income from water sold to South Africa :

a) The Jordane project would divert water from the Upper Senqunyane, a source of the Orange, toward the capital providing 36 Mw (180 Gwhy) and 5 m^3s of water for export.

b) A dam at Oxbow (N) at the sources of the Malibamatso could produce at Lala 100 Gwhy from 120 Mw installed capacity.

c) The complete ($2B) Eastern Highlands project provides for damming over 20 years the Malibamatso at Pelaneng, Soaï and then Oxbow, followed by the Upper Orange (Senqu) at Polihali and Taung.

Installed capacity would thus rise ultimately to 190 Mw and energy to 510 Gwhy. Finally,

d) a dozen mini-hydro plants might provide 10-20 Gwhy.

8 SOUTH AFRICA

Although mineral commodities contribute 60% of export revenue, their share declined to less than 20% of GDP. The Republic is, nonetheless, the leading producer of half of the commodities in Africa : it is a class in itself. Conversely, there are flows : the scarcity of oil, gas, coking coal and bauxite. These and the necessity of exporting her products and importing capital and labor oblige South Africa to rely heavily on the world.

81 ORGANIZATION and PERFORMANCE

South African mining is oriented towards the precious, which represent two thirds of the total of Table 108, processed materials excluded. In terms of value, 70% of production is concentrated in the Witwatersrand basin, 15% in the Bushveld. The remainder spreads over eastern Transvaal, the north-western and eastern Cape. Nominally independent, Bophutatswana shares 12 (60) % of PGM, 19 (60) of cobalt, 7 (50) nickel, 30 (4) vanadium and 25 (50) % of chromite reserves (and production). It has vast copper resources ; manganese, iron ore and asbestos deposits skirt its fragmented borders. There is also amosite in Lebowa, chrysotile asbestos in Ka Ngwane and coal in Venda and KwaZulu. There is otherwise little employment in the homelands and the number of foreign migrant workers has been reduced to less than a million.

> GOALS : Equal education and opportunity for all. Development of small mines. Further expansion of coal and platinum mining, return to higher production levels for others. Conservation of gold resources. Hydrocarbon exploration.

More than 700,000 people earn four billion rands in the minerals industry - sensu stricto -. Investments generally reach two billion rands per year or R6B along with mining supplies.

The rand has declined from almost a dollar a few years ago to 40 cents in 1985. Table 108 therefore indicates sales data in rand. - Figures for PGM, uranium, cobalt and some other products are rough estimates.

Table 108 South African Production in Million Rands and Reserves.

Commodity	Value RM	Production 80s	1963	Resvs.
Precious minerals and metals				
Gold	11000	t 700	8000	kt 20
Diamond gem	500	Mct 4	1.6	Mct
" industrial	20	Mct 6	2.4	370
Platinum	600	t 50	6	kt 25
Palladium	200	t 25	2.5	kt 15
Other PGM	100	t 15	1.5	kt 10
Silver	90	t 200	90	kt 10
Emeralds, &	1	kg 600	400	
Tiger's Eye	0.1	t 100	130	
Sources of energy				
Coal	3500	Mt 150	45	Bt 58
Anthracite	200	Mt 4	1	Bt 1
Uranium	300	kt 6	5	190
Gas				Tcf > 1
Condensate resc.				Bb 30
Ferrous metals				
Iron ore	350	Mt 20	17	Bt 6
Manganese	200	Mt 3.5	1.6	Bt 13
Chromite	130	Mt 3	1	Bt 3
Vanadium	70	kt 20	2.5	Mt 8
Nickel	60	kt 20	3	Mt 9
Cobalt	5	t 300		kt 30
Non-ferrous metals				
Copper	400	kt 200	65	Mt 6
Zinc	50	kt 60	-	Mt 12
Lead	40	kt 100	0.2	Mt 5
Zircon min.	40	kt 170	3	Mt 7
Antimony	40	kt 10	20	250
Tin	30	kt 2	1	25
Rutile	10	kt 20		Mt 2
Monazite	10	kt 10	2	Mt 1
Industrial minerals				
Lime	120	Mt 2	1	
Phosphate P_2O_5	100	Mt 1	0.2	kt 180
Sulfur	50	kt 600	500	
Salt	30	kt 700	200	
Titanium	20	kt 250	30	Mt 30
Bentonite	1.5	kt 20	9	Mt 2
Barite	0.2	kt 2	3	Mt 30
Attapulgite	0.2	kt 5	51	24
Diatomite	0.1	kt 1	0.2	kt 100
Corundum		t 30		

Commodity	Value RM	Production 80s	1963	Resvs.
Processed materials				
Petroleum		Mb 80		
Oil fr. coal		Mb 50		
Coke		Mt 6		
Steel		Mt 8		
Ferrochromium		kt 800		
Ferromanganese		kt 400		
Ferrosilicon		kt 100		
FeSiMn		kt 20		
FeSiCr		kt 30		
Silicomanganese		kt 50		
Silicon		kt 30		
Manganese		kt 30		
Aluminum		kt 180		
N in ammon.		kt 650		
P_2O_5in fertil.		Mt 1		
Cement		Mt 8		
Building materials				
Sand, &	300	Mt 40	0.3	
Asbestos	100	kt 200	230	Mt 8
Limestone	80	Mt 15	7	Bt 2
'Granite'	30	kt 200	3M	
Vermiculite	20	kt 160	100	Mt 73
Gypsum	7	kt 550	200	Mt 90
Kaolin	5	kt 120	40	Mt 20
Slate	5	kt 40		
Shale	2	kt 500	0.2	
Mica	1	kt 3	2	
Other stone	1	kt 1	5	
Talc	0.7	kt 10	8	
Wonderstone	0.5	kt 4	2	
Pigments	0.4	kt 1.5	4	
Perlite				Mt 3
Fluxes and ceramics				
Fluorite	60	kt 350	60	Mt 90
Silica	20	kt 1.5	0.3	Mt 900
Andalusite	20	kt 150	11	Mt 50
Flint clay	6	kt 120	60	Mt 30
Feldspar	4	kt 50	46	
Fire clay	2	kt 150	250	
Magnesite	2	kt 30	110	Mt 4
Sillimanite	0.5	kt 3	62	Mt 1
Wollastonite				kt 100

The Republic uses only 15% of the commodities it produces, or 25% if the precious are excluded. Coal matches 60% of domestic sales, followed by aggregate and sand, 7% ; limestone and dolomite, 6% ; copper, 6% and iron ore, 4%.

Benefitiation ratios of such commodities as copper and nickel are normal at roughly 80%; with only 60%, they are even low for zinc. In contrast, antimony concentrates, 60% of chromite, 40% of the iron ore and a quarter of the manganese ore are upgraded.

Exports

Sales have doubled in five years in rand terms. While they have declined by 20% since their peak of 1980 in constant dollars, they are growing again. It should be remembered though that quality and exports fetch higher prices than local sales and the difference is considerable in some instances as shown in table 109. Traditionally, gold represents more than 60% of mineral commodity exports, coal 12%, ferro-alloys, chrome and manganese ore 6%, steel and iron ore 4, PGM 4, diamonds 3%, fertilizers, uranium, aluminum and copper 4% combined and all other products together another 4%.

Uranium sales have declined along with demand, but aluminum exports have trebled in seven years time, even though South Africa has no bauxite and little hydropower by African standards. Similarly, Foskor works one of the lowest grade phosphate deposits in the world, but the value of exported fertilizer, matches total sales of rock. South Africa even exports 90% of the 'granite', 70% of the wonderstone, 60% of the slate, 40% of the flint clay sold. Yet exports of the two latter represent only 30 and 15% of the weight produced.

Table 109 Exports as a) weight % and b) value % of total sales and destinations as % of exports.

Commodity	Pt	Pd	Rh	Ru	Fe	Mn	Cr	FeCr	FeMn	all	V	Ni	Co	Cu	Pb	Zn	Sb	Zr	And	Asb	Fl	Ver
Weight%					50		30	95			95			60	100	30	60		50	95	90	95
Value%		99			65	80	50	90	70	90	95	70	90	60	100	30	60	100	60	95	95	95
US	45	60	80			10	30	15	70			40	50			5	60			10	50	
Japan	50	30	20	100	40	40	35	35	5	30	80		45	30	30	20	15		30	40	25	
Germany	5				15	10	10	30		15		10		50	60	70	5		40		15	10
Other EEC						10	15		15	40	15	10		20	10		5		20	20		30

Ruthenium includes Osmium and Iridium,all. = other alloys including silicon, And.alusite, Asb.estos, Ver.miculite.

Switzerland imports a third of gold, the UK much of the remainder along with most silver. The CSO directs from London most diamond sales, notably to Belgium, smaller gems to India and industrials to Ireland. Coal exports have doubled in four years time in terms of value ; tonnage grew by 50%. Europe takes 60% and Japan a quarter. Most rutile and three quarters of titanium slag are shipped abroad and exports of phosphate concentrates are expected to increase from the current 25% of sales. Finally, the United States now take a half to two thirds of the uranium.

8 2 STRUCTURE OF THE INDUSTRY

Tables 110-113 sum up the structure and activities of the 'houses' which co-own, manage and service mining and other companies.

Table 110 Percentage control of commodities by the 'Houses'.

Σ Group	Au	U	Pt Ni	Ag	Diamd	V	coal	Fe	P	Mn	Cr	Asbe	Cu	Verm	Pb	F	Sb	Sn	Ti
53 AAC &[a]	47	70	60	25	95	65	23			15	5						100	5	
20 Gencor	17	20	30	8			28			30	30	95				40			30
17 GfSA	20	5		60			4						15		100			85	
11 State							23	95	95		10								
9 Barrand	10			4			17				30					25			
5 Anglovaal	5			2						35									
3 Lonrho	10						1					5							
2 RTZ								2					85	100					

Diam(onds), Asbe(stos), Verm(iculite), Sb Antimony.

If processing is included, the share of the state and Gencor increases though.

Table 111 Share (%) of equities and earnings of mining groups in South Africa.

AAC &[a]	AAC	De Beers	JCL	State	Gencor	CGf	Barrand	A.Vaal	Lonrho	RTZ
Equity	24	24	2	22	11	9	8	2	1.5	1
Earnings	28	20	5	19	12	5	10	2	1.5	1

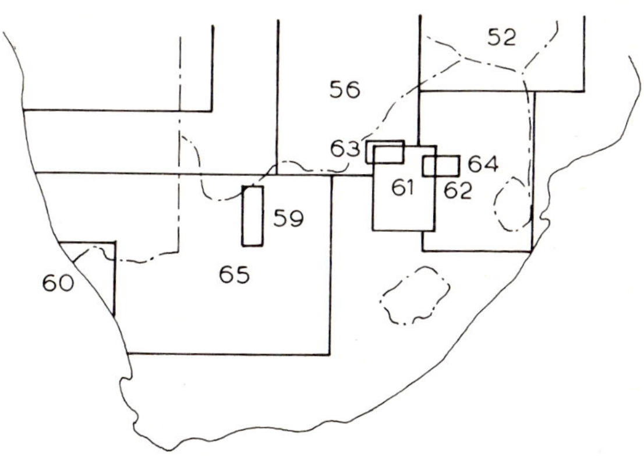

Assemblage of South African figures.

Table 112 Structure of South African groups, their holdings % and activities.

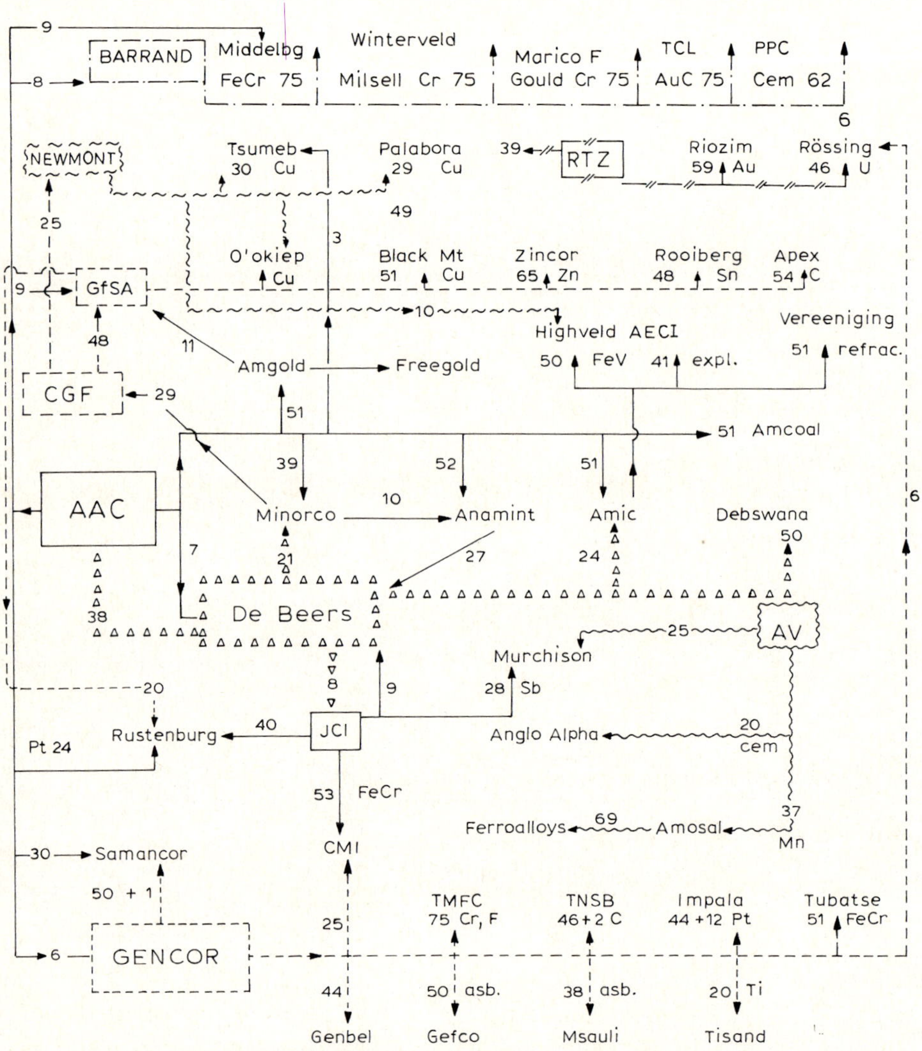

These groups diversified into finance, service and other industry.

Table 113 **Breakdown in % of group market value and income.**

	Market value			Income		
	gold + U	other mining	non mining	gold + U	other mining	non mining
Anglo American	50	20	30	40	25	35
De Beers	30	10	60	30	15	55
Johannesburg C.	15	50	35	15	40	45
Gencor	45	30	25	30	20	50
Goldfields	80	10	10	85	5	5
Anglo Vaal	40	20	40	30	15	50
Barrand	10	5	85	15	10	75

The mining finance houses

De Beers and AAC are each other's principal shareholders. The latter controls JCI and a third of Goldfields shares. Their investments (a) and income (b) are allocated as follows:

Group	Gold		Industry		Finance		Coal		Platinum		Diamonds		Other mining	
	a	b	a	b	a	b	a	b	a	b	a	b	a	b
AAC	50	42	17	20	10	11	7	10	5	4	5	7	3	-
De Beers	30	30	30	32	20	17	7	7	3	2	4	5	3	1
JCI	13	12	33	22	20	17	15	14	20	15	7	5	2	2

Industry comprises trade above, and finance insurance. AAC & Amgold merged their OFS mines into Freegold and they owned the following mine shares : I Administered by AAC. Elandsrand 48 + 23, Vaal Reefs 10 + 17, Western Deeps 29 + 13, Welkom 14 + 19, II by Goldfields : Deelkraal - 9, Doornfontein - 7, Driefontein 3 + 11, Kloof - 8, III by Anglovaal : Hartebeestfontein 7 + 16, Lorrain - 12, IV Barrand : Blyvooruitzicht - 20, V by Gencor : St Helena - 19%.

Vaal Reefs contributes a third of the profit of the AAC administered mines, Western Deeps 20% and Western Holdings 15%, Johnny's owns 24% of Randfontein, 33% of Rustenburg Holdings and dominates expanding PGM markets ; it also administers Murchison antimony. Coal plays an important role at JCI too with the Tavistock Collieries and Arthur Taylor, a joint venture with Total. Amcoal has 10 Bt of reserves, its 15 collieries produce 40 Mt, 2/3 of it bought by the state. Finally Amic controls the world's leading boart, vanadium and explosives firms.

GfSA has 70,000 employees ; gold mines are 80% of its business and it owns 30-34% of their capital (except for Deelkraal). Driefontein and Kloof alone represent 55 and 20% of market capitalization and profit. CGf owns another 10%

of Kloof and GfSA is the leading zinc, lead and tin producer.

Consumer products are the mainstay of Barlow RAND's activities, but mining ferro-alloys and cement (with lime) represented 15, 7 and 8% of assets, 6, 2 and 4% of turnover and 26, 1, and 13% of profit. 75% held TCL manages Harmony and the other goldmines of which it owns less than 30%. Nonetheless, gold accounts for 33% of income against 40% for the less costly collieries, which include 53% of Witbank and Welgedacht, 37% of Rietspruit (a joint venture with Shell) - and 4 Bt of export allocation. Further, 75% held Rand Mines manages the Middelburg coal consortium led by BP. Similarly, Barrand plays an important role in chrome, ferro-alloys and fluorspar.

In contrast, Anglovaal is equally involved in food and mining. The latter and building materials provided close to half of profits led by Hartebeestfontein. Furthermore, Anglovaal controls the second manganese producer ; it has shares in a dozen non-managed goldmines and a stake in antimony.

Gencor and the state

At the Afrikaner house, Gencor, owned by Federale Mijnbouw and controlled by Sanlam insurance, shareholders interests (and income) broke down thus : gold and uranium 45 (28)%, platinum 15 (10)%, coal 7 (7)%, others 8 (3)%. Although Buffelsfontein contributed two thirds of the profit of its gold mines ; Gencor held 100% of preference but only 9 + 2 of ordinary shares, compared to 25% of St. Helena, 21 + 21% of Winkelhaak, 32 + 6% of Unisel and 36 + 22% of Kinross and 69% of unprofitable Beatrix. The second figure refers to holdings of the Genbel subsidiary. Impala is the second PGM producer and Quebec Iron and Steel has a majority in the titanium smelter. In contrast, Gefco , Samancor and TMFC dominate asbestos and strategic ferro-alloys, manganese and chrome businesses. Finally TNSK has 13 collieries including two at its Blinkpan subsidiary (35% owned by AAC). The Government controls the power industry through Escom which allocates coal mining contracts and can buy 60% of all coal mined. The transport authority SATS shares out freight cars and port facilities. To these should be added the Yskor, Foskor and Sasoil triad. Yskor (or Iscor) controls steel , dolomite and 30% of coking coal capacity ; it mines quartzite and sand and in Namibia zinc, lead and tin. The korporasie employs 60,000 people, half of them white, and almost breaks even on R2-2.5B of turnover and R2B capital.

Sasol was founded in 1950 at Sasolburg, development of Sasol II at Secunda started in 1974 and Sasol III in 1979. Following its sisters, the government's 50% stake in Sasol III will also be divested. With a capital of R2B Sasol has a higher turnover and its net earnings have doubled to R400 M. Employment reached 30,000 in six collieries, which now supply 23% of coal output, the oil from coal, chemical fertilizer and explosives plants and marketing divisions.

The Uranium Enrichment Corporation and the Diamond Cutting Industry are other state enterprises like IDC's Alluvial Diamond Diggings on the Atlantic coast, Foskor and Soekor the oil company. Moreover, the state owns 60% of Premier and 70% of Finsch the leading diamond producers which it leases to De Beers. Last but not least taxes total 45-55% of earnings at the mining companies and 15-20% at the finance houses.

83 GOLD AND DIAMONDS

Diamonds and precious stones

The 1869 discovery of an 82 carat gem by a Griqua gave impetus to the development of what became South Africa. In the seventies diggers switched their search from ancient rivers to the pipes of Kimberley and by 1888 De Beers amalgamated their tiny workings. Production reached a million carats during World War II and grew from 2.5 Mct in the fifties to 7-8 Mct in the seventies. Sales were higher than R300M in 1981 and 1982 but lower prior to 1977 in current rands ; but they were in excess of 500 million rands in 1979, 1980 and 1983-85. The largest stones, the Cullinan and the Excelsior, some 3,000 and 1,000 carats, were found around the turn of the century, and another fifteen gems of 300 to 700 carats have been recovered in a hundred years.

There are three principal diamondiferous areas.

I. Gems show over 60 km of the lower Orange terraces. At Alexander Bay the state produces 100,000 carats averaging 0.7 ct per stone. It also leases some thirty sea concessions which extend 200 km into the ocean at Oranjemund. South of the Orange gems provided half of all South African diamond revenue. A number of workings cluster around Langhoogte and Kamaggas 40 km inland and three dot the shore over 250 km south to Hondekklipbaai (The 'dog's cliff'), Richtersveld, Tweepad, Bonte Koe (the 'checkered cow'). To these should be added Annex Kleinzee, Buffels Marine and Brazil. Strykloof is another 200 km to the south.

II. Some twenty mineralized kimberlites spread over 400 km between Postmasburg (E Cape) and Lesotho. Finsch mine, the western most of these was inaugurated in 1967. It is the principal producer of small gems. Extraction reaches ever deeper here, below 600 m, but grades appear to decline. Conversely, around Kimberley only remnants or dumps are worked. Rovic and the Star are other pipes in the Free State and

III. alluvial diggings cluster in three areas : around Kimberley, Beitsbank and Barkly West in the south, in the Schweizer Reineke district, west of Klerksdorp and further north in the Lichtenburg district near Botswana.

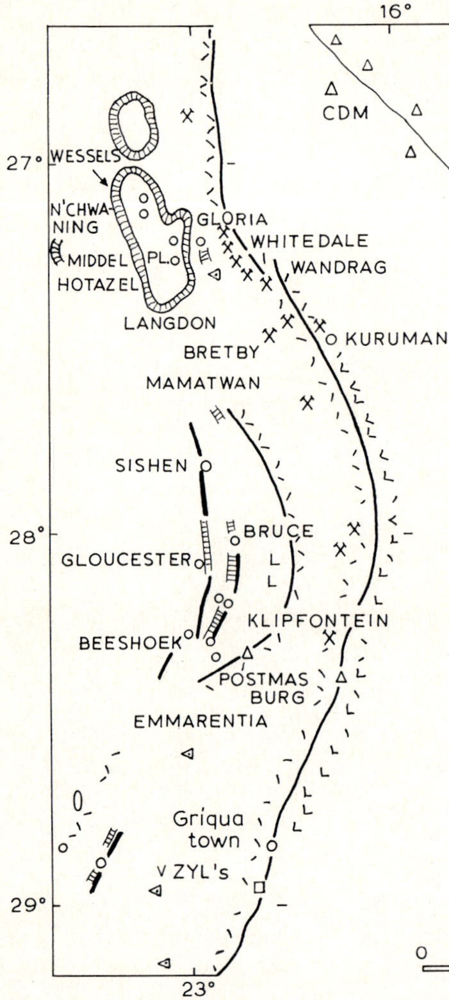

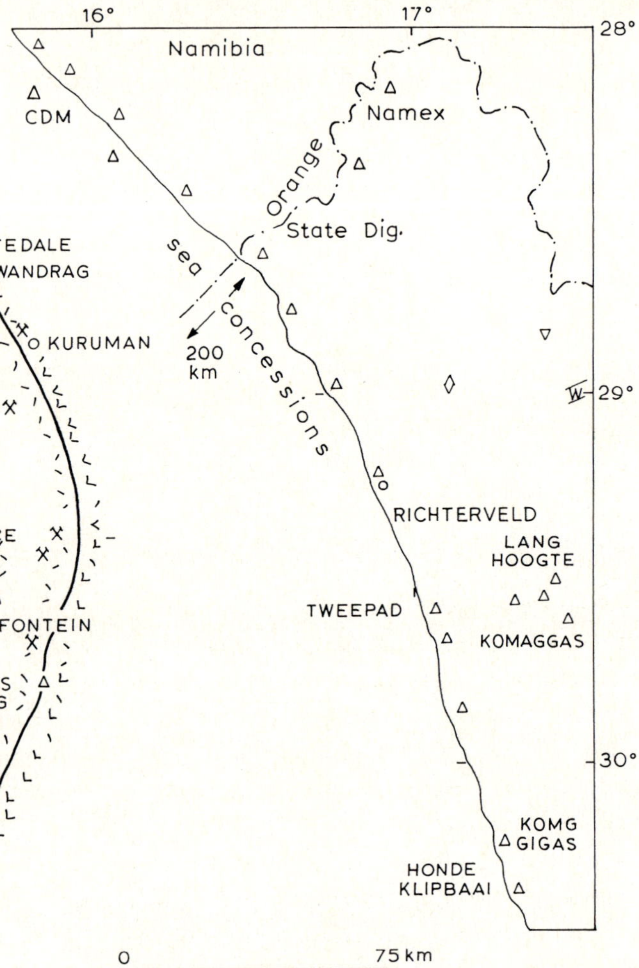

Fig. 59. (Left) Manganese, iron and asbestos. (Legend on page 13.)
Fig. 60. (Right) Diamonds.

IV. Premier, the world's largest underground diamond mine (S Tv) has been closed and reopened a couple of times. There are also other pipes in the Transvaal even far north at Venetia in the extension of the Orapa trend.

Emeralds from Gravelotte (NE , IV) are the principal precious stones, followed by Tiger's Eye. This quartz pseudomorph after crocidolite occurs at the southern end of the Cape blue asbestos belt. The best shows spread over 40 km at Niekerkshoop and the principal workings include van Zyl's, Sandfontein, Blouboskuil and Seekoenuis. Jasper is widespread further north and some other occurrences include :

Almandite and aquamarine : Jackalswater (NW Cape), amazonite : Honeydew (N), Kenhardt (W Cape), chrysoberyl : Kenhardt and Daberas, garnet : Bergview (N), heliodor : Soutpansberg (N), jade garnet : Rustenburg (W Transvaal), morganite : Steyns Puts (W Cape), verdite : Barberton (E Tv).

In effect, Namaqualand, Griqualand (NW Cape) and the northeastern Transvaal are a lapidary's paradise and collectors are referred for details to pages 613-7 of MRA.

Table 114 Ore, diamonds extracted (1000 t, 1000 ct) and grade (ct/100 t) in South Africa.

Mine	1000 tons	1000 ct	ct/ 100t	Mine	1000 tons	1000 ct	ct/ 100t	Complex tons	1000 ct	ct/ 100 t
Premier mine	5500	2000	35	Bultfontein mine	1200	350	30	Koingnaas 1500	650	40
Premier dump	1600	650	40	Bultfontein dump	670	300	35	Buffels 3400	200	6
Finsch	5000	5000	1000	De Beers	650	150	20	Longhoogte 150	30	20
Wesselton	1400	300	20	Dutoitspan	600	100	15			

Gold

Africans have worked gold for centuries, yet Walker and Harrison discovered the Rand only in 1886. Production reached 100 t in 1898, it plummeted in 1901 and grew from 200 to 300 ty between 1908 and 1923; another period of growth, up to 400 t in 1941 and 1958, was interrupted between these dates. Output peaked with 1,000 t in 1970, whence it declined to 650 t in 1981. It has been increasing ever since and 40,000 tons, more than half of the world's gold, have been produced to date.

The milling rate reached 70 Mty in 1960 from which it rose slightly to 80 Mty in 1980. But out of the current 150 Mty milled, 50 Mt are retreated slime dumps. Yields ranged around 7 g/t until 1954, peaked with 13 g/t between 1965 and 1971 and reverted to 7 g/t by 1980. Their decline to 6 g/t continues with higher gold prices and better technology in ever deeper shafts.

These mines are, indeed, a living tribute to the managerial and financial talents of half a million people engaged in this enterprise : 8-10 m wide shafts are sunk to more than 2,000 m depth. They are, likewise, at the forefront of excavation techniques including ventilation, dewatering, and refrigeration. In the mills machines crush and grind the ore, and up to 98% of the gold is recovered by gravity concentration, cyanidation and the carbon in pulp process. The bullion so produced contains 86% gold and 4.5% base metals.

Up to ten reefs, seams and channels have been mined in five gold fields along the northern and western rim of the Witwatersrand basin where they do not lie too deep. Eleven mines operate on the West Wits Line, 4 at Klerksdorp along with a retreatment plant, 10 plus 1 in the Free State, 12 plus 9 on the Rand and 4 in the Evander field.

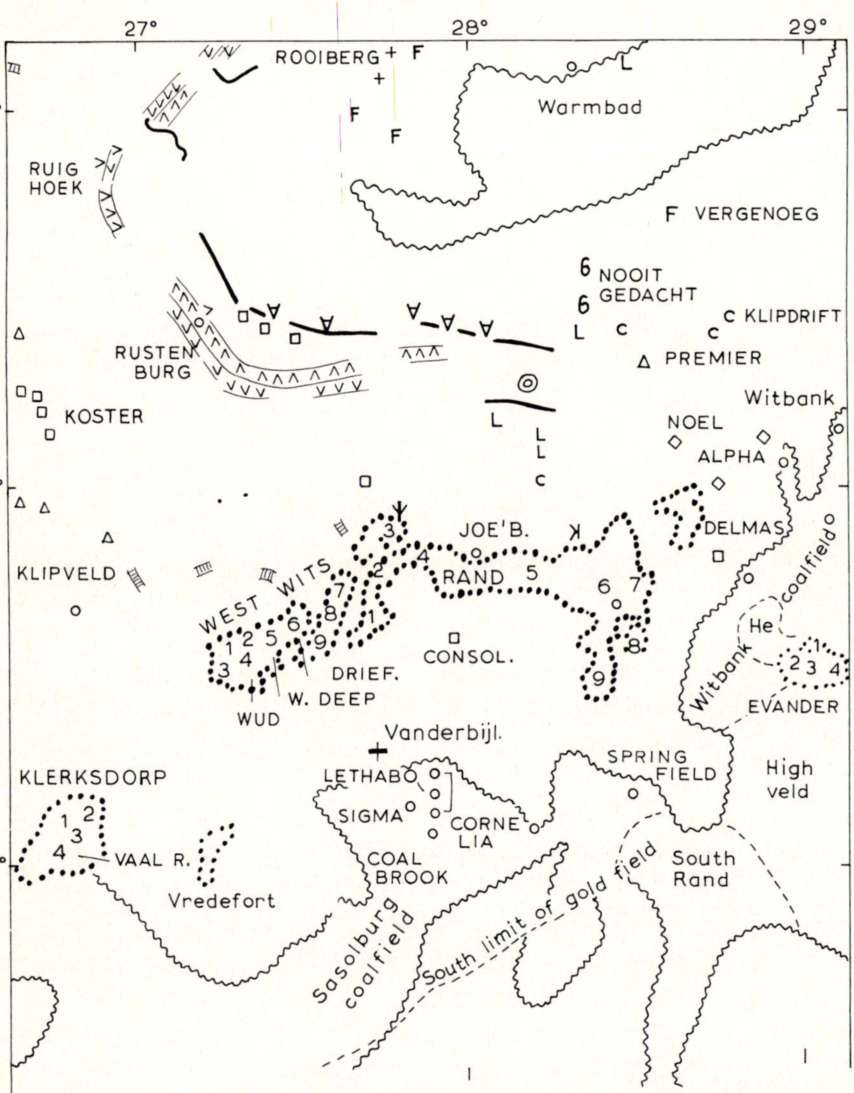

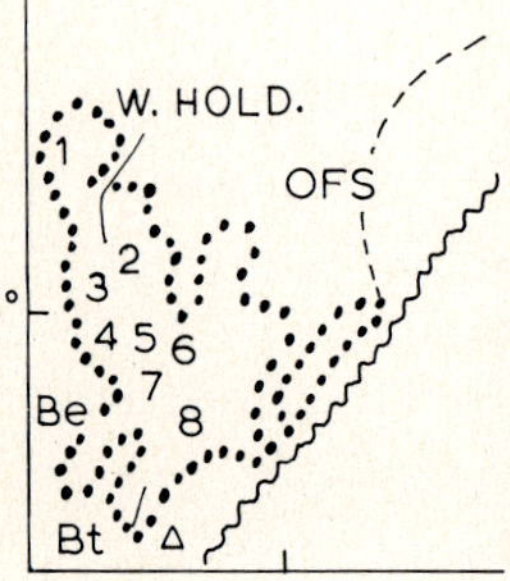

Fig. 61 The Witwatersrand and the southern Bushveld

OFS 1 Loraine, 2 F.S. Geduld, 3 Western Holdings, 4 St Helena, 5 Pres. Brand, 6 Pr. Steyn, 7 Unisel, 8 Harmony, Be(isa), Bt Beatrix. KLERKSDORP 1 Hartebeestfontein, 2 Stillfontein, 3 Buffelsfontein, 4 Vaal Reefs. WEST WITS 1 Doornfontein, 2 Blyvooruitzicht, 3 Deelkraal, 4 Elansrand, 5 Western Deep Levels, 6 Driefontein, 7 Venterspost, 8 Libanon, 9 Kloop. WITWATERSRAND 1 Western Areas, 2 Randfontein, 3 West Rand Consolidated, 4 Durban Deep, 5 East Rand Proprietary, 6 Ergo, 7 Grootvlei, 8 Marievale, 9 Witwatersrand Nigel. 1 EVANDER, 2 Leslie, 3 Bracken, 4 Winkelhaak.

Table 115 Production and grade of the principal Witwatersrand basin gold mines.

N°	Mine	t	g/t	N°	Mine	t	g/t	N°	Mine	t	g/t
A3	W. Holdings	40	4	F6	Driefontein	73	13	A4	Vaal Reefs	80	8.5
R8	Harmony	30	4	A5	W Deep Levels	40	11	V1	Hartebeestfont.	30	10
A2	F.S. Geduld	27	6	F9	Kloof	36	10	G3	Buffelsfontein	30	8.5
A6	Prd. Steyn	25	6.5	R2	Blyvooruitzicht	16	7	G2	Stilfontein	10	6
A5	Prd. Brand	23	6.5	F1	Doornfontein	10	6.5	Klerksdorp		150	8
G4	St. Helena	12	6	F8	Libanon	10	5				
V1	Loraine	8	5	A4	Elandsrand	10	5.5	J2	Randfontein	30	5
G7	Unisel	9	7	F3	Deelkraal	7	5	J1	Western Areas	18	5
A	Joint Metalg.Sch.	4	10	F7	Venterspost	6	4	R5	E R Proprietary	10	4
Orange Free State		180	5	West Wits		210	9	R4	Durban Deep	8	3
								G7	Grootvlei	7	4
G4	Winkelhaak	15	6	Abbreviations				A6	Ergo	6	0.3
G1	Kinross	13	6	A	AAC			G3	W.R.Consolid.	4	2
G2	Leslie	4	3.5	G	Gencor			G8	Marievale	1	3
G3	Brackon	3	3.5	R	Barrand			9	Witw. Nigel	1	3
Evander field		35	5.5	F	Goldfields			Rand		85	4
				J	JCI						
				V	Anglovaal			Numbers refer to Fig. 63			

The share of total earnings of the five fields is : West Wits almost 40%, Klerksdorp and OFS 24% each, and Evander and Rand 7% each.

The gold fields

The Orange Free State field is the preserve of AAC. In its southeastern part, Western Holdings, a combination of Welkom, Saaiplaats and Erfdeel, will contribute 27 ty of gold in the nineties; 30% of this $700M project is financed by Lonrho. Further, President Brand's expansion into Du Preez, Leger and Adamsonsvlei adds a third more capacity in 1986/7. AAC also retreats slimes in the JMS and explores east of Western Holdings and south of the field with GfSA. So does Gencor who operate Beatrix under the auspices of Buffelsfontein and mill Unisel's ore at St. Helena, where a new shaft extends minelife by eight years. South of Beatrix, Randfontein, JCI (45%) and AAC (10%), own valuable ground.

From 10 Mt tons of ore, 5 tons of additional reserves on Afrikander, and 4 Mt of residues and 3 carbon-in-pulp plants, the world's largest, Vaal Reefs, produces more gold than any country except for the U.S.S.R. One of its shafts reaches a depth of 2,340 m and its ultimate hoisting capacity will be 33 Mty. Also in the Klerksdorp field, Gencor's Stilfontein acquired ground from Hartebeestfontein where Anglovaal will treat lowgrade ore defined by radiometric sorting. South of the West Wits mines, AAC drilled Western Ultradeep Levels, JCI explored south of Western Areas and GfSA extends Kloof towards Fochville. Profitable West Wits is the fief of Goldfields whose Driefontein expanded into Texasgulf leases and ranks first in earnings and second in production. Second, at AAC's Elandsrand 200,000 tMo of ore will be mined in 1988. JCI's Randfontein extracts 450,000 tMo and the Doornkop section's capacity will increase from

500,000 ty to 3 Mty. Unfortunately though Western Areas requires intense refrigeration.
The Witwatersrand produced more gold than any other field, but most of the remaining mines are marginal. ERPM is to sink a shaft deeper, Grootvlei is better off, but most of the earnings came from uranium, i.e. Ergo's slime treatment which AAC expanded to Daggafontein, east of which Ergoli plans a new mine. Likewise, at City Deep Barrand reworks 200,000 tMo out of 45 Mt of dumps which assay 0.7 g/t and plans a new mine near Alberta south of City Deep and Crown. Further, interest focuses on the potentials of reefs which might be located lower than previously worked reefs in the mines.

Gencor started the Evander field, where higher prices added to Winkelhaak's reserves. Together with Kinross milling capacity exceeds 360,000 t and developing the 4 g/t Poplar deposit to the north will cost $280M.

The Vaal Reef shows in the west and the Dominion Reef has been worked in the Ottosdal basin west of Klerksdorp. Similar outcrops are widespread in Transvaal, Natal (Goedehoop) and KwaZulu. Gold also shows at Schweitzer Reineke (W Tv), in the Klein Letaba district and Pilgrim's Rest (NE Tv) including Manville and Malati. More important is Barberton (E Tv) : Mynwerk Beperk's (Gencor) Fairview with 1.5 ty and Anglovaal's Sheba, Agnew and New Consort. However, Centurion and Connat are smaller. Another 6 Mt of ore have been delineated on Makonjiwan and Imperial (grades >6 g/t). These mines produce 5 ty combined and the Bushveld platinum mines, Palabora and Murchison should be credited with another 3 tons.

84 ENERGY

Uranium and byproducts

Recovery ratios average 2500 pyrite (in situ 4000) : 25 uranium : 1 gold : 1/10 silver : 1/4500 osmiridium. The East Rand has been the world's leading producer of osmiridium which averaged 38% osmium, 35% iridium, 14% ruthenium, 12% platinum and 1% rhodium. Although hardly more than half of the osmiridium was recovered, production now reaches 300 kgy. Gold contains 9.5% silver and a third of the output comes from bullion. Demand for pyrite has increased until 1983 along with uranium extraction and production trebled in 20 years. AAC alone can produce 500,000 t of sulfuric acid in Ergo and 350,000 t in the JMS. In addition to these Sasol recovers acid and Palabora produced 100-150,000 ty (see 'copper'), but suspended recovery of 200 t of uranium concentrates out of 8000 t of resources which have been originally assigned to carbonatites. The

Dominion Reef has been credited with 70,000 t and higher in the sequence, the Witwatersrand 'Supergroup' with 410,000.
Each 200 m of the Pilanesberg in Bophutatswana contains 50 kt uranium which might become economical in the XXIst century. Likewise, Karroo sandstones and coal have been credited with 40 kt of reserves and 70 kt of additional resources. Other low grade materials include the blacksands of Richards Bay and Springbok Flats (see titanium), granites of the northwestern Cape, and the Mozaan formations in KwaZulu.

Table 116 South African uranium grades (g/t), production and capacity in 1000 t.

Co.	Producer	g/t	Prod.	Cap.	Co.	Producer	g/t	Prod.	Cap.
AAC	Jt. Metallurgical	0.15	700	1000	JCI	Randfontein	0.17	600	600
BR	Harmony+	0.09	500	1000	JCI	Western Areas	0.45	300	500
GM	Beisa+	0.56	500	500	AAC	Western Deeps	0.09	150	500
AAC	Vaal Reefs	0.23	2000	2000	BR	Blyvooruitzicht+	0.14	250	500
GM	Buffelsfontein	0.2	600	1000	Gf	Driefontein	0.14	150	500
GM	Chemwes	0.15	500	500	AAC	Ergo		150	250
AV	Hartebeestfont.	0.14	400	2000	GM	W Rand Consolid.			500

Out of the remaining 460,000 tons of resources, 190,000 tons are considered as low cost reserves ; including 2,000 t at Phalaborwa, 4,000 t in Karroo sand - stones and coal and the remainder in the Witwatersrand basin.

Since 1950, the Witwatersrand basin produced 100,000 tons of uranium. Half of the production now comes from the Klerksdorp field and 20% from the Free State. Production has reached 5,000 t in 1957, declined to 3-4,000 ty between 1964 and 1977, it oscillated around 7,000 ty between 1980 and 1984 when it started its current decline.

Oil, gas, and power

Even though the capacity of the continent's first nuclear power plant, Koeberg (SW Cape) can be doubled to 3.6 Gw, it meets only a small segment of needs. South Africa imports oil, it ships in contrast low cost coal and exports power to most of its neighbors. Indeed, coal contributes 85% of energy and 92% of power requirements. With the addition of Kendal (Kuthala) 4 Gw, Matimba, Lethabo, Duvha and Matla, 3.6 to 3.7 each, and the Tutuka and Majuba projects, 2 x 4 Gw, Escom's coal fired capacity is doubling to 50 gw. To these should be added the pumped storages of the Drakensberg and Palmiet (1 + 0.4 Gw).

Rather than relying on vast but intermittent supply from Cabora Bassa in Moçambique, South Africa harnessed more of its modest hydropower resources than any other country on the continent. Part of the Orange River has even been diverted into the Fish River through the Ruitlage-Steynsburg canal, simultaneously providing water for irrigation.

Table 117 Initial and ultimate capacity of South African power plants (Mw).

Orange River	Initial	Ultimate	Orange River	In.	Ult.	Fish River	Ult.	Fish River	Ult.
De Aar	11	45	Kakamas	1	1.5	Bedford	18	Cradock	9
Petrusville	12	18			1.1	Steyns-		Hofmeyr	18
Douglas	6	13			4.5	burg	14	Sondags R.	9
Augabries	4	7	Vioolsdrif	0.8	0.9	Middle-	6	Graaf Reinet	4
In(itial), Ult(imate)						burg	14		

Out of eighty, ten wells tested 3-50 Mcfd low methane gas in Mossel and Still bay (S Cape), two had 800 and 1200 bd of associated oil and the E-S4 bore flowed 2,600 bd of 36° API oil. According to Soekcr the condensate potential of the basin reaches 30 Bb and the FA-EM field has gas reserves of >1 Tcf out of which 150 Mcfd could provide 20,000 bd of fuel over 15 to 20 years.

The goldfields have produced more gas, however. Average composition of the Evander and OFS gases was respectively 96 and 66-90% methane, 3 and 7-27% nitrogen and 0.6 and 2-13% helium. In 1963 600,000 cfd and 40 Mcfd may have escaped from these fields, to which should be added 2 Mcfd of helium. The methane was used to heat boilers in the Virginia (OFS) uranium plant... and at Bongwan, southwest of Durban, up to 10 Mcfdy of CO_2 were compressed between 1926 and 1956.

Occurrences of pseudocoal, probably a bitumen, are scattered over 130 km near Merweville (CS Cape) and up to 1.50 m thick seams of torbanite, a shale, averaging 50 l/t oil are interbedded with lowgrade shale and coal, mainly along the eastern and northwestern rim of the East Transvaal basin, e.g. near Mooifontein, Ermelo, Wakkerstroom, Utrecht and Witbank. At Lichtenburg (SW Tv) peat has been burned and some lignite was quarried at Knysha (S Cape). South Africa is a leader in synfuels of which Sasol I plant produced 5,000 bd plus 35 Mcfd town gas out of 2 + 2 Mt coal. By now, Sasol I uses 6.5 Mt high ash coal and consumption of Sasol II and III is in excess of 30 Mty indicating a total potential of 100,000 bd. AECI considered expanding its methanol capacity from 400 to 2,700 bd. Likewise, Anglovaal and Utah Mining (30%, a subsidiary of BHP of Australia) had synfuels plans based on 1 Bt strippable reserves at Delmas.

Coal

Africans have used coal for centuries, but industrial production only started in 1859. Output reached 15 Mt in 1927, 20 Mt in 1940, 50 Mt in 1963, 60 M in 1972 and might pass the mark of 200 Mt. Exports will exceed 40 Mt by 1988, and should reach their ultimate limit of 80 Mt in the nineties with the expansion of the Richards Bay terminal, upgrading of Durban and Maputo facilities and additions of rolling stock. Export coal for power stations, all Waterberg (W Tv) coal and metallurgical coal is washed; all other raw coal is only screened.

A third is mined opencast, the remainder underground, mostly by the board and pillar method. Productivity averages 300 ton/man/months and three times as much in stripmining. South African coal is therefore among the cheapest in the world. Secunda contributes a fifth of the output, ten collieries, more than half and 40 collieries out of 90 supply 80% of production. Although thick seams underlie vast areas, 90 collieries mine them only along the shallow rims of the basin, which arch from north of Durban beyond Johannesburg. Out of 110 Bt of in situ reserves 40% are though in the Waterberg field on the Botswana border.

Table 118 South African coal reserves, in billion tonnes.

Field	in situ	rec.	saleable	Field	in situ	rec.	saleable	Waterberg	proved	ind.	20 MJ/kg prov.	20 MJ/kg ind.
Witbank 5	0.8	0.6	0.3	Klip River	0.7	0.5	0.3	Upper	18.5	8	17.5	-
all other	15.4	11.8	6.1	Vrijheid	0.1	0.1		Lower	14.5	8.8	8.2	4.8
Highveld 5	1.3	0.9	0.5	South Rand	3	0.7		Springbok		in situ		saleable
all other	15.6	10.1	0.9	Sasolburg	4.8	2.2		SW		2.5		0.6
Utrecht	0.7	0.5	0.3	Orange FS	8.9	4.9		NE		4		1

rec(overable), ind(icated), 5 : five seams

We might add the 30 Mt of the Guba seam of the Molteno-Indwe field (NE Cape) which are not yet commercial. Washed power station smalls average 25-31 MJ/kg, 9-21% ash, 20-12% volatiles and 1-4% moisture.

The quality of coal increases towards the east, where most recoverable anthracite reserves are :

- 280 Mt of $<$12% ash at Utrecht, Klip River and Vryheid, plus 200 Mt containing 12-20%.
- 150 Mt of the same category in KaNgwane (E Tv).
- 40 t (beneficiated) $<$ 12% ash Nongoma (KwaZulu, N Natal). Gencor's new colliery exports from here 875,000 ty with a lifespan of 22 years.

There is little straight coking coal and recoverable, beneficiated metallurgical in the bituminous coal can be summed up as follows : Highveld N°5 230 Mt, Witbank N°5 220 Mt, Witbank N°2 160 Mt, Klip River 160 Mt, Utrecht 100 Mt, Vryheid 40 Mt. Further, 150 Mt have been indicated in the Springbok Flats (C Tv) field and seven blocs of the Soutpansberg Field (N Tv) total 270 Mt ; they heat $>$ 27 MJ/kg.

More important, in situ resources of the Upper Zone of the Waterberg field might reach 3 Bt.

The collieries

Iscor's Grootgeluk, near Ellisras (Fig. 56), the first colliery in this Waterberg field, produces 2-3 Mt steam coal and 1 Mt coking coal to which another 1.3 Mt can be added at Durnacol. The latest acquisition, Hlobane, has a capacity of 1 Mt. Sasol's Secunda, Brandspruit, Middelbult and Twistdraai at Bosjesspruit, one of the world's largest underground collieries, produce more than 30 Mty and 70% of in situ reserves are recovered. Sigma is also expanding beyond its 6 Mty rate.

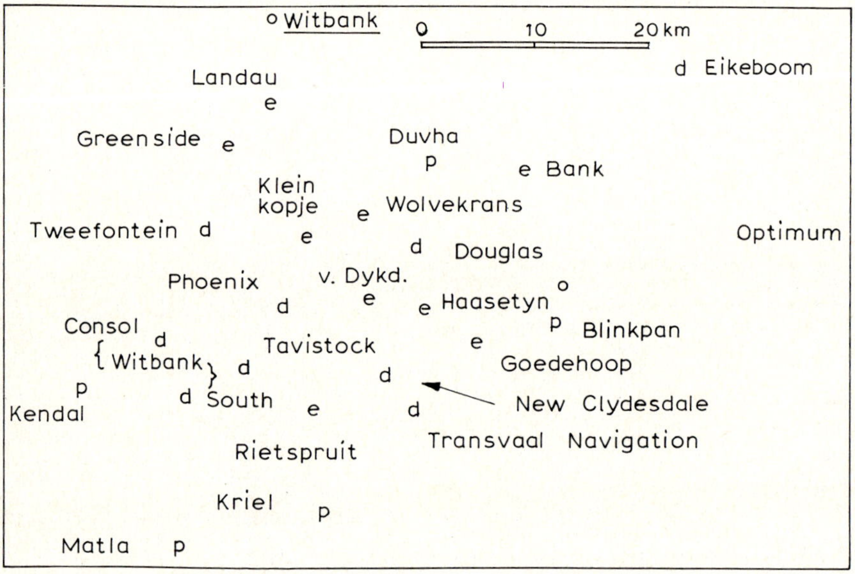

Fig. 62. Witbank collieries selling to e: export, d: domestic and p: power markets

Gencor's TNSK has 8 collieries in the Transvaal, 2 in Natal and jointly operate a ninth in the Transvaal (see MRA pp 184-8, 520-3). Matla is to expand to 10 Mt, the Optimum colliery will increase exports to 5 Mt and is to supply 8 Mt lower grades to Escom.

Barlow Rand's Duvha has also increased deliveries to Escom near Witbank. The opencast Rietspruit colliery, a joint venture of TCL and Shell, is one of the most mechanized in the country. Middelburg, another export venture with BP, has started up in 1983 and two new collieries will serve Escom's Kendal and Majuba stations by 1988 and 1989. To continue, Goldfields' Apex (2 Mt) is merging with Clydesdales' two collieries in Transvaal and Natal. Amcoal's (AAC) 13 collieries produce 35 Mt out of which 22 are sold to Escom, with Kriel shipping 8 Mt, Arnot 5, Springfield 3.5 and New Largo 0.7 Mt. New Denmark and Vaal start deliveries to the Tutuka and Lethabo plants with an ultimate capacity of another 25 Mt, but opening of the South Cornelia colliery has been postponed into the nineties. Other collieries, such as Kleinkopje (2-4.5 Mt), Goedhoop (2 Mt), Bank, Landau, Vryheid Coronation, and Natal Anthracite supply the trade or anthracite. Further, JCL's production is roughly 4 Mt at Tavistock, Phenix and Witbank.

85 THE METALS OF THE BUSHVELD

Platinum group, nickel,cobalt

The Bushveld holds much of the world resources of a half a dozen strategic metals. Most of these can be mined along its rims where the reefs crop out in 2 to 6 km long stretches which add up to 100 km within distances of 250 km at Rustenburg (SW), 150 km at Lydenburg (E) and 60 km at Potgietersrus (N). Resources are vast and platinum (palladium) and other PGM reserves have been estimated as follows in the Merensky reef : demonstrated 6,200 (2,800) 1,600 t , identified 10,000 (4,000) 2,600 t ; UG2 reef proved 15,000 (6,000) and 4,000 t. There are 31,000 tons PGM in three reefs to a vertical depth of 600 m, but the ore has been drilled as deep as 1,700 m. As shown in table 119, the proportion of metals varies :

Table 119 Average PGM content (g/t) and composition (%).

Reef/complex	g/t	Platinum	Palladium	Ruthenium	Rhodium	Iridium	Osmium	Gold
Merensky	5.3	61	25	8	3	1	0.8	4
UG2	6	41	34	12	9	2	1.7	
Platreef	2.9	43	48	4	3	0.8	0.6	
Bushveld		47	32	5	7	2	1	3

The Merensky and UG2 reefs are a meter thick each and Ni/Cu/Co/Pt/Pd/other precious metal ratios have increased to 2000:1200:20:3:1.5. Most of the latter are gold and ruthenium. Moreover the UG2 reef contains 41% Cr_2O_3 over 80 cm of thickness. Lonrho and Falconbridge have succeeded in dressing this seam at Westplats, where it contains 5.6 g/t PGM against the Merensky's 4.8 g/t. Since palladium costs a third or a quarter of platinum, RPM reopened Klipsspruit on this seam and GfSA planned to recover at Amandelbult through flotation 79% of 7.5 g/t, i.e. 7.5 ty platinum and 2.5 t palladium from 2.5 Mty of ore.

Production has increased rapidly from 3 t in the fourties to 16 ty in 1960, 40 ty in 1970 and should reach 100 ty soon, mostly on the Merensky reef. Although yields average only 3 g/t at Rustenburg grades of 9 g/t persist over stoping width of 70 cm and 100 m along the dips. The seams within the reefs are, however, only a mm to a few cm thick.

From north to south workings include RPM at Amandelbult, Elandsfontein, the Union section at Turfbult and Zwartkop, Impala's Bafokeng and Wildebeestfontein, RPM at Rustenburg and Westplat at Middelkraal. Impala's mines are in Bophutatswana and the others straddle the border. Finally Atok, an RPM subsidiary, operates in the northeastern Bushveld. All RPM concentrates are smelted and

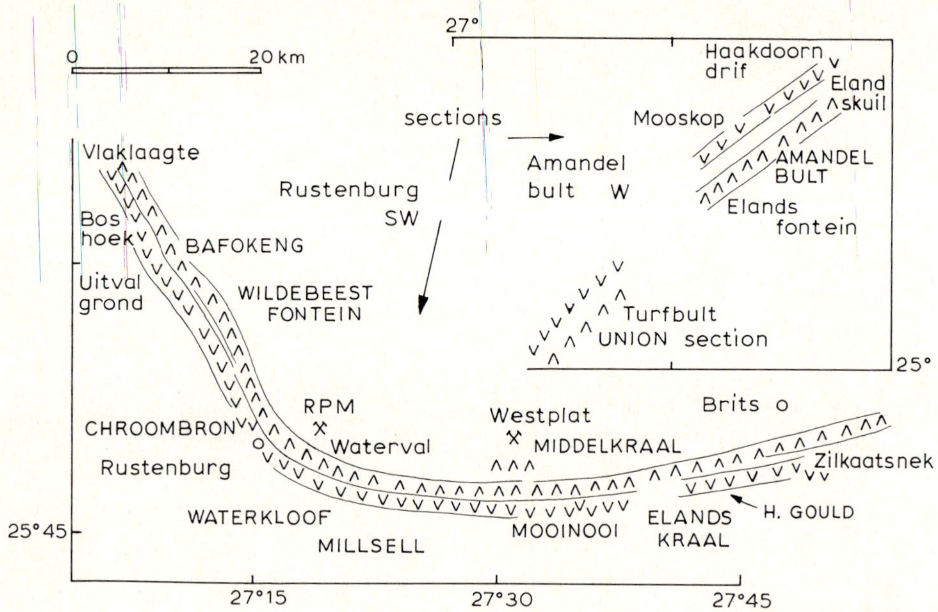

Fig. 63. Western Bushveld. (Legend on page 13.)

refined at Rustenburg, Impala at Springs on the Rand and Westplat in Norway.

Cobalt, nickel, vanadium

As a byproduct of platinum mining, cobalt follows the same pattern. The Matthey-Rustenburg refinery has a capacity of 500 t metal in sulfate, Impala produces 99.8% pure powder and Westplat whose cobalt-nickel sulfate plant operates near Rustenburg supplies 99% pure cathodes. The reefs might contain at least 150,000 t cobalt, but official reserves are a fraction of these. Further, attemps have even been made to work the Kuilsrivier gold-copper deposit in the southeastern Bushveld.

Based on milling grades of 0.19% Ni and 0.11% copper, the Merensky Reef alone contains 4.7 Mt nickel and 2.7 Mt copper to a depth of 1,000 m. To these should be added UG2 and the Plat Reef which contains the highest nickel grades. The Pilanesberg pipes in the western Bushveld is more accessible, but smaller. Different deposits are also known to occur at Vlakfontein and Volspruit in the northern Bushveld, and in a 40 km radius around Mt Ayliff between Lesotho and the sea.

Chromitites of the Bushveld average 0.2% V_2O_5, but the world's principal vanadium deposits are the titaniferous magnetite seams which dip 15-30° and crop out over 120 km out of 175 km in the west between Doornkop and Middeldrift

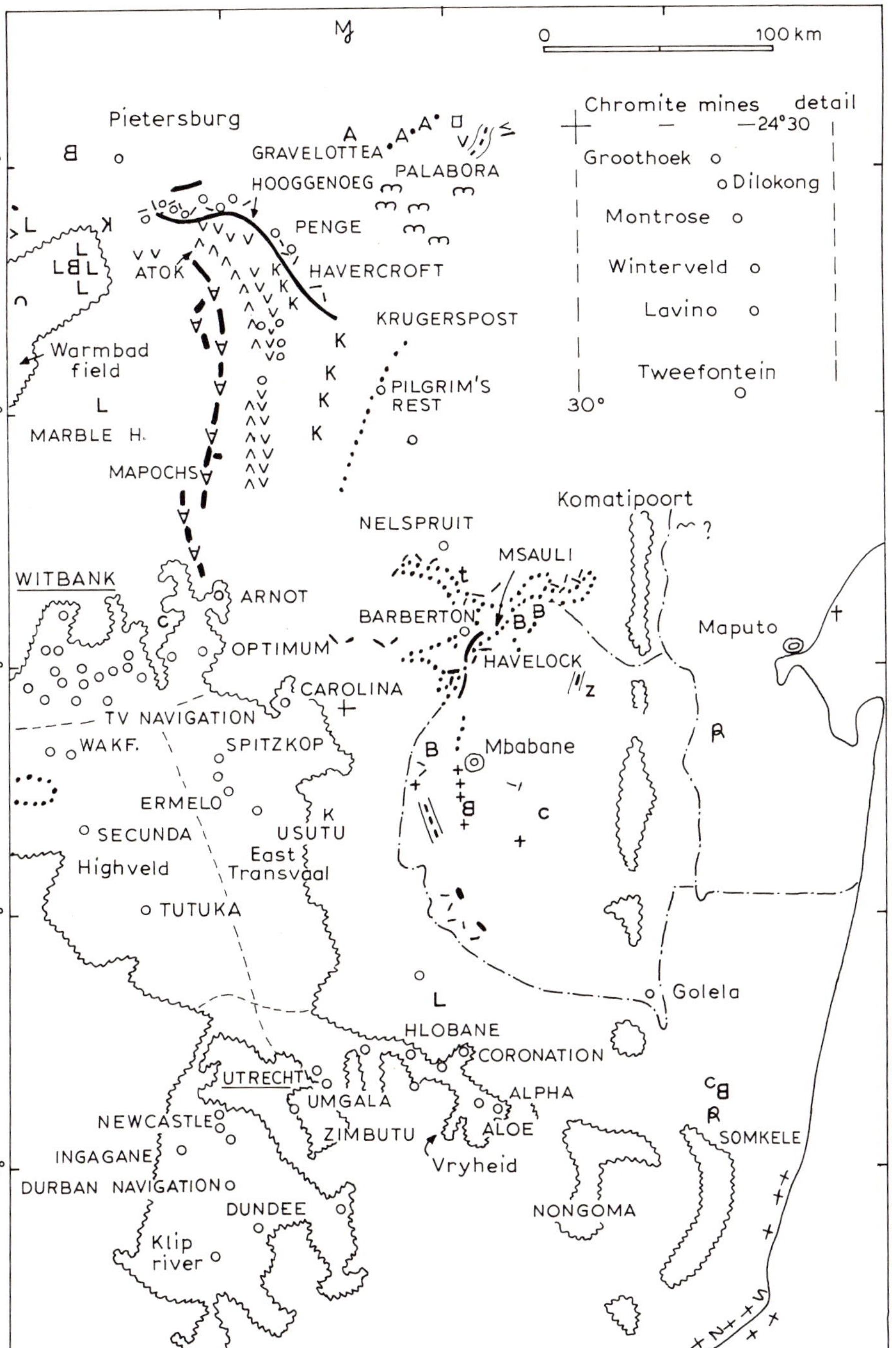

Fig. 64. South Africa, east. (Legend on page 13.)

(S) and over 150 km in the east. The latter are 1.2 m thick at Stoffberg and 2.7 m at Hoeraroep (NE) ; grade is stable at 1.6-2%. In the southwest, the upper seam contains only 1.65% V_2O_5 but it is 80-150 cm thick and in the north-west the two upper seams are 0.9-3.4 m thick and assay 1.2 to 2% V_2O_5. Their reserves to only 30 m depth represent a half of the world's total. There is also overall 50% Fe, 20% TiO_2 and 0.01% Ni. At Kennedy Vale (E), Highveld has mined a plug type deposit and the magnetites of Tugela (south of Swaziland) contain 120,000 t V_2O_5 (grade 0.56%).

Sales started in 1957, passed the mark of 20,000 t in 1970, 50,000 t in 1980, bottomed out with 30,000 t in 1983 and have been in excess of 40,000 tons. Highveld's Mapochs mine at Roossenekal (E) is the largest. Transvaal Alloys is further south at Wapadskloof and Vametco, the former Ucar, operates at Brits, Krokodilkraal and Uitvalgrond.

Chromite

Bophutatswana and the Western Bushveld (Fig. 63) also hold three quarters of the world's high chromium (Cr:Fe 2/1) resources to which should be added two thirds of the high iron (Cr:Fe 1.5-2/1) resources in the east, mostly in the Steelport sector (De Young et al. 1984) as shown in table 120.

Table 120 Exploitable (R1E), subeconomic (R1S) and extension (R2E) resources of the Bushveld in Mt with Cr_2O_3 grades (%) as produced.

	Area	Prod.%	R1E	R1S	R2E	other
Rustenburg-Zwartkop	W	43	390	920[a]	390	Pt
Nietverdiend-Lehurutshe	W	48	1	55	70	Fe
Nietverdiend-Marico	W	45	1	55	70	
Mankwe-Bafokeng	W	42	230	860	230	Fe
Steelpoort	E	44	370	1600[b]	370	Pt, Fe
Potgietersons	N	48		20[c]	14	metallurgical
UG2	E	marginal 38% Cr_2O_3				Pt, Fe
UG2	W		R1M 5420			Pt, Fe

Resources average 40-45% Cr_2O_3 except for a) : 30-40%, b) : 35-45% and c) 42-48%. Seams slope to more than 1,000 m of depth. Among a dozen, the 80-130 cm thick Magazine (LG6) seam is the most important in the western belt, which stretches from Marikana beyond Rustenburg and then bends towards the north. Generally, the 30 cm thick Leader seam is stoped together with LG6. Up to 30 mines worked underground the gently dipping seams ; they open and close according to demand and currently include from north to south : Rustenburg Minerals and Bathloko near Ruighoek (Samancor), Chroombronne, Waterkloof and Milsell (Barrand), Mooinooi (Gencor), Elandskraal (Chrome Chemicals) and Henry Gould. Further west, the high chrome ores of the opencast operations Nietverdiend ('not deserved') and AOM's Zeerust are often used in refractories. Samancor's Grasvally mine in

the northern Bushveld even exhibits a Cr/Fe ratio of 2.3/1.
The eastern arc extends from Deelkraal 120 km to Scheiding and supplies more than half of the output. There are only two, rarely three, important seams in the 1 to 1.8 m thick Main Zone south of Steelpoort. Mines include from north to south TMFC's Groothoek and Montrose, Dilokong inbetween; the largest chrome mine, Barrand's Winterveld; IMC (AAC) - Lavino's Grootboom and TMFC's Tweefontein. Earlier mines comprise TMFC's Kroondaal (S), Hendriksplaats, Swatkop, Tweelaagte, Bokfontein ; TCLE's Kroondaal (S), Driekop (E), Chrome Corp.'s Jagdlust (Union Carbide), to which might be added Metallurg (W) and Vereeniging Refractories.

Regular production started in 1924, reached 500,000 t in 1950, 800,000 t in 1960, 1.4 Mt in 1970, peaked with 3.3 Mt between 1977 and 1981, a level it should exceed in 1986 again, with increased exports of lump ore and ferroalloys.

86 OTHER METALS

Iron ore, tin

Even though tin has been mined from time to time at Van Rooi's Vley and Bokseput (NW Cape) at Kuilsrivier, Stellenbosch and Langverwacht (near Cape Town), and at Vlaklaagte (N of Pretoria) and Elandsfontein in the central Bushveld, the principal deposits are to the north. Rooiberg has been worked by the ancients and its production has doubled between 1960 and 1970. The country produced 6-700 tons between 1891 and 1910 and 5-600 tons yearly up to 1969. Tin in concentrate oscillates since then between 2000 and 3000 t. 1975 to 1981 averaged 2950 t and some 90,000 tons have been produced to date. Rooiberg smelts its own tin, so does smaller Zaaiplaats. Union at Nabomspruit exports concentrates and the Vanderbijlpark smelter reduces Namibian concentrates.
Production of iron ore increased from 3 Mt in 1960 to 7 Mt in 1970, reached 16 Mt in 1976, 31.5 Mt in 1981, bottomed out in 1983 with 17 Mt and ranged 24 and 28 Mt inbetween. Yskor's Sishen mine (N Cape) has a capacity of 25 Mty of three grades of hematite : high 66-70, medium 63-66 and low 63-60% Fe ; there are only traces of P and S and no Ti. The stripping ratio is 4:1 and opencast reserves based on a 60% cutoff attain 1 Bt. To these can be added iron ores next to the Postmasburg manganese deposits, 300 Mt on Sims, Gamagara and Sacha and similar underground reserves on the same farms plus Woon and Fritz. Yskor has also mined Amosal's Gamagara Hills deposit near Bruce and Lelyveld and several hundred million tons are available along the same trend between Demaneng and Beeshoek, at Welgevonden with smaller quantities at Carintoul and Boven-

zeekoubaart. All in all, Cape Province has been credited with 4.2 Bt of ore reserves to a depth of 30 m.

Secondly, Thabazimbi (W Tv) has a capacity of 5 Mty upgraded to 62% Fe and reserves of 100 Mt with a cutoff grade of 57%. The southern extension between Kwaggahoek and Donkerfontein is, however, discontinuous. Thirdly, Mapochs (see 'vanadium') has a capacity of 3 Mt with proven ore reserves of 100 Mt. Indeed, titanomagnetites of the Bushveld grade 55-57% Fe, 8-15% TiO_2 and 1.4-1.9 % V_2O_5 and 16 Bt of chromitite assay 20-27% ferrous oxide, yet Palabora has stockpiled 150 Mt titanomagnetite and sells only small amounts of this. Lastly, smaller deposits include Syferfontein (S Tv) 0.5 Mt, Deelkraal-Paardekop (SW Tv) 13 Mt, 43-52% Fe, and Elandsriver-Slaaihoek (E Tv) 46% Fe. Lastly, the Vergenoeg fluorspar mine operates for Dunswart the small Kroomdraai workings which have 4 Mt of reserves.

Manganese ore

Manganese ore output reached 500,000 t in 1940, 0.8 Mt in 1950, 1.2 Mt in 1960, 3 Mt in 1970, peaked with almost 6 Mt in 1975 and averaged 3 Mt since 1983 out of a capacity of 9 Mt. Eightyfive % of the production come from Samancor's six and Amosal's five mines in the Kalahari field (N Cape, Fig. 59) which holds most of the world's resources (De Young et al. 1984) : a) 15 Bt R1E,30-48% Mn ; b) 300 Mt R1E, $>$ 40% Mn ; 8.3 Bt R1E, $\leqslant$ 40% Mn ; 6 Bt R2E $>$ 40% Mn and 3.9 Bt R2E $\leqslant$ 40%. To the south of the scattered ore bodies of Dikgathlong, Sapsford and Leinster, a 6 m thick bed of 1 Bt averages 40-48% Mn and 10-20% Fe at Black Rock. Bottom ore grading 44-65% Mn and 18-6% Fe from Samancor's underground mine at Wessels is used for blending ; Hotazel is similar. Amosal works Gloria and N'Chwaning and the eastern belt extends to Annex Langdon and Lehating. In the stretch towards Mamatwan (SW) grades range only from 20 to 40%, where self-fluxing, calcium and magnesium rich ores average 38.5% Mn 4.5% Fe under a thin cover. This trend extends to Middelplaats, the deepest mine, and at least 1200 m along the dip. Only lower grade ores are left to the south in the 65 km connecting Sishen and Postmasburg with 15 Mt reserves grading 30-48% Mn and one mine : Gloucester. Ore in the eastern Klipfontein Hills is also inconsistent, but Samancor works the detritals of Lohathla in the western , Mokaning-Beesthoek belt. There are furthermore 5 Mt R1E and 10 Mt R2E assaying 35-39% at Bronkhorstfontein and Baden (NW Tv). Minor occurrences include the Krugerdorp area, and Dwarsberg (W Tv), the western Transvaal. Rand London is the principal producer of chemical grade ore here, to which should be added Roodepan, Klipveld and Viljoen. Samancor produces chemical grade ore at Smart in the Kalahari field. The only battery grade mine works the 1-5 m thick orebody 6 km NE of Gopane in Bophutatswana and the current breakdown of South African production

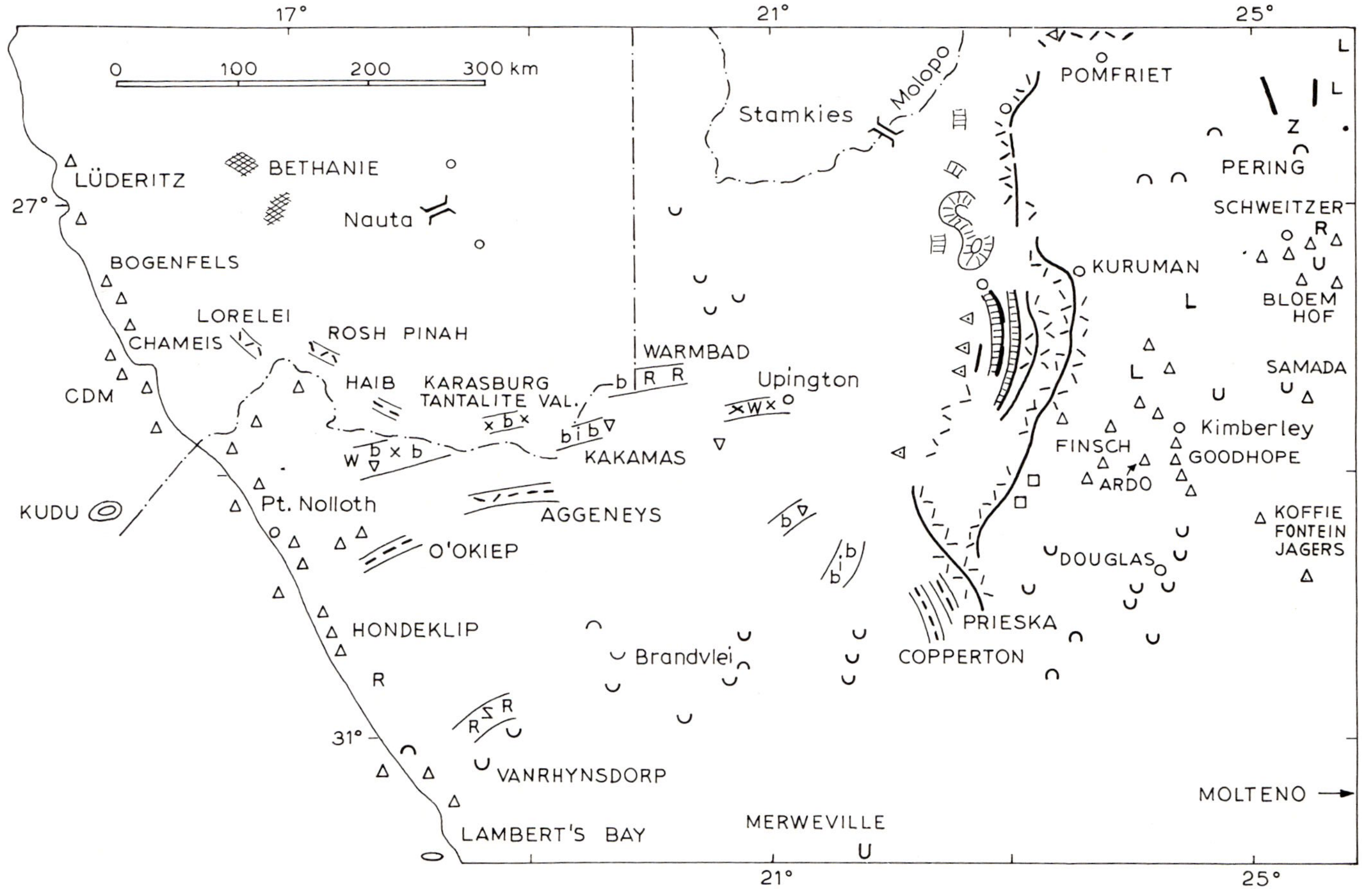

Fig. 65. Northern Cape and southern Namibia. (Legend on page 13.)

is as follows :

		Chemical grade	
Mn %	proportion %	Mn %	proportion %
30-40%	40	35	2
40-45	14	35-65	4
45-48	15		
48	25		

Base metals, silver, barite

The three reefs of the Bushveld average 0.1% copper, but the platinum mattes supply only 6,000 ty. Phalaborwa is thus the principal copper producer, the Gamsberg (W Cape) holds the largest zinc reserves and the Black Mountain group at Aggeneys (W Cape) comprises the largest lead-silver deposits.

Table 121 South African base metal (Mt), barite (Mt) and silver (kt) reserves.

	Phalaborwa	Gams	Broken H. Black	Broken H. Mt.	Big Syncl.	Prieska	O'okiep	Pering	Messina	Areacheap	Platveld	Maranda	Rand
Copper Mt	2.7		0.1	0.1		0.8	0.4		0.07			0.02	
Zinc Mt		10	0.8	0.9	1.7	0.2		1		0.07	0.06	0.15	
Lead Mt		0.7	1.7	1.8	1			0.2					
Silver kt			1.2	3	1	0.4						0.04	1.5
Barite Mt		3	24										

At Phalaborwa mineralization is persistent to 1000 m depth or more and PMC quarries 100 Mt, treats 30 Mt, mills 300,000 t and smelts 120,000 t copper. It also produced 130,000t sulfuric acid, 100,000 t magnetite, 5,000 t baddeleyite (ZrO_2) 150-200,000 t vermiculite, 20 t silver, some platinoids and has extracted in the past 150 ty uranium. Even though PMC mills ore grading only 0.5% Cu, it has a recovery ratio of 84% and makes 1.5% profit. Costs of underground mining contribute to the problems of the others who have better grades.
Both O'okiep and the Black Mountain group struggle with debt. Nonetheless, O'okiep provided 20,000 t copper and 5 t silver mostly from Carolusberg. Other mines of the 1500 km^2 field include Hoist and Spektakel. Orebodies of the Aggeneys group may extend even deeper and Black Mountain provides 80,000 t lead, 20,000 t zinc, 4,000 t copper and 100 t silver. The balance of silver comes from the Rand. Substantial reserves are left at Copperton (C Cape) following the closing of Prieska in 1986. Palabora smelts what is left at Messina (N Tv) and Shell starts mining Pering (NE Cape) opencast in 1986/7 and selling 33,000 ty of concentrates to Zincor. In addition to these, the Richterveld deposit (NW Cape) is said to hold 1 Mt copper (grade 1%) and the Argent mine has been occasionally worked. Finally, the Goudmyn (NW Tv) has been closed and most barite was mined opencast in the Barberton district (E Tv), followed by Blauberg (NW Tv).

	Gams	Black Mtn.	Broken H.	Big Syncline	Prieska	O'okiep	Pering
Zinc %	7.41	0.6	2.9-3.9	2	3.7	1	
Lead %	0.5	4.2	8.4	1			0.04
Copper %		0.7	0.4-0.6		1.8	1.84	
Silver g/t		26	60-100	10	7		

Antimony, titanium, zirconium, &

Between Gravelotte and Monarch Kop handspalled or cobbed and floated ore grades 2.8 Sb and production ranged between 10,000 and 15,000 t antimony in 60% concentrates, peaked in 1979 with 20,000 t, bottomed out in 1983 with 6,000 t and is expected to increase after 1987, when the Monarch East shaft will reach its ultimate depth. Some other occurrences of the Murchison Range (NE Tv), e.g. Monarch (East), United Jack, Weigel, and Shabari have been worked occasionally at ever increasing depth. Dumps contain 2% Sb and gold is recovered by press leaching and from arsenical tailings.

Some 20,000 tons of rare earth-thorium monazite is left at Steenkampskraal (W Cape). Similarly Rodewal nearby contained zircon and the Bothaville (NW OFS) - Wolmeransstad sandstones include 300 kt monazite and 15 Mt TiO_2. The Upper Zone seams of the Bushveld averaging 19 and 13% TiO_2 and 0.3 and 1.5% V_2O_5 also represent 320 Mt TiO_2. At Santa Lucia Bay (KwaZulu) there are 200 Mt of heavy sand and perhaps 100 Mt at Sandy Point and Umgagaba (Morgan Bay, Transkei).

The 40 Mt of resources of the old dunes north of Richards Bay (N Natal) are dredged in 2 km wide and 30 m thick stretches in artificial lagoons. Although maximum thickness may attain 100 m, the well bedded lower layer is only 5 m thick. The dredge serves the floating wet plant which rejects light quartz,&. Gravity, magnetic and electrostatic separation produces the concentrate. The sands average in effect 5% ilmenite, transformed into slag, 0.3% rutile, 0.65% zircon and 0.2% monazite plus magnetite which is stockpiled. Tisand ships premium, white 66% ZrO_2 ceramic zircon and a 65% ZrO_2 refractory product with a maximum of 0.3% Fe_2O_3 and TiO_2 for foundry sands. The world's only producer of baddeleyite, Phalaborwa, supplies 15% of the mass of zirconium minerals.Palabora produces 97-98 and 99% ZrO_2 grades. Further, FOSKOR sells 96,98,99 and 99.5% ZrO_2 concentrates.

Lastly, other occurrences include :
Aluminum clays Ngome, Weza and Bauxite (27% Al_2O_3, 20% Fe_2O_3, 5-8% reactive SiO_2, all in Natal), Arsenic Barberton (E Tv), Groenvly with Mo (W Tv), Beryl (NE Tv) : Gravelotte, Leeuwspruits with Nb-Ta, Selati, Lithium (NW Cape) : Noumas with Be, Ta, Norrabees, Spodumenkop with Ta, Columbite-tantalite Kokerboomsrand (NW Cape), Witkop (SE Tv), Niobium : Glenover (W Tv), Tungsten : Nababeep (N Cape), Mc Taggart's Camp (NW Cape), Zaaiplaats tin (C Bushveld).

87 BUILDING MATERIALS

Following the precious and sources of energy, building materials rank third among mineral industry products and among them cement contributes 40%, aggregate and sand a third, asbestos and limestone combined 10% of the value.

Cement and limestone

Table 122 South African cement capacity (Mt).

Group	Plant	Area	Pc	Mt	Group	Plant	Area	Pc	Mt
Pretoria Portland Cement	Slurry	(W Tv)	d	1.4	Anglo Alpha Cement	Dudfield	W Tv	d	1.8
	Hercules	(Pretoria)	d+sd	0.6		Roodepoort	Tv	w	0.32
	Jupiter	(Jo'burg)	d	0.4		Ulco	E Cape	w	1.2
Cape Ptd.	De Hoek	(W Cape)	d	0.9	Blue Circle	Lichtenburg	W Tv	w	2.2
E. Province	Pt. Elisabeth	(E Cape)	d	0.26		Whites	OFS	w	0.4
					National	Durban	Natal	w	0.26

Pc : process, d(ry),s(emi)d(ry),w(et).

Its northern Transvaal mill will operate fully in 1987, but Barrand's PPC group already constitutes 40% of total capacity, followed by Swiss dominated Anglo-Alpha managed by Anglovaal. The three companies share the 0.5 Mt clinker plant at Simana (S Natal) and Slagment near Durban. Further, Blue Circle UK has reduced to 42% its stake in its South African subsidiary. Obviously, all plants are coal fired. In terms of mass the cement industry is the principal user of limestone ; only 20% of the volume are burnt as lime, but they provide 60% of the value. A breakdown of the remainder is as follows, according to mass (and value) : cement 40 (70)%, metallurgy 10 (20)%, agriculture 7 (12)%.

The weight of limestone quarried doubled between 1965 and 1980, the peak year and its value increased sixfold. By 1983, output reverted to the 13.7 Mt of 1975, but the value trebled with inflation since the price increased by 20% last year. Growth is also expected to resume.

The best limestone deposits are in Cape Province at Steinkopf (NW) and Vanrhynsdorp, with marble nearby at Namaque, Piketberg (W), Saldanha Bay, De Hoek, Bredsdorp (SW) Oudtshoorn (S), and Alexandria (SE). The 240 km long Bromley-Danielskuil (NE) trend is, of course, nearer to the industrial hartland ; the Lime Acres, Ouplaas and Ulco quarries operate here. In the Transvaal limestones belong to the Chuniespoort series, with Lichtenburg, Slurry and Dudfield (SW), Glen Douglas (S), De Hoek, and Marble Hall (SC). PPC quarries 55% of all limestone at Loerie (SE Cape), Pienaarsrivier and Beestekraal (SC Tv).

Stone, aggregate, gypsum

While 'granite' and slate output changed little since 1975, their value increased fivefold in the sixties and threefold during the seventies. Marble producers saw their heyday in 1975 when they produced 20,000 tons. The same year, 300,000 tons of 'granite' and 50,000 tons of slate were sold in the whole country. In 1978, following the creation of Bophutatswana, the Republic supplied only 5,000 tons marble, 200,000 'granite' and 30,000 tons slate. A breakdown in tons (and value), in the Republic alone, shows 'granite 30 (70), shale 60 (4) and slate 6 (20)%. Most raw granite blocs and a third of slate is exported.

The best 'granites', i.e. gabbros, are quarried at Belfast and Colorado (SC Tv). They contribute 30% of the mass. Boschpoort, Impala, Marikana and Marlin supplied 65% in the southern Bushveld and Kakamas (CW Cape), Paarl and Clifts (W Cape), the remainder. Slate comes from Elephant, Mazista, Daradak, Keeloy, Scherman, and Sterling (WSW Tv). Twelve quarries of the Swartruggens contribute 70%, 3 at Marico 12%, and Koster the remainder.

Wonderstone, a metamorphosed ash, is exported to the U.S. from Gestoptefontein and Driekuil further west and the yellow ochre of Riverdale (Cape) is shipped to the U.K. Lastly the Buffalo River (Natal) provides most of the red pigment. On the other hand, other materials comprise :

Carbonatites : Phalaborwa (NE Tv), Spitzkop (CNE Tv), Cementrock: Hankie (SW Cape), Rietfontein (W OFS), Limestone : Warmbad (C Tv), Marble :Hennemann (OFS), Metallurgical dolomite : Glen Douglas, Mooiplaats (SW Tv), Quartzite : Pompey (NE Tv), Schweizer Reineke (SW Tv), Lekkersing (W Cape), Shale : Delmas (S Tv).

Natural gypsum output more than doubled in the sixties, increased by 20% in the seventies and by 40% between 1980 and 1985, to reach some 600,000 tons, when the value sold doubled. The cement industry takes three quarters of the production. Granaatbostolk (W Cape) can be expanded to 250,000 ty, Knersvlakte, De Duine, Zandveld, Bergplaas and Trekkersdraai are nearby. Others include Groensdoorns and Aries (NW Cape), Ysterfontein (C Cape), Gipsvlakte (E Cape), Olivenput, Windsorton and Rietpan (W OFS). To these should be added phosphogypsum, 4.5 t for each ton of P_2O_5. Reserves (in Mt) are distributed as follows : Bushmanland 46, Vanrhynsdorp-Vrededal 7, Cape West Coast 1.8, Prieska-Hopetown (C Cape) 1, Kalahari (N Cape) 24, Kimberley-Boshof 5 (E Cape, W OFS), Springbok Flats 1 (C Tv) and Calvinia-Karoopoort, Baroe-Wolvefontein, Bitterfontein and Cradock 0.6 combined.

Asbestos, vermiculite, mica

Reserves of extractable fiber have been reduced by 50% to 8 Mt but crocidolite resources are abundant indeed. Only 10% of this brittle, superior fiber is spinnable, but it attains up to 100 mm length in the east Transvaal field and Cape blue is up to 150 mm long. Amosite is unique, it regularly attains a length of 100 mm at Penge and more than 38 mm long white fiber is the most expensive. Production (in 1000 t) and sales (in million rands) developed as follows :

Weight kt	1940	1950	1960	1970	1980	peak	year	value	1960	1970	1980	1984
Amosite	20	50	65	100	60	100	1970	RM	6	11	21	24
Crocidolite	10	30	70	140	100	200	1977	RM	12	19	60	100
Chrysotile		15	30	50	110	111	1977	RM	3	4	20	25

The east Transvaal crocidolite field extends ENE over 90 km from Chuniespoort to Steelpoort, but at Penge, the largest mine, and to the southeast only amosite is left. Smaller workings include Gemini, Cork, Dalton, Tubex, Rooiwal, Kromdraai and Haenerstsburg. Further south, the principal chrysotile mines of the Barberton district were Msauli in KaNgwane, Kaapsehoop, Stella. Koedoe, Tubex, Mundt's, Kalkkloof and Joubertsdal were smaller or older and 300,000 tons of fiber were left at Stolzburg. - The Cape blue belt of Gefco (Gencor) extends over 400 km from Pomfriet (N) to Whitedale and thence to Wandrag, Klipfontein and Emmarentia, even south of the Orange. Older workings comprised Koegas, Breckenridge, Carrington, Strelley, Mt. Vera etc. At Wandrag Lonrho produced 10,000 t and other occurrences included Noumas (NW Cape), Umtwalumi (Ciskei) and Melmoth south of Swaziland.

At Palabora, the world's second producer, vermiculite is a byproduct of copper. Out of 100 Mt of material, 2 Mt are selected, half of these are milled and a ton of concentrate grading 90% is winnowed out of 5-7 tons of feed. Mica smaller than 35 mesh is eliminated. While golden yellow vermiculite expands 26 fold, western Loolekop material has a coefficient of 11. Production doubled between 1964 and 1974, peaked with 220,000 t in 1976, bottomed out with 150,000 t in 1983 and is increasing again. Additional reserves are available in the northern orebody and some vermiculite has been recovered at Nat Kruit in the corundum fields (NE Tv). To these might be added minor shows in the Glenover and Schiel carbonatites.

The 70 km long mica belt extends from the Macloutsie to Malelane via Phalaborwa (NE Tv). Pegmatite swarms are abundant here at Square, Morelag, Inyoku and Excellence, and also Noumas (NW Cape), Umtwalumi (Ciskei) and at Melmoth south of Swaziland. At Nxwala, nearby, 3 Mt of perlite expand 20 fold.

88 REFRACTORIES AND CERAMICS

Silica and clay

Sales of silica trebled in the sixties and again in the seventies to 1.5 Mt, a level to which they should revert this decade, following their 1982 low of 1.2 Mt. At any rate, the value of silica sold already increased by 50% in inflated rands between 1980 and 1985. The principal quarries are concentrated south of the Bushveld at Ladonna (SW), De Rust, Noel Lancaster and Spitzkop (SE) on 300 Mt of high grade glass quartzite stretching from Silkaatsnek to Breedtsnek. To these can be added another 40 Mt east of Delmas. Filter Sands operated nearby on the Rand and there are 9 Mt of foundry sand at Nylstrom (C Tv), Vanderbijlpark, Forstia and Witfontein (S Tv). Quartzites crop out further north at Potgietersrus (NC Tv); there are some 2.5 Mt of reserves at Weltevreden and silicon is produced at Potgietersrus (NC Tv).

At Viljonskroon 20 Mt of silica can be recovered from the only major deposit of the Free State and the best glass sands of KwaZulu and Natal are at Nseleni, Fernwood, M'bazwana and KwaNibela. Excellent silica bricks are made from the quartzites of Mossel Bay and Riversdale (S Cape) and O'okiep copper also ships some sand. Last but not least, at Philippi (SW Cape) 68 Mt of pure sand have been measured to a depth of 2 m.

A breakdown of clays according to weight (and value) is the following : flint clay 25 (42)%, kaolin 30 (35)% ; bentonite 4 (12), fire clay 40 (10) and fuller's earth 1 (1)%.

Sales of refractory flint clay of a bulk density of > 2.3 t/m^3 peaked with some 230,000 t in 1970, 90% of which were exported. They declined to 200,000 t between 1975 and 1981. Although higher than 100,000 tons, hardly 20% are exported now. Earlier estimates attributed 7 Mt to the Pretoria-Belfast belt where the principal pits include Premier, Nooitgedacht, Bronx, Klipdrift and Kopermyn. Bronkhorstspruit contributed 70% of the production, Pretoria 15% and Cullinan 10%. To these might be added 35 Mt of reserves and 200 Mt of resources of plastic and semi-flint clays on the East Rand.

Kaolin output has trebled in the seventies, dipped to 100,000 t in 1982 and should soon revert to its 1981 peak of 130,000 t. Strowan, Glenhoek, Crous, Collingham and Rainbow contributed in the Albany-Grahamstown area (SE Cape) 60% of this mass. Plastic clays at Albertinia and Riversdale are of better quality than in Mossel Bay (S Cape) and Serina near Simonstown (SW Cape) provided almost 15% of the output. These accumulations extend from Brakkloof and Nordhoek to

Stellenbosch. Brackenfell, Fishhoek, Nordhoek, Hendriksvlei, Kommetjie, and Somerset are in the western Cape, and there are also many shows in Namaqualand. Lastly, west of Johannesburg Premier Bricks and Haaskraal supply more than 15% of sales.

Ventersdorp has been credited nearby with a million tons of hydromuscovite vitrifier and Burgersdorp (ESE Cape) with 40 Mt of ceramic illite.

Bentonite reserves are smaller than those of other clays, but sales have doubled in the seventies to 40,000 t and have hardly exceeded 15,000 ty since 1982. Oceaan in the Koppies (N OFS) supplies some 70% of the highgrade output with the remainder coming from Swellendam and Cape Bentonite (WSW Cape) nearby. These shows extend as far as Heidelberg, Knysna and Pronksberg (SE Cape). Brown bentonitic fire clay accumulates at Lawley and Boksburg in the Transvaal and with some fuller's earth at Dwaalboom (NW Tv). Lastly, high grade attapulgite occurs in the Springbok Flats coalfield (C Tv).

Fluxes and industrial materials

Fluorite shipments increased by 70% in the sixties and trebled by 1980 to their peak of 520,000 t ; they bottomed out with 270,000 t and are expanding again. 89% of the output is of acid grade, 10% metallurgical and 1% ceramic. Fluorite is shipped in lumps and flotation concentrates are briquetted. Country reserves reach 90 Mt CaF_2 and the principal areas include :

I The 60 km long belt (SW Tv) with 90 Mt ore grading 15% CaF_2 comprising Barrand at Zeerust, Phelps Dodge at Rhenosterfontein and U.S. Steel at Marico. Others include Buffelshoek, Witkop, Wintershoek and Van den Heever.

II The Warmbad district (SC Tv) with its 60 Mt of reserves assaying 16% CaF_2 ; the largest mine is Gencor's Buffalo, followed by Morgenza, Tooyskraal, Zwartkloof (5 Mt, 14%) &

III Vergenoeg's (Bayer) Kroomdraai (C Tv) 20 Mt assaying 30% and the Rust de Winter hematites with similar grades

IV Rietfontein-Ruigtepoort in the Rooiberg tinfield (C Tv), 5 m thick veins at Wyndhoek (NC Tv)

V Kenhardt-Dyasonklip (NW Cape) and

VI Hlabisa, KwaZulu, 400,000 t, 35% CaF_2.

Andalusite production increased sixfold to the sixties, five fold in the seventies and peaked with 200,000 t in 1980. It bottomed out with 116,000 t in 1983 and by now exceeds again 150,000 t. Prices plummeted during the decline and have doubled since 1981 in rand terms. The principal deposits on Annesley, Streatham, Hoogenoeg, Havercroft , Holfontein, Morgenzoon and Rand London's in the northeastern Bushveld reemerge at Krugerspost (E Tv). Proved and probable (including possible) reserves attained per 3 m interval with a cutoff of 8% :

I Eastern Transvaal Magaliesberg stage 130 kt (700 kt), Daspoort 700 kt (1Mt), Timebal Hill 500 kt (550 kt).

II Western Transvaal Daaspoort, Andrafax, (60 Mt).

III Thabazimbi Timeball Hill stage (150 kt), Weedon Mines.

Sillimanite (and corundum) production mostly from Pella (SW Cape) peaked with 80,000 t in 1960, it has hovered between 15,000 t and 30,000 t during the seventies and declined to a 1,000 tons since 1983. There are also 2.5 Mt further south at Kamiesberg ; Niemöller's is to the NW. Moreover, the Ndandla deposit includes in KwaZulu 50 Mt ore averaging 43% kyanite.

Wollastonite resources at Lelie and Wilgehout fonteins (W Cape) and Keimoes (NW Cape) are in excess of 100,000 t and Apiesboomen alone has several million tons of magnesite with other occurrences in the Lydenburg belt (NE Bushveld) at Letaba (NE Tv), Soutpansberg (NW Tv) and Kaapmuiden (E Tv). Pyrophyllite occurs in the Khaki shales of the Witwatersrand and at Masala (SE Tv). Lastly, talc was quarried at Nqutu (KwaZulu) and steatite at Ngoweni.

Mtubatuba has 20 Mt of graphite reserves in KwaZulu grading 63°C and 30% ash along with anthracite. There is also graphite in the Soutpansberg field (NNE Tv), flake at Gumbu and amorphous material at Mutabe.

Corundum shipments have declined to 20 t out of 50-100,000 tons of resources between Messina and Mogalekwana, Blauberg and the Kruger Park and Leydsdorp and the Olifants river (NNE Tv).

Occasionally, diatomaceous peat is recovered at Bankplaats (SE Tv) out of 200,000 t of reserves, iceland spar at Kenhardt (N Cape) and soda at Zoutpan (C Cape).

89 FOOD AND PROCESSING

Water and food

The north western half of Cape Province has little water and the Karroo has few continuous aquifers. Originally, the Transvaal dolomites provided enough water for industry,yet by now even a canal bringing Zambezi water has been considered (see 'Botswana'). Meanwhile, the Bloemhof station pumps 1270 Mm^3 and Sterkfontein on the Tugela (Natal) might double pumping to 2700 Mm^3. Rerouting of the Mkomaas into the Mgemi has been surveyed, nearby. The Great Usutu's flow might be redirected towards the Vaal, the flow of the Komati has been regulated and Orange-Vaal diversion (850 Mm^3y) has been discussed under 'Energy'.

A series of 110 salt pans crosses diagonally the dry areas of Cape Province as far as its northeastern corner. Salt is also ponded at Port Elizabeth; production doubled in the sixties and increased by 20% in the seventies.

Small resources include Glenover (NW Tv) 2-3Mt P_2O_5 (grade 20-30%), Schiel (NE Tv) 1.5 Mt (5%), and the Saldanha peninsula (SW Cape) 3.5 Mt (9%) plus 1.7 Mt P_2O_5 (grade 6%) sold as Langfos. More important, Foskor exports a 40% P_2O_5 concentrate from Phalaborwa, where each 30 m of foskorite contains 1.7 Mt P_2O_5 in assays of 8.6% (cutoff 6%) or with an average of 7% 150 kt P_2O_5 to 100 m, 730 Mt from 100 to 600 m and 600 Mt down to 1000 m. The southern foskorite, to be mined from 1990 on, holds 80 Mt to a depth of 300 m, grade 6.5%, cutoff 5%, and the surrounding pyroxenite 34 Mt P_2O_5 per 30 m depth in assays of 6.9%. Moreover, apatite and tailings from PCL add 500,000 t capacity.

In 1980, the peak year, 20 Mt of ore provided 3.3 Mt concentrates containing 1.15 Mt P_2O_5 plus 11,000 t copper and 8,000 t baddeleyite, a zircon abrasive. Production of phosphatic fertilizer has doubled in the sixties and grew 50% in the eighties to a capacity of 480,000 t P_2O_5 at Fedmis' Phalaborwa facility, 400,000 t and 50,000 t at Triomf's Richards Bay and Potchefstrom plants, 40,000 and 20,000 t at Knoch's Modderfontein and Somerset West and 30,000 t at Omnia's Rustenburg operation. Exports reached 480,000 t in 1979 and are increasing. Triomf even reexports. AECI's ammonia capacity has expanded at Modderfontein to 1,000 td, based on 3,000 td of coal and Sasol also offers a full range of NPK fertilizers.

Processing

Power, petroleum refining, oil-from-coal, cement and fertilizers have been discussed under the appropriate headings. Nucor converts ammonium diuranate into uranium hexafluoride feedstock and UCOR is to enrich 300 tons from 1987 on.

The Rand Refinery supplies much of the world's 995°/oo pure gold. The South African Mint has struck some 1,500 tons of this into Krugerrands, mostly one ounce coins, sold by Intergold for the Chamber of Mines. Considering a 3-5% markup on coins over bars and the 22 carat fineness of the coins, minting of 100 ty or 3 million pieces ultimately adds \$ 60-90 M to their value, that is more than the mineral sales of thirty African countries. Platinum group metals are refined at Germiston, Springs and Brakpan on the East Rand. Minting of platinumrands started in 1984 and a fifth of the nickel recovered in the refineries are used by Southern Cross, Yskor and Union in the production of steel and alloys. Similarly, some cobalt is used domestically, 60% in cemented carbides and 25% in paints and ink. Palabora produces zirconium sulfate tetrahydrate from which hafnium can be extracted. Half of the copper it refines is processed into 70-80,000 t of rods, but casting has plummeted from 20,000 to 1,000 t. There are also small copper, zinc and tin smelters at O'okiep ; Springs, Vanderbijlpark ; Rooiberg and Zaaiplaats.

Richards Bay Iron and Titanium Ltd. processes and smelts ilmenite dredged by Tisand Ltd. into 85% titanium slag. Tioxide Ltd converts roughly a third of this into ten grades of pigment. Alusaf's smelter nearby imports alumina since 1971. Its production has doubled in 1983 to 170,000 t. More than half is exported and another expansion to 300,000 ty has been considered. Alustang transforms liquid aluminum, a third of the output, into electric rods, Hulett Metals produces aluminum alloys and secondary products along with Alumex, Non-ferrous Metal Works and Metal and Chemical Industries Ltd.

Antimony Products Ltd. transforms much of Murchison's concentrates into trioxide.

Production of lime more than doubled in the sixties and doubled in the seventies, stayed level afterwards at some 2 Mty and is increasing again. Unslaked lumps match two thirds of shipments, unslaked ground lime 20% and slaked 10-12%. Barrand's PPC is the leading producer of lime at Marble Hall. About 40% of locally sold asbestos is incorporated into buildings and flooring and a third into sewerage and drainage. Stucco manufacturers bought 20% of natural gypsum. Cullinan prepares a broad range of refractories and insulators and AECI, a joint venture of AAC and British ICI, is the world largest private explosives firm. Sasol also manufactures explosives and both make a variety of chemicals, Coalplex, a joint venture of AECI and Sentrachem, produces PVC from calcium carbide and coal and its Newcastle (Natal) subsidiary, Karbochem, is the world's biggest synthetic rubber-from-coal plant. Another project involves the production of activated carbon.

Steel and ferroalloys

A half to a third of iron ore produced is sold locally and converted into steel; further, a quarter of the steel so produced is exported. Steel output doubled in the seventies, ranged from 8 to 8.4 Mt between 1979 and 1982 and from 7.2 to 7.5 Mt since then. Exports peaked in 1979 with 2.7 Mt and exceeded 2 Mt ever since, and their value averaged 500 MR. Yskor's Vanderbijlpark (S Tv), Newcastle and old Pretoria works dominate the business. Dunswart is smaller and USCO, Scaw Metals and Mac Willaw treat scrap. Middelburg considered expanding its capacity of corrosion resistant 3CR12 containing 12% chromium to 2 Mt and numerous works supplied secondary steel shapes, rolled and cast products.

Highveld Steel and Vanadium Corp. smelts vanadium slag for export at Witbank (S Tv). Production ranged from 60,000 t in 1981 to 35,000 t in 1983 and some 50,000 by now. Transvaal Alloys at Wapaadskloof salt leaches vanadium pentoxide for export and Highveld's Vantra division reduces it into ferro-vanadium, also produced by Vametco (formerly Ucar) at Brits along with Nitrovan and Carvan.

South Africa is the leading producer and exporter of ferroalloys and Gencor's Samancor dominates this activity.

Table 123 Ferroalloy capacity and peak production.

	Steel	Ferrochromium	Ferromanganese	Capacity	kt
Capacity	kt 10000	970	750	Ferrosilicon	80
Peak production	kt 9100	900	660	Silicon	40
Peak year	1980	1985	1979	Manganese	40

Table 124 South African ferroalloy, manganese and silicon production.

Name	Group	Share %	Plant	Product	capacity kt
Ferrometals	Samancor	100	Witbank	high CFeCr	240
			Witbank	highgrade FeSi	35
Transalloy			Witbank	medium C FeMn, FeSiMn	
Silicon Smelters	Samancor		Pietersburg	silicon	42
Metalloys	Samancor	100	Meyerton	high C FeCr and FeMn, FeSiMn	40
			(Kookfontein)	milled and atomized FeSi	
Tubatse	Gencor	51	Steelpoort	high CFeCr	140
	Vametco	49			
Middelburg	Barlow	100	Middelburg	high CFeCr	200
Steel and Alloys	-Rand	100	Krugersdorp	also low CFeCr, FeSiCr	100
Electrolytic Metal			Krugersdorp	Mn metal	20
Ferralloys	Amosal	69	Machadodorp	high and low CFeCr, FeSiCr	110
	US Steel	31	Cato Ridge	high CFeMn, FeSiMn	
CMI	JCI	53	Lydenburg	high CFeCr	140
	AAC	25			
Rand Carbide	Highveld	100	Witbank	FeSi	55
Highveld	AAC		Witbank	SiCr, SiMn	
Delta Manganese			Nelspruit	Mn metal	20
				dioxide, oxide and chemicals	

The production of ferrochromium absorbed only a third of chromite sales until 1975 and was at par with exports between 1980 and 1982. The 1980 peak of 710,000 tons ferrochromium has been matched in 1983 and shipments may approach a million tons in 1985 with the installation of the new 20 MVA plasma arc furnace which treats fines.

Although ferrosilicomanganese sales are increasing, ferromanganese production lingers at 70% of capacity. The same holds for manganese metal where South Africa dominates export markets. Finally, ferrosilicon and silicon metal plants will hopefully reach full capacity.

9 ISLANDS AND THE SEA

There is more nickel, cobalt and manganese at the bottom of the seas than on land and oil exploration of the shelf may still surprise us ; the Repoblika Malagasy has graphite, and a sixth of the African hydropotential. There are, in contrast, few minerals on the smaller islands, but sophisticated populations enjoy a better standard of living.

9 1 MALAGASY REPUBLIC

Madagascar has such a variety of underutilized minerals that a wit called it a mineralogical museum .

By 1960 the Malagasy monopoly of crucible grade flake graphite has been lost and electronic technology reduced the premium on phlogopite mica.

Most sales declined since 1964 when the writer suggested some of the urgent

> GOALS : Agressive marketing of gems, graphite, mica and of the ferro-chromium project. Reopening of abandoned mines. Incentives for shale development, foreign investment and promising oil exploration. Drilling for water in the south-west.

Apart from the cement plant proposed in 1964, few products are made for local use. Infrastructure mostly supports the Highlands, Airmad serves the rest of the island.

The state owns Kraomita Malagasy and military Omnis controls energy projects. Since petroleum burdens a third of imports, international companies prospect for oil. French mined mica, Swedes mica and graphite, of which they imported 25%. The EEC took 50-70% of the commodities, France chromite, the U.K. and Germany graphite and the U.S. imported 20% of each.

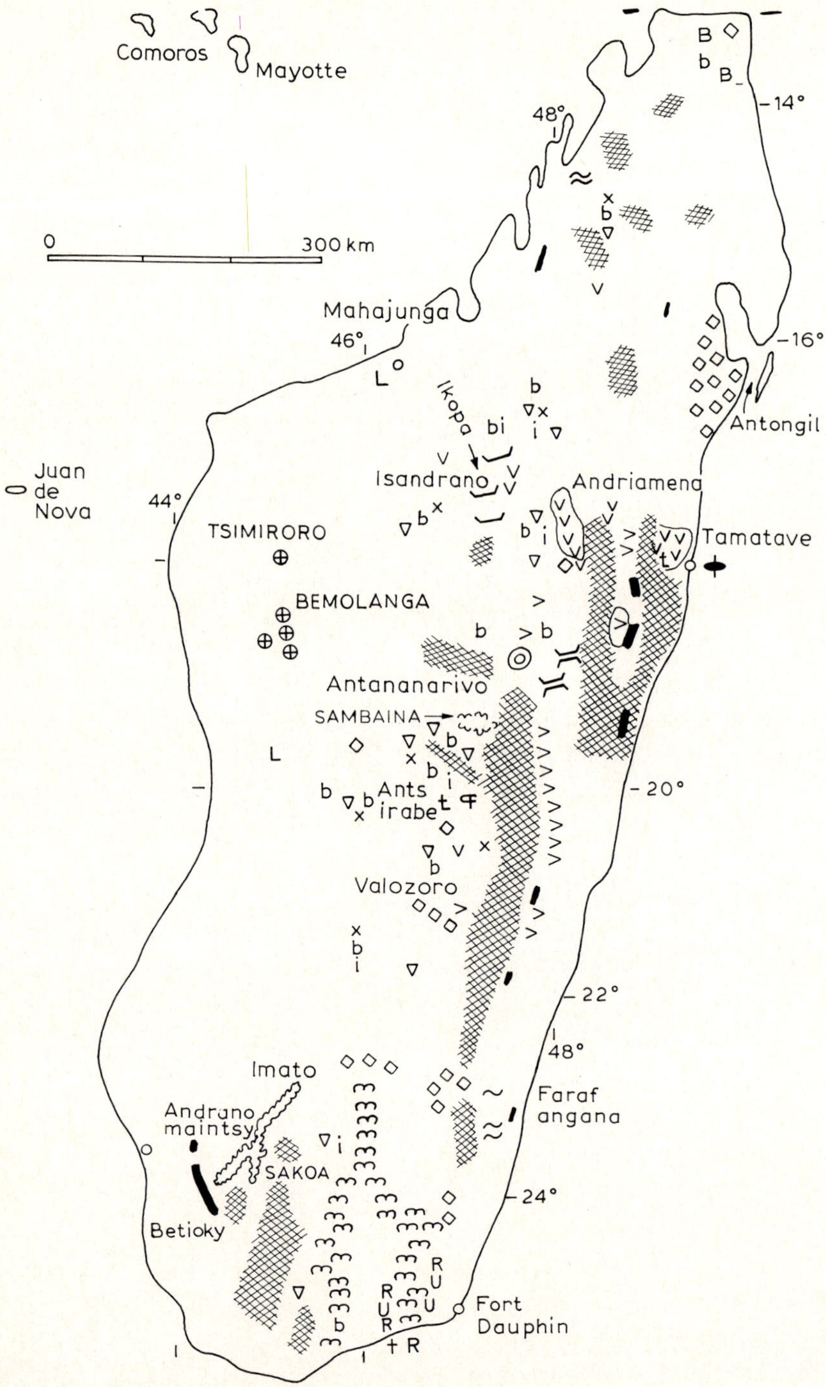

Fig. 66. Malagasy Republic. (Legend on page 13.)

Table 125 Malagasy production, reserves and resources.

Production	kg
Gold	4
Amethyst	10-30
Apatite	30-500
Calcite	0-2
Citrine	7-50
Cordierite	150-350
Gem garnet	5-60
Quartz geodes	50-60
	tons
Agate	10-40
Amazonite	0.7-3
Amethyst geodes	4
Aragonite	1
Beryl	5-65

Production		
Petroleum	Mb	3
Phlogopite	t	300
Beryl conc.	kt	10
Feldspar	kt	2
Salt	kt	30
Cement	kt	30
Calcite		1-2
Kaolin		2-3
		tons
Beryl		5-65
Celestine		20-30
Cipolino		10-400
Garnet ornam.		0.5-2
abrasive		5
Jasper		2-20
Labradorite		5-25
Quartz ornam.		1-20
Rhodonite		0.3-15
Rose quartz		40-80
Tourmaline		1-2

Production			Reserves	
Hydropower	Mw	90	Gw	64
Graphite	kt	15	Mt	5
Chromite	kt	50	Mt	44
Nickel	t	50	kt	100
Bastnäsite	t	20	kt	20
Piezo quartz	kg	50	kt	5
Uranium			kt	5
Heavy oil			Bb	4.8
Tarsands			Bt	3
Coal in situ			Mt	500
Lignite "			Mt	100
Bauxite rsc.			Mt	150
Barite			kt	500
Gypsum			kt	40
Resources				
Monazite			kt	80
Zircon			kt	160
Ilmenite			Mt	4
Iron ore			Mt	550
Copper			kt	6

Energy

The Malagasy energy problems epitomize the contrasts of underdevelopment. Hydropower potential ranks second in Africa, but only 0.15% has been harnessed including the mini- and microhydros now under consideration.

Table 126 Hydropower in Madagascar.

Developed	Mw	Ikopa	Mw	Site	River	Mw
Andekaleka (CE)	58	Mahavola	400	Firingilava	id.	540
Mandrakaka	12	Antanandava	300	Marokoloy	Andriamena	270
Antelomita	10	Antafofo	100	Ambondiroka	Betsiboka	210
Volobé (S)	5	Vohitsera	320	Vohombohitra	Betsiboka	120
Namorona	2	Isandrano	144	Manara	Betoafo	100
Manondona	2			Mania	Tsinjorano	100

Major sites are well located on the Ikopa (CNW), 160 km from the capital, but only Antelomita has been developed. Further, capacity of the Andekaleka plant can be doubled if required.

To continue, hot springs are aligned between the capital and Namoromona, at Tsarimpioky (SW), Ranomofana (SE),etc.

Between 1955 and 1963 CEA has recovered 4000 t of uranothorianite north of Ft. Dauphin (SE) and Amboasara in a 80 x 10 to 30 km area of the Mandrare valley. Some ore within these phlogopite rocks grades indeed 0.3-0.4% uranium and thorium, with a ratio of 1:2 to 1:12 and at Belafa the thorianite contained 13-26% U. Unfortunately, at Sakoatelo, the largest deposit, the proportions were unfavorable and few low-cost reserves were left in the smaller deposits such as Ambindandrakemba and Kotovelo.

A subsidiary of CEA and Péchiney (now PUK) has also recovered 1,750 tons of monazite between Antete, Vohibarika, and Ialanamainty (SE, see titanium), where dune sands contain 2-3% heavy minerals which include 3% monazite.

Bastnäsite, another rare earth-thorium mineral, occurs at Ambatofinadrahana and at Itremo (E). There are 120 t uranium between Antsirabe and Mandoto south of Antananarivo and the sandstones of Folakara (NW) are also radioactive. Finally, uranium accompanies the rare metals, e.g. at Ampasindava in the far north.

Coal and oilshale

The 3700 km^2 Onilahy coal basin stretches over 200 km east of Toléari, including from south to north the Beroy, Beomby, Ianapera (10 km, 7-10 m), Sakamena (10 km), Sakoa (10 km, 3 m), Savasy, Vohibory (12 km, 12 m) and Imaloto (10 km, two seams 3 + 2.5 m) fields. Figures in brackets indicate the length and thickness of seams; their potential reaches 1 Bt, but only the 3 to 6 m thick seam IV of Sakoa has been included in workable reserves of 60 Mt to 400 m depth because of the high ash content, i.e. 15% when washed with a 75-80% recovery. There are, further, 25% volatiles and 0.4-1% S ; coal heats, washed, 5700 cal/g and BP surveyed a 4-5 Mty $500M project. Earlier, between 1941 and 1957 when the Andranomaintsy colliery closed down some 30,000 t of coal have been extracted. By contrast, the Antanifotsy-Ianaborona lignite probably heats, nearer to the capital, 2900-3000 cal/g. There are 15-20% ash, 0.6-1% tars and 5-15% volatiles and reserves range from 30 Mt upwards. Neighboring shows include I. Anjoma, Antoby, Tanato, Ankararana, Tsimahabeharona, Andakandrano and Ambodivona II. The Ankazondrano, Mandrosohasina, Andranokely and Analambelatra seams surround the Sambaina plain nearby. Fossile peats are particularly abundant at Pilanona and Antanifotsy and 26 Mt bituminous shales contain 9.4-15% of light, low paraffin oil, mostly at Soamandrariny-Anjoma and Mandrosohasina.

The Isalo formation covers much of the Mahajunga (NW) and the larger Morondava basins (W) and along the Tsimiroro and Maroboaly highs (40 km southeast of Ankazobe, ESE of Maintiriano) 4.8 Bb of heavy oil (5-8° API) have already been delineated. Exploration continues in these basins, further north and in the Mozambique channel. 55 km SSW of Tsimiroro, at Bemolanga 200 m thick sandstones cover 360 km^2 ; 130 m of this column average a permeability of 430 millidarcys and 5.3% bitumen (1-8%). Highest grades, 5.7-15%, concentrate in the top quarter, but the composition of derivates hardly varies : lamp oil (kerosene) 25%, light tar 24%, heavy tar 21%, coke 8%, and gasoline 7%. The same sedimentary basins surround most of Madagascar and oil companies favor the area between Morondava and Toliara (SSW) for exploration. Meanwhile, the product mix of the

Tamatave refinery (E) has been further expanded.

Graphite and Phlogopite

For centuries, the Malagasy used the black pigment, graphite ; industrial exploitation started in 1907, and reached its peak of 40,000 tons by 1917. The grades were low and only since 1927, when De Robillard designed his first flotation cell, have miners produced graphite containing 85% carbon. Before that date and later, there were many fluctuations caused by overproduction, lack of labour and small scale operations. By 1947, miners had invested, mechanized and concentrated their operations near Tamatave harbour and thus reduced the cost of exploitation and freight. Yet, by 1960, sales declined, out of more than 30, only eight operators remained in business in 1964 and three firms, Gallois, Swedish SMGI and Izouard now supply 50, 30 and 10% of the output. Meanwhile, resources remain vast, not only west and south west of Tamatave, but in the whole Graphite System which is the backbone of the Eastern Highlands, covering several hundred kilometers south of Antananarivo and near Ampanihy (S). Indeed, groups of 3 m thick layers containing 3-6% graphite extend for kilometers, but miners can only quarry weathered, soft rock, e.g. in the Sahanavo Valley, at Ambatomitamba, Marovintsy, Sahamamy, Antsirakambo (Gallois), Ampangadiatany and Faliarano (Izouard).

How do they exploit this mineral ? Water carries the ore, excavated by bulldozers, towards the plant where it is washed, crushed, floated, screened and finally dried. For each ton of ore containing 3-6% graphite, miners need ten times as much water, and they install for each ton per month of graphite 1-2 kw of power. In contrast to graphite, mica is restricted to a narrow triangular area of the south . Rossi discovered the first phlogopite deposit in 1912, but only the founding of SMGI (and of the Union des Micas) in 1926 gave impetus to the industry. SMGI now supplies some 90% of the output, STO and lower grade BSI from Benato, the largest deposit, and VSP, DSP and FP grades from Ampandandrava. The Hilon brothers work Sahamasy' and older workings included Mafilefy, Marovala, Morarano, Sakasoa and Vohitramboa.

Chrome, nickel, iron, bauxite

In 1976, 200,000 t of chromite fines have been trucked by Ugine (now PUK) from Andriamena to the Vehidiala spur of the Tamatave railway. Following nationalization output declined to a quarter and stabilized in 1982 at some 80,000 t. Even though hydropower could be generated at Vohombohitra and plasma technology facilitates the use of fines, the Moramanga ferrochromium project has been in abeyance for 20 odd years. Some 300 pods show over 70 km between Bemanevika, Andriafia, Antsairana, Amboasary, Telomita and Andranomadio to Bepilopilo in the

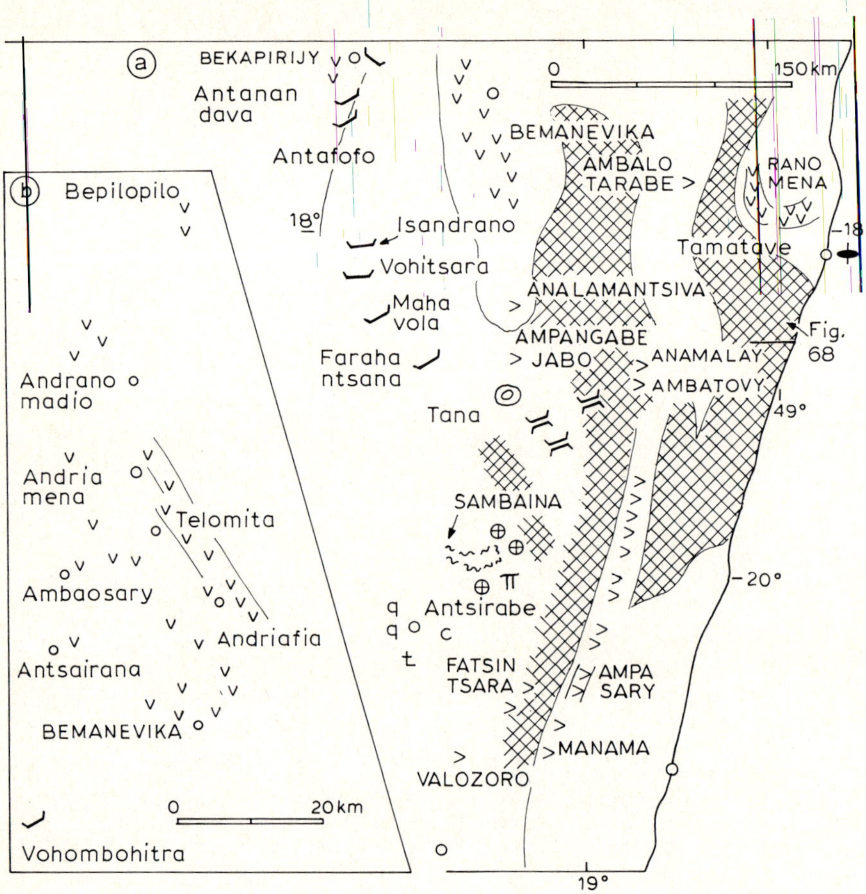

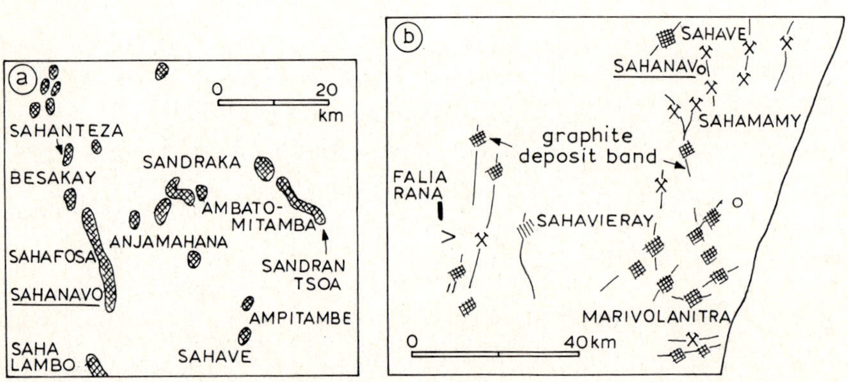

Fig. 67. (a) Malagasy Plateau. (b) Chromite. (Legend on page 13.)

Fig. 68. (a) Graphite. (b) Southwest of Tamatave. (Legend on page 13.)

north and the belt is up to 30 km wide. The small Bekapirijy deposit is further west and there are 250,000 tons in the Ranomena-Befandriana belt 30 km north of Tamatave.

Table 127 Malagasy chromites.

	Reserves Mt proved	probable	Cr_2O_3 % mined	benefitiated	ratio Cr/Fe
Bemanevika	1.3	1.6	42	50-52	2.2-2.75
Remainder	-	2	37	48.5	2

	Bekapirijy	Andriamena	Maevatanana	Ranomena
Cr_2O_3%	30-60	38	26-30	42-48
Cr/Fe	0.5	2	0.7	1.1-2.2

The cutoff grade is here 30%, grain size averages 0.3-0.4 mm and the 'remainder' contains 30% silica which can be reduced through dressing to 2%. Recovery was expected to reach 70%. Out of a total of 1.5 Mt of resources, the northern nickel deposits comprise nearby :

Ambalotarabe 3,000 t metal, 600 x 300 m, laterite 1.7%, serpentines 1.2-8% Ni,

Analamantsiva 12 x 2.5 km, 0.2-0.3% Ni in laterite.

Ampangabe-Jabo 30 x 60 m, 0.2-0.3% Ni in hard peridotite.

Ambatovy-Anamalay 500,000 t Ni, 0.16% Ni in clays, 0.5-2.5% in hard rock ; 40 Mt iron ore, 20-30% Fe.

Valozoro (E) 60-70,000 t Ni, Ø 700 m, 1.2% Ni in lateritic clay, 1.5-2.5% in hard harzburgite.

The Ampasiry belt of nickel deposits connects Moramanga with Manama and Valozoro with some 30 nickeliferous bodies. Lastly, the Fatita hydropower site is at a distance of 60 km.

Nearby, the Fasintsara deposit contains nickel and 30-60 Mt of iron ore grading 36 and 34% Fe respectively in the lower and upper zone. At Soalala (NW) 3-400 Mt of ore grading 48% Fe can be upgraded to 60%. There are also 100 Mt of iron ore at Betioky (S) near the coal field and up to 5 Mty of bauxite could be shipped out of 70 Mt of laterite on the Manantenina-Esama plateau (SE, 40% Al_2O_3), and we can add to these 40 Mt on Ankarana and Ampefivato, 120 km to the north. There are another 40 Mt at Manakara nearby (38% Al_2O_3) but the Marangoka and Analavory laterites (NW) are subeconomic.

Other resources

The west and southwest are sub-arid, but at Ankatrafay, water rises from a depth of 100 m, at Erada from 110 m and at Ambatolany from 30 m. Toléari (SW) and Bezaha are also supplied by artesian water; south of Antsohihy water was reached at 120 m depth in sandstones, but the rest of the Karroo only yields 1-2 m^3h. In contrast, the Morondava limestones (W) can provide 250 m^3h, artesian pressures build up at Mahajunga and the Cretaceous sandstones of Marovoay-Boboka (NW) also contain abundant water. Unfortunately, south of Befandriana, the Santonian sandstones are at a depth of 900 m. Lastly, the wells of Sambava (N) and Antalaha (NE) are in alluvial sands and coastal sands yield 2-5 m^3h.

Strongly weathered porous soils and greybrown soils of the grasslands surround east and west the laterite plateaus of the Red Island. There is also ample rain in the highlands and Madagascar had a surplus of rice in the past...

Salt is recovered through ponding mainly at Diego Suarez (N), where limestone is also widespread. These layers extend towards Soalara (see 'iron') and Amboina (NW, Mahajunga), where a cement plant has operated since 1956. There is also pozzuolana here and near the Ibity cement plant (C), which operates at less than capacity, but could be expanded to 180,000 ty.

Moreover, 50,000 and 12,000 t of calcite and cipolino were quarried in 1977. In contrast, recovery of some ornamental and gemstones has increased and the reader is referred to pp 208 and 614-7 of MRA. Let us only mention here the beryls, aquamarine at Marijao and Tongafeno, sapphire at Ankaratra, morganite at Sahatany and Anjanobonoina and with emeralds, indigolite and rubellite at Androfia. The Malazialina (C), Analava and Berere fields produced the famous 7 m long beryls. Together with Vorondolo, they were also the principal source of tantalite-columbite and Befanamo (Berere) supplied 30 kg scandium. Yet, because of Brazilian competition recovery of piezoelectric quartz plummeted at Ramartina, Antambohelibe (C), Antongil (NE), Harombe, Tsivory (S) &. Beforona and Ejeda provided most corundum and Besosa, Ianatrata and Ampandramaika hold most garnet.

Most gold -3 t - was washed in 1909 ; Andavokoera (N) was the principal producer and other occurrences include

Barite Andavokoera, Ampandrana, Bismuth Ampopihaonana, Andranagoaka, Belalala, Copper Vohibory, Vohémar, Dolerite Mahatsinjo (E), Glauconite (K) Analevalona (SW), Granite Ambatomara (C), Gypsum Ankay (NW), Kaolin and alum Ampanihy, Kyanite, sillimanite Manajary (SE), Andavabato, Lead and zinc Besakay, Manganese Bepeha, Latona (S), Mineral water Antsirabe, Namarona, Molybdenum Mafilefy, Talc Sahatany, Titanium, Zircon Tamatave (E), Marombe (SW), Wollastonite Androy (S), Behara (N).

92 SMALLER ISLANDS

Since most islands (Fig. 1,69) are hardly more than volcanic cones, resources of five million people packed into 18,000 km^2 are mostly restricted to pumice, coral, building stone, gravel, ponded salt, guano and some hydropower even in

Table 127a The smaller islands and presidios.

Islands	Area 1000 km^2	Population 1000	Products and resources
Comoro	1.8	400	geothermal steam, pumice
Mayotte	0.4	60	stone pumice
Réunion	2.5	600	hydropower 450 Gwh, clinker plant 200 kt
Mauritius	2	1000	hydropower 100 Gwh, NPK 10 kt, salt 6 kt, lime 7 kt, basalt 1 Mt
Seychelles	0.4	70	hydropower 1 Mt, guano 5 kt (Astove)
Suqutra	3	20	guano on K'al Fir'wan
Italian	0.2	30	Pantelleria, Pelagians
Spanish	0.03	140	Melilla and Ceuta presidios and islets -
Madeira	0.7	350	clinker capacity 70 kt
Canary	7	1600	clinker 1 Mt, pozzolana 1 Mt, petroleum 40 Mb, 20 kt salt+
Cape Verde	4	400	clinker project 60kt,pumice 10kt, salt 7 kt, geothermal potential
Bioko, E.G.	2	60	gas, hydropower 2 Gwh
Saõ Thomé	1	90	hydropower 3 Gwh, clay 20 t, phosphate
British	0.2	10	manganese St. Helena, phosphate Gough.

Réunion and Mauritius

Sugar and rum are the key products of Réunion, a 'département' of France, which imports 2 Mb of petroleum products. Electricity requirements, which also increased fourfold to 450 Gwh, are met by Passerelle, where 130 m head on the Langevin stream provides 36 Mw, which cover all electricity needs along with other micro-hydros. To continue, Ciments de Bourbon's grinding plant (51% Macore S.A., 25% Blue Circle group) has a capacity of 220,000 ty of clinker at Le Port, west of St. Denis. Loams and clayey soils are shallow in the Mascarenes and little water stays in the permeable volcanic rocks. The steep slopes of Réunion are eroded because of deforestation and water on the coast is brackish.

> GOALS : Development of hydropower resources and substitution of petroleum imports by coal. Regional cooperation including the Malagasy Republic in cement, fertilizer and other supplies.

There is more water and soil on the gentler relief of Mauritius and Rodrigues, the easternmost of the Mascarenes, even receives a meter of rain. Birds deposit guano here and on the Cargados islands north of Mauritius. Moreover, a 30 td plant produces fertilizers in St. Louis harbor for the sugar cane plantations. Even gold shows in the Fressanges and Midland veins, but by 1990, Mauritius will import 2 Mb of petroleum products and perhaps 80,000 tons of cheaper coal, since little hydropower is available in the dry season from October to January at :

Site	dry season	total Mw	Site	dry season	total Mw
Champagne	26	30	Magenta	0.9	0.01
Tamarind	4.6	7	Chute Cécile	0.4	1
Ferney	0.9	10	Eau bleue	-	1

Fig. 69. The Islands.

Champagne generates 40 Gwh and microhydros provide 60-80 Gwh out of 450 Gwh which have doubled during the seventies.

Smaller islands in the Indian Ocean and Mediterranean

Large colonies of penguins deposit guano on the Prince Edward islands which belong to South Africa 1500 km south east of Durban, but less is known in this respect concerning the French Crozet island, 1300 km to the east, halfway to the windswept Kerguélen.

Guano has also been recovered from the Juan de Nova islet, a French possession in the Mozambique channel. The easternmost of the Comores, Mayotte (Mahoré) is a French 'territorial collectivity'. It lives like the République fédérale islamique des Comores, from cloves, ylang-ylang, and vanilla, planted on soil distinguished by rapid humus build-up. The geothermal potential of Anjouan and the Grande Comore have also been surveyed since electricity demand trebled to 10 Gwh in a decade.

France also controls Tromelin island north of Réunion, but the Agalega group, further north, is Mauritian.

> GOALS : Oil exploration on the Fortuna, Saya de Malha and Nazareth banks between Mauritius and the Seychelles and on the Medina bank of Malta. Development of hydropower and water resources respectively on the Seychelles and Suqutra.

Ninety % of the population of the Seychelles lives on Mahé, where there are six minihydro sites with a total capacity of a Mw, viz. Cascade, Mare aux Cochons, Grand St. Louis, Islette and Mamelles. The most attractive seems to be Grande Anse, with a storage capacity of 700,000 m^3, but currently all energy needs are met by petroleum imports which will exceed half a million barrels before 1990. Electricity demand has more than doubled in ten years to 50 Gwh. Fuel is even more expensive on the outlying island groups north of Madagascar. Among these, Astove Island is the principal source of high phosphate guano, 20,000 t of which were produced in 1957. Other deposits include Picard and West islands in the Aldabra Ring, St. Pierre in the Amirantes, Bird and Dennis isles in the Seychelles.

The Veddaic people of Suqutra in the Yemen Democratic Republic live from fishing, perling, and 'dragon's blood'. Their mountain soils are poor and they have little water. Birds drop guano though on Ka'l Fir'awn which rises steeply from the sea 20 km from Abd el Kuri, west of Suqutra, and 170 km from the Horn of Africa (Fig. 18c).

There is more gas around the Pelagians, Linosa and Lampedusa , which has brought fame to the 'Leopard' than there is water on the islands. Pantelleria in the straits of Sicily even has to import sweet water.

The principal livelihood of the Presidios are the Spanish troops, the harbors, official and unofficial trade, but there are no mineral resources in the mainland enclaves of Ceuta and Melilla, the fort of Peñon de Velez, conquered by Philipp II three hundred years ago, and on the Alhucemas and Chafarinas islands.

93 ATLANTIC ISLANDS

Even though most Açores are nearer to Cape Mogador, these islands are generally attributed to Europe. The Madeira archipelago consists of the Dezertas, flat

Porto Santo (40 km^2) and Madeira proper, the original model of the plantation economy, based on the volcanic soils of the south and east and on abundant water of the Picos da Vara and Arieiro distributed by privately owned canals. Bananas, wine and sugarcane are still the mainstays of the economy along with tourism, which prompted the opening of a clinker grinding plant in Funchal, the third city of Portugal.

Canary Islands

The combined GDP of the two Canary provinces ranks twelfth in Africa and Las Palmas is, by far, the biggest harbor; Santa Cruz is among the top ten and the three airports are the busiest on the continent. There is also one car or telephone per 3 inhabitants. Yet only 8% of the population is active in industry.

Lanzarote, Teneriffe and La Palma have a good geothermal potential, but one of the largest refineries in Africa, Cepsa's 160,000 bd facility at Santa Cruz de Teneriffe risked closure. Similarly, the Islands export cement to the peninsula, i.e. Spain proper, but 2% growth per year has slowed down to half a million tons in Teneriffe. Conversely, consumption has doubled to 600,000 t in the province of Las Palmas. The two plants, La Auxiliar de la Construccion and Cementos Especiales, each have a clinker grinding capacity of 750,000 ty and add 74% value on imported limestone. The islands are the dominant producer of pozzuolana in Africa and light concrete blocs made with tuff are the main building material.

The 'Afortunadas' are a gift of the trade winds, which bring the rains. Red soil developed in the rainiest spots of the northern slopes and it is quarried e.g. at Puerto de Erjos in western Teneriffe at 1000 m altitude and trucked to the coastal banana plantations of the sunny south, where there is little torric soil. Rich soil costs thus \$2.50 in situ and 5 m^3 of water cost a dollar.

> GOALS : Maintenance of the refinery. Further development of water resources, particularly on the smaller isles and of pozzolanic cements on the main islands.

Usage of water is particularly high : 490 m^3 per inhabitant in the province of Sta Cruz and 220 m^3 in Las Palmas province, which works out respectively as 330 and 150 Hm^3y. But the recharge on Gran Canaria alone is 110 Mm^3y. Surface waters are a tiny part of this as confirmed by the hydraulic balance (Hm^3) on the best endowed islands :

Island	mmy	Total	Evapotranspiration	Run off	Infiltration
Teneriffe	450	920	530	130	260
La Palma	660	480	240	90	150

Where galleries are the principal resource :

Table 128 Water origin and demand in the Canary islands (Hm^3y).

	Gr.Canaria	Teneriffe	La Palma	Gomera	Hierro	Fuerteventure	Lanzarote
Galleries	20	200	40	1			0.5
Wells	100	30	30	2	2	5	
Sources	3	5	15	8			
Produced[a]	155	235	85	15	2	7	1.5
Demand 2000	165	235	100	17	2	7	2.5
Available	200	390	240	65	25	35	1.5

[a]including purification and desalination, more of which will be required on Lanzarote

Cape Verde

There are almost as many Capeverdians in the U.S. as in the islands where a quarter of the population is concentrated on Santo Antaõ and Saõ Vicente and a half in Santiago, where Praia, the capital, is located. Fishing and bunkering are the principal occupations, but 7% of exports were mineral products. Maio next to Santiago supplied 15,000 ty of pumice until 1980, and on Santo Antaõ, the mixing of 600,000 ty of pozzolanic cement has been mooted.

The Republica do Cabo Verde is a member of the Sahel group and mountain soils and water are so scarce that tourism has not developed yet. While industry requires only 1000 m^3d on Santiago and Saõ Vicente, total usage is already in excess of 70,000 m^3d. Although recharge might reach 300,000 m^3d, water has to be pumped from 100 m deep wells in the farming areas of Santiago. As consumption increases ever deeper wells would have to be dug. Since hillsides are steep, each gallery and dam serves 10-15,000 people and irrigation schemes use water from the galleries. Furthermore, water is also piped to the valleys. Conversely, Mindelo and Praia desalinate respectively 750 m^3d and 1200 m^3d. Yet, while water by gravity costs only $40/$m^3$, desalted water which is pumped on Fogo to an altitude of 600 m costs the double.

GOALS : Rational exploitatipn of water resources for farming and tourism. Development of the light cement industry.

Saõ Vicente and Sal even used 40,000 b of fuel to desalinate 550,000 and 10,000 b - a ratio of 15:1. No wonder that this water is expensive. As its name says, salt is also recovered on Sal from the Crater of Pedra Lume, just 200 m from the sea, when tides wet the 60 m thick layer. Shipments declined though from their 1980 peak of 22,000 t. Net petroleum imports are growing rapidly from 500,000 by. So does electricity generation which has increased 50% in ten years to 12 Gwhy. Lastly, steam rising from the Fogo caldera (west of Santiago) might have a temperature of $\geqslant 300°C$. Little power is required, however, on this isolated island.

South Atlantic islands

Iron ore and rutile occur near its borders, but little is known about the resources of Equatorial Guinea (Fig. 30-32). The Kribi gas field is just 50 km north of the border and the oil indicating salt layers extend over 30 km along the coast of land based Rio Muni. Drilling started therefore off Bata. Further north, the Rio del Rey basin extends to Bioko (Fernando Poo) island. Gepsa, a joint venture of the state and Hispanoil has struck subcommercial gas 36 km northwest of Malabo, the capital at 3830 m depth, south of Cameroonese water. While lava flows cover oil sands here, further east, a stretch of the favorable shale 'diapirs' (domes) are in Guinean waters.

> GOALS : Further oil and mineral exploration, above all around Bioko, utilization of fertilizer here and in Sao Thomé. Development of microhydros.

Until such time that commercial oil is found, petroleum imports will require a third of export (coffee) earnings. The major trading partner is Spain and 25-30 Gwh are generated from 7-10 Mw capacity out of which 2 Gwh were hydro. Lastly, stone, sand and perhaps salt are used even on volcanic Annobon island south of São Thomé.

The Republica democratica de São Thomé e Principe can produce 20 ty of earthenware for export. The Carregerdo generates hydropower and thermal electricity consumption increased 60% in 10 years to 8 Gwh. Nearby, on the Ilheu de Cabras trachyte alters to aluminum phosphate. There is, moreover, guano on Boatswain Island near Ascension, on Gough (Fig. 71), south of Tristan da Cunha, and on Egg Island near St. Helena. Some manganese occurs here too in Prosperous Bay, and thermal power generating capacity has been doubled in 1979 to 2 Gwh.

94 THE SEA

The density of seawater reaches 1025-1026 kg/m^3 a) between East London (S) and Maputo (SE), b) from Nouadhibou (NW) to Tangier, where salinity attains 35-36% and c) in the Mediterranean where it exceeds 37%. Conversely, between Calabar and Gabon (W), salinity is less than 33% in February. Seawater is also rich in other elements, e.g. :

element	g/l		mg/l		mg/m^3		mg/m^3
Chlore	19	Potash	380	Phosphorus	70	Cobalt	0.5
Sodium	10	Bromium	65	Nickel	2	Gold	0.004
Magnesium	1.8	Boron		Manganese	2		

The coast, shelf, seafloor nodules, brines and marine energy are the other key resource categories : firstly, tidal waves have deposited ilmenite ; less, but more valuable rutile, zircon and monazite along the beaches of West Africa as far as Ghana and on the Indian Ocean from the Horn to Natal including Madagascar as well as in the Nile delta. Africa is, therefore, a leading producer of both titanium minerals. Diamonds shone over 1000 km off Namibia and South Africa, but the sea hardly provides enough salt for African needs.

II Nonetheless, salt horizons define the oil fields between Gabon and Angola, where offshore production predominates. In Nigeria, a quarter of oil and up to a half of gas reserves might be in shallow seas. The shelf bulges out off Gabon where gas finds are 70 km off the coast and 105 km off Namibia, and the East African and Malagasy coast. The Mascarene-Seychelles highs have a similar potential, so does the Medina bank which straddles the economic zone of Libya and Malta. Gas fields on the maritime border of Tunisia and Libya are, indeed, 200 km from the coast.

III Phoscorite pebbles, muds, sands and 'pavements' continue to develop along the Moroccan, Senegalese, Namib and Cape coast, on the Sierra Leone seamount and the Aguilhas Bank where samples average 10-25% P_2O_5 with peaks of > 33%. Glauconite is also abundant and resources might reach

	Morocco	Namibia	Aguilhas	other South Africa
to depth, m	500	50-200	20-55	200
Billion t ore	1.5	4	0.5	2

IV On the Ivory Coast the height of the tide could generate energy. Waterfalls supply, however, cheaper hydropower. South of the equator the belt of high temperature gradients stretches eastward from Tanzania, yet the rigth OTEC site with a difference of 20°C over more than 1000 m depth has not been located in the Seychelles.

Red Sea brines

V The top ten meters of hot brines in the Atlantis II Deep reportedly contain \$5B worth metals : 80 t gold, 4000 t silver, 2 Mt zinc, 500,000 t copper, some lead and traces of Co, Ni, Cd, Sb, Sn, Ge, Bi, In and Hg. Some sulfide layers originally contained as much as 20% zinc. Atlantis II brines are still active and exhibit the highest chlorinity. Eighteen other deeps include those explored by the 'Chain' and 'Discovery' for Sedco, financed by Saudi Arabia, which spent \$300M in 1981.

Seven of these show concentrations of zinc and except for the northernmost two, Oceanographer and Kebrit, all deeps have extensive hydrothermal beds. Layers of the 300 km^2 Chain deep even grade 30-40% manganese and brine pools nestle in the deepest parts of long local basins within the rift. Water is some 1000 m deep and the metalliferous muds are at a depth of 2-3000 m. Agitation tests have already been made and a pilot plant is planned.

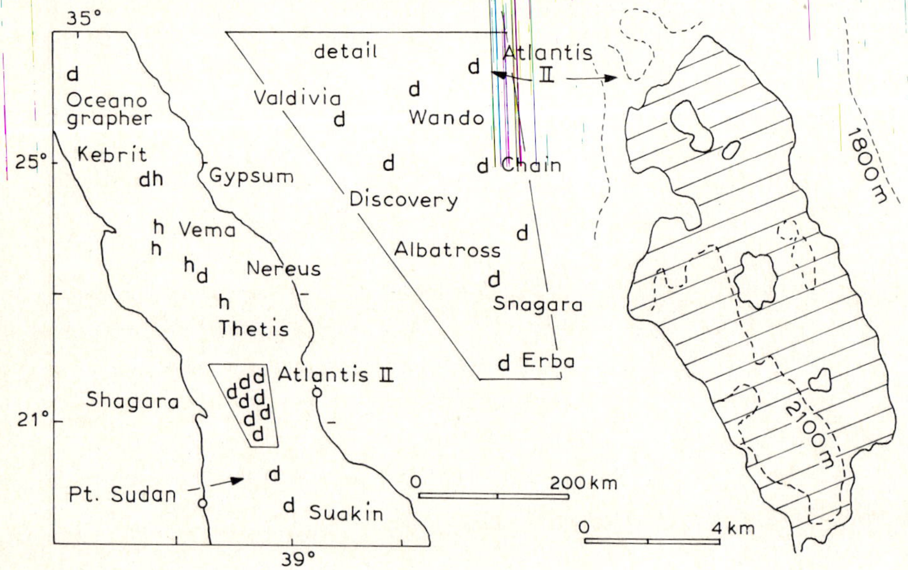

Fig. 70. Red Sea brines (*d*), sediments (*h*).

These are undoubtedly the largest metallic resources of Sudan and Saudi Arabia. Meanwhile, reserves of the Mid-Atlantic brines have not been estimated yet, nor have perhaps extensive methane hydrate deposits been quantified.

Table 129 Zinc content °/oo, maximum depth of deeps, maximum thickness, surface and temperature of brines.

Deep	Zn °/oo	depth m	thick m	surface km^2	heat C°
Nereus	0.5	2460	40	20	30
Valdivia	0.4	1670	120	4	30
Atlantis II	5.4	2060	180	55	60
Discovery	0.8	2050	210	10	45
Erba	0.36	2395	20	7	30
Pt. Sudan	0.17	2835	320	5	36
Suakin NE	0.04	2830	55	10	25
SW	0.05	2850	75	2.5	23

Nodules (Fig. 71-72)

VI Sediments contain manganese along broad belts centered on the mid-Atlantic ridges (south). Nodules also grade 20% manganese a) between St. Helena and Brazil and b) southeast of the Walvis Ridge which blocks Antarctic currents - along with an assay of >25% in the southern Cape Basin. The mean nickel content is 0.5%.

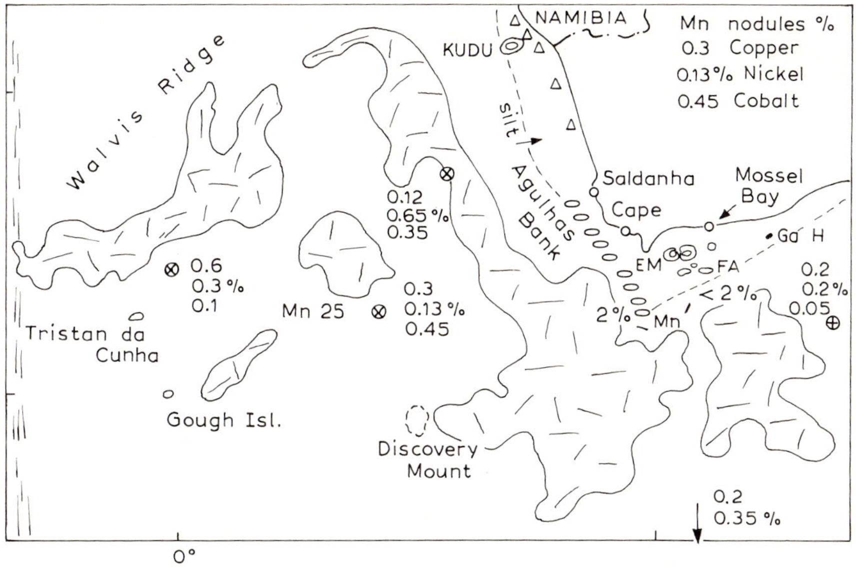

Fig. 71. The Cape Basin. (Legend on page 13.)

Cobalt values increase on the other hand from the Cape (0.2%) to Tristan da Cunha (0.6%) and level out to 0.4% towards Brazil. Copper assays are more erratic with an ill-defined opposite trend from 0.1 to 0.4%.

A broad belt of manganiferous sediments also spreads along the Mid-Indian Ridge, from the Antarctic Sea to the equator, but nodules contain little manganese south of 20°S east of 70°E, with an unusual low extending south of Maputo and east of Durban. In contrast, south of 12°N, nodules generally grade >20% with extensive deposits a) between Soqutra and Réunion b) southeast of Diego Garcia and c) south of Rodrigues. Cobalt assays range from 0.15% to 0.3% with the famous Coetivy peak of 0.95%, most nickel values spread between 0.25 and 0.4% and the best copper grades, 0.25 and 0.5%, are southeast of Diego Garcia and Madagascar. - Since land based reserves are adequate these resources will remain the heritage of future generations.

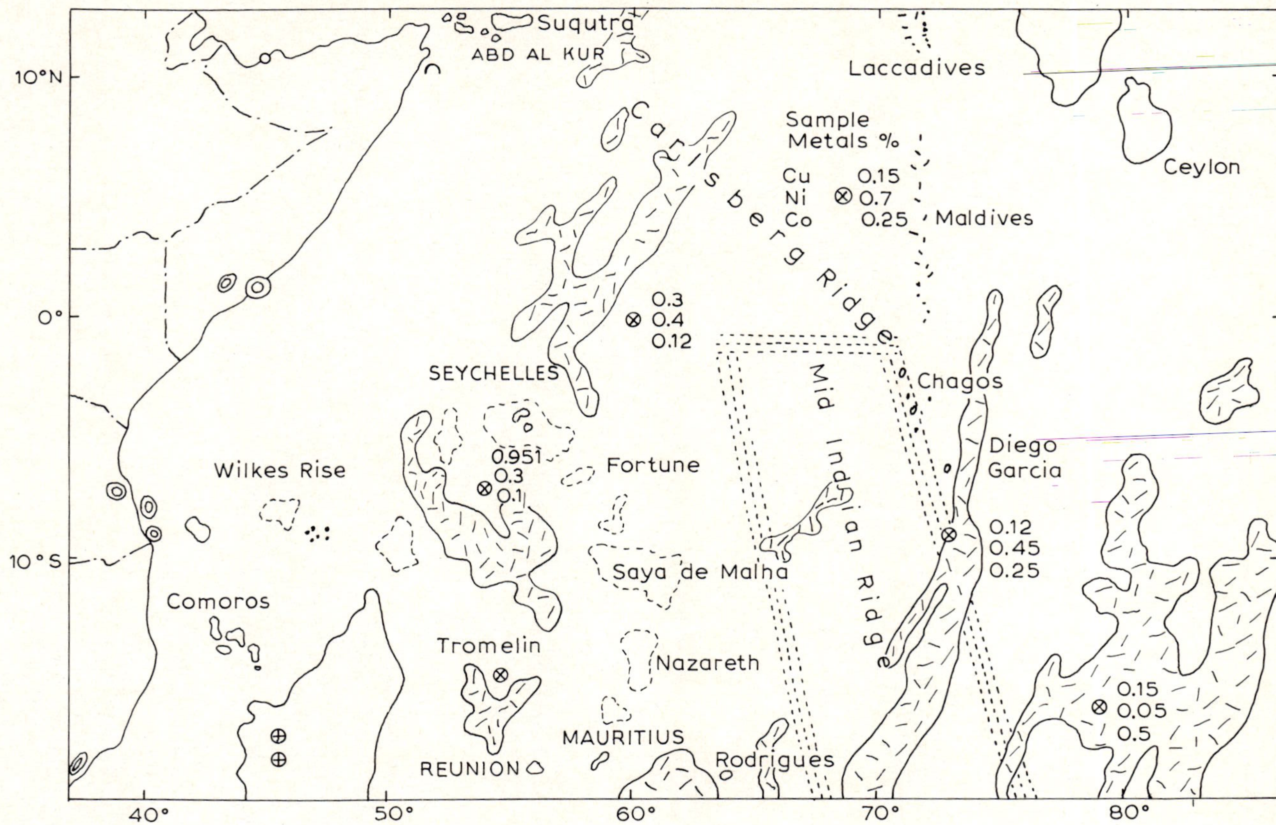

Fig. 72. The Indian Ocean, west. (Legend on page 13.)

LIST OF ACRONYMS

AAC : Anglo American Corp., South Africa.
ACM : African Chrome Mines, Zimbabwe.
ADB : African Development Bank, Abidjan.
Adfaed : Abu Dhabi Fund for Arab Economic Development.
ADMC : Agricultural Development and Marketing Corporation, Malawi.
AECI : African Explosives and Chemical Industries, South Africa.
Afesd : Arab Fund for Economic and Social Development.
AGC : Ashanti Goldfields Corp., Ghana.
Ageco : Arab Gulf Exploration Co., Libya.
Agiba : EGPC + Aegyptco.
Alcan : Aluminium Ltd, Canada.
Alcoa : Aluminum Co. of America, U.S.A.
Alucam : Compagnie Camerounaise de l'Aluminium.
Amaprop : Anglo-American Properties Ltd, South Africa.
AMAX : formerly American Metal Climax, U.S.A.
AMC : African Manganese Co., South Africa.
Amcoal : Anglo-American Coal Corp., South Africa.
Amgold : Anglo-American Gold Investment Co., South Africa.
Ammosal : Associated Manganese Mines of South Africa.
ANGLOVAAL : Anglo-Transvaal Consolidated Investments Co.
AOM : Associated Ore and Metals, Ltd., South Africa.
ARC : Agip Recherchas Congo.
Arédor : Association pour la Recherche et l'Exploitation du Diamant et de l'Or S.A., Guinea.
Armico : Arab Mining Company, Jordan.
Asaic : American South African Investment Co.
ASED : Austrian Economic Development Agency.
ATMN : Amalgamated Tin Mines of Nigeria Ltd.
Asmar : Soc. des Ciments de Marrakech, Morocco.
AV : Anglovaal, South Africa.

BAC : British Aluminium Company.
Barrand : Barlow Rand Ltd, South Africa.
BGM : Bureau géologique et minier, Burkina.
BGR : Bundesanstalt für Geowissenschaften und Rohstoffe , Germany.
BMC : Bong Mining Co., Liberia.
BP : British Petroleum Ltd.
BOTREST : Botswana RST, Ltd.
BR : Barlow Rand Ltd, South Africa.
BRGM : Bureau de Recherches Géologiques et Minières, France.
BRPM : Bureau de Recherches et Participations Minières,Morocco.
BTK : Banque Tuniso-Koweïtienne.
BTKD : Banque Tuniso-Kuwaïtienne de Développement.

Cadei : Centre d'achat et de développement de la région minière du Tafilalet, Morocco.
Cadem : Société des Ciments Artificiels de Meknès, Morocco.
Cadetaf : Centrale d'Achat et de Développement de la Région Minière du Tafilalet et Figurg, Morocco.

Carbomoç : Companhia Carbonifera de Moçambique.
CAST : Consolidated African Selection Trust, U.K.
CBG : Cie. des Bauxites de Guinée.
CCCD : Caisse Centrale de Coopération au Développement, France.
CDG : Caisse de Dépôt et de Gestion, France.
CDM : Consolidated Diamond Mines, Namibia.
CEA : Commissariat à l'Energie Atomique, France.
CEAO : Comité Economique de l'Afrique Occidentale, Ouagadougou.
Cemo : Cie d'Exploitation des Mines de 'l'Ouarsenis, Morocco.
CEPGL : Communauté Economique des Pays des Grands Lacs, Gisenyi, Rwanda.
CFA : Communauté Financière Africaine (French-speaking Africa).
CFM : Compagnie Française des Mines, Zaïre.
CFMU : Compagnie Française des Minerais d'Uranium.
CMSA : Chrome Mines of South Africa.
CGf : Consolidated Gold Fields, U.K.
CIDOLOU : Cimenterie Domeniale de Loutete, Congo.
Cie : Compagnie.
Cimangola : Empresa do Cimentos do Angola.
Cimat : Société de Ciments et Matériaux, Burkina Faso.
CIZA : Ciments du Zaïre.
CMA : Cementos Marroquies.
CMI : Consolidated Metallurgical Industries Ltd, South Africa.
CMOO : Compagnie Minière de l'Oubangui Oriental, Central African Republic.
CMT : Compagnie Minière de Touissit, Morocco.
CND : Comptoir National du Diamant, Central African Republic.
Coger : Compagnie Gabonaise Elf Raffinage.
Comabar : Cie Marocaine des Barites.
Comici : Compagnie Minière de la Côte d'Ivoire.
Comferci : Compagnie des Mines de Fer de la Côte d'Ivoire.
Comiki : Compagnie Minière de Kiéré, Burkina Faso.
COMILOG : Compagnie Minière de l'Ogooué, Gabon.
Comimat : Compagnie des Mines et Matériaux, Ivory Coast.
Comina : Compagnie minière d'Audriamena, Malagasy Republic.
Comincor : Commercial and Industrial Corp. of Sierra Leone.
Cominor : Compagnie minière du Nord , Mauritania.
Comuf : Compagnie minière d'Uranium de Franceville, Gabon.
Cotomib : Compagnie togolaise des phosphates du Bénin.
CPG : Compagnie des Phosphates de Gafsa, Tunisia.
CPT : Compagnie des Phosphates du Togo.
CSM : Compagnie Saline du Maroc.
CSO : Central Selling Organisation (De Beers), London.
CSPT : Compagnie Sénégalaise des Phosphates de Taïba.
CTS : Compagnie des Tourbières du Sénégal.
CTT : Cie de Tiforout Tiranamine, Morocco.

Delimco : Deutsche-Liberian Mining Co.
Diamang : Companhia de Diamantes de Angola.
Dicorwaf : Diamond Corporation of West Africa, Namibia.
Diminco : National Diamond Mining Co., Sierra Leone.
Disc : consortium of Deminex, Shell, BP, Egypt.

Ecowas : Economic Community of West-African States, Lagos.
EDF : European Development Fund, Brussels.
EDM : Electricité du Mali.
EEC : European Economic Community, Brussels.
EECI : Eau et électricité de la Côte d'Ivoire.
EG : Engrais de Gabès, Tunisia.
EGL : Energie des Pays des Grands Lacs, Bujumbura.
EGPC : Egyptian Petroleum Co.
EMZ : Entreprise Minière du Zaïre.

Encim : Empresa Nacional do Cimentos, Angola.
Endiama : Empresa Nacional de Diamantes de Angola.
ENI : Ente Nazionale Idrocarburi, Italy.
Enura : Empresa Nacional del Uranio (Spain).
ERC : Entreprise de Ciments et Dérivés, Algeria.
ERGO : East Rand Gold and Uranium Co., South Africa.
ESCOM : Electricity Supply Commission, South Africa.
ETAP : Entreprise Tunisienne d'Activité Pétrolière.
EVL : Eurocan Ventures of Canada, Ltd, Sierra Leone.

FED : Fonds Européen de Développement, Brussels.
Fedmis : Federale Kunsmis, South Africa.
Fosfang : Empresa Mineiral dos Fosfatos do Zaïre, Angola.
Fosbucraa : Fosfatos de Bou Craa, Western Sahara - Morocco.
FPZ : Soc. des Fonderies de Plomb de Zellidja, Morocco.
Freegold : Orange Free State Consolidated Goldmines (AAC), South Africa.
Friguia : Société d'Economie mixte Friguia, Guinea.
FREDDIES : Free State Development and Investment Corp.
Ftcic : Foreign trading, contracting and investment Co., Kuwait.

GBC : Ghana Bauxite Co.
GCD : Ghana Consolidated Diamonds Ltd.
GCM = Gécamines
Gécamines : La Générale des Carrières et des Mines E.P., Zaïre.
Gencor : General Mining Union Corp. Ltd., South Africa.
GEPSA : Empresa Guineo-Espanola de Petroleos S.A., Equatorial Guinea.
GfSA : Gold Fields of South Africa, Ltd.
Ghacem : Ghana Cement Works Ltd.
GM : General Mining Union Corp. or Gencor, South Africa.
GNM : Ghana National Manganese Corp.
GPC : General Petroleum Corporation, Sudan.
Gupco : Gulf Petroleum Co., Egypt.

Hidrominas : Empresa Nacional de Aguas subterraneas (Angola).

ICG : Industries Chimiques de Gafsa, Tunisia.
ICM : Industries Chimiques Maghrebines, Tunisia.
ICS : Industries Chimiques du Sénégal.
IDA : International Development Association (World Bank), Washington.
IDB : Islamic Development Bank.
IDC : Industrial Development Corp., Zimbabwe.
IEOC : International Egyptian Oil Co. (Agip).
IFC : International Finance Corporation (World Bank), Washington D.C..
IMC : International Minerals and Chemicals Co. (U.S.).
INCO : International Nickel Corp., Canada.
INI : Instituto Nacional de Industria, Spain.
ISC : Iron and Steel Company, Egypt. '
ISMI : International Strategic Minerals Inventory (USBM, USGS).
Iscor : Iron and Steel Corp., South Africa.

JCI : Johannesburg Consolidated Investment Co., South Africa.
JICA : Japanese International Cooperation Agency.
JMS : Joint Metallurgical Scheme, South Africa.

KBO : Kagera Basin Organisation, Kigali.
Kfed : Kuwaiti Fund for Economic Development.
Kftcic : Kuwait foreign trading, constructing and investment company.
KfW : Kreditanstalt für Wiederaufbau, Germany.

Lamco : Liberian-American Swedish Minerals Co.

LEB : Liberia Electric Corp.
Libeth : Liberian Bethlehem Iron Mines Co.
LISCO : Liberian Iron and Steel Corporation.
LJV : Lamco Joint Venture Operating Co., Liberia.
LMC : Liberia Mining Company.
LME : London Metal Exchange.
Lioc : Liberia Iron Ore Co. Ltd.
LONRHO : London Rhodesian Investment Co.

MC : Maroc Chimie.
MDC : Malaŵi Development Corporation.
MEMACO : Metal Marketing Corp. of Zambia.
MHO : Société Métallurgique de Hoboken - Overpelt.
Miba : Société Minière de Bakwanga, Zaïre.
Mifergui : Soc. des Mines de Fer de Guinée.
Miferma : Société des Mines de Fer de Mauritanie.
Minaquartzo : Empresa Mineira de Quartzo, Angola.
Minorco : Minerals and Resources Co., Ltd, Bermuda.
Miomco : Marampa Iron Ore Mining Co., Sierra Leone.
MMA : Mine Management Associates, Canada.
MP : Maroc Phosphore.
MTD : (Mangula) Ltd, Zimbabwe.

Namico : National Mining Co., Uganda.
NCC : National Chemical Corporation, Ethiopia.
NCC : Nigerian Coal Corp.
NEPA : Nigerian Electric Power Authority.
NMC : Nigerian Mining Corp.
NNPC : Nigerian National Petroleum Corp.
NSDA : Nigerian Steel Development Authority.
NTMC : Nigeria Tin Mining Co.

OBK : Office des Bauxites de Kindia, Guinea.
OCP : Office Chérifien des Phosphates, Morocco.
ODI : Office pour le Développement Industriel, Morocco.
ÖEG : Österreichische Erdölgesellschaft, Austria.
Okimo : Office des mines de Kilo-Moto, Zaïre.
Omnis : Office militaire national des industries stratégiques, Malagasy Rep.
OMRG : Office Mauritanien de Recherches Géologiques.
OMT : Office du Manganèse de Tambao, Burkina Faso.
OMVG : Organisation pour la Mise en Valeur du Fleuve Gambie; Senegal, Gambia.
OMVS : Organisation pour la Mise en Valeur du Fleuve Sénégal; Mali, Mauritania, Senegal.
ONA : Omnium Nord-Africain, Morocco.
Onarem : Office national des ressources minières, Niger.
Onarep : Office national de recherche et d'exploitation pétrolières.
Opcia : Omnium des Produits Chimiques pour l'Industrie et l'Agriculture, France.
OPG : Office des Phosphates de Gafsa, Tunisia.
OTP : Office Togolais des Phosphates.
OURD : Overseas Uranium Research and Development Corp., Japan.

PCIAC : Petro Canada International Assistance Corp.
Petrangol : Companhia de Combustiveís do Lobito, Angola.
Pétroci : Société Nationale d'Opération Pétrolière de la Côte d'Ivoire.
PNC : Power Research and Nuclear Fuel Development Corporation.
PPC : Pretoria Portland Cement, South Africa.
PUK : Péchiney-Ugine-Kuhlmann, France.

Roremina : Empresa Nacional de Rochas Ornamentales, Angola.
RPH : Rustenburg Holding Co.
RPM : Platinum Mines.

RTZ : Rio Tinto Zinc Co. Ltd, U.K.
RTZ : Rio Tinto Zimbabwe Ltd.
RUL : Rössing Uranium Ltd, Namibia.

S.A. : Société Anonyme (Ltd. Co.)
Sacem : S.A. Chérifienne d'Etudes Minières, Morocco.
Sadcc : Southern African Development Coordination Conference.
Saepa : Société Arabe des Engrais Phosphatés et Azotés, Tunisia.
Safa : Société Arabe du Fer et de l'Acier, Kuwait-Mauritania.
SAG : Société Aurifère de Guinée.
SAI : Société Arabe d'Investissement.
SAM : Société Auxiliaire du Manganèse de Franceville, Gabon.
Samaes : Société Africaine des Métaux et Alliages blancs, Morocco.
Samine : S.A. d'Entreprises Minières.
SAP : Société Africaine des Pétroles, Senegal.
SAR : Société Africaine de Raffinage, Senegal.
SATS : South African Transport Services.
SCB : Société des Ciments du Bénin.
SEBACAU : Syndicat des Bauxites du Cameroun.
SCN : Société Nigérienne de Cimenterie.
SCO : Société des Ciments du Bénin.
SCPA : Société Commerciale des Phosphates et de l'Azote, France.
SCS : Soc. Chérifienne des Sels, Morocco.
Sdics : Soc. de Développement des Industries Chimiques du Sud, Tunisia.
SDIM : Soc. de Développement Industriel et Minier, Morocco.
Sébacam : Société d'Etude des Bauxites du Cameroun.
Secpt : Société Egyptienne de Ciment Portland.
Sedco : Saudi-Sudan Commission for the Exploitation of Red Sea Resources.
SEEG : Société d'Energie et Eau du Gabon.
Sefercam : Société d'Etudes de Fer du Cameroun.
Segma : Soc. d'Etat du Gaz Marin, Tunisia.
Sémipso : Société d'Etudes des Mines de Phosphates de Stra Ouertane, Tunisia.
SEPH : Société d'exploitation du potasse de Hollé, Congo.
SET : Société des Engrais du Togo.
SFD : Saudi Fund for Development.
SIBEKA : Société d'Entreprise et d'Investissement, Belgium.
Sicovam : Soc. Interprofessionnelle pour la Compensation des Valeurs Mobilières.
SFI : Société Financière Internationale, Morocco.
Sfmrm : Syndicat franco-malien de recherches minières.
SGM : Société Générale des Minerais, Belgium.
SGMC : State Gold Mining Corp., Ghana.
Sgrem : Soc. Gén. de Recherches et d'Exploitation Minières, France, Niger.
Shell-BP : Shell-BP Petroleum Development Co. of Nigeria.
Siape : Société des Industries d'Acide Phosphorique et d'Engrais, Tunisia.
SICF : Soc. des Industries Chimiques du Fluor, Tunisie.
Sieromco : Sierra Leone Ore and Metals Co.
SIOP : Société Ivoirienne d'Opérations Pétrolières.
SIR : Société Ivoirienne de Raffinage.
Sitep : Société Italo-Tunisienne d'Exploitation Pétrolière (Agip).
SLDC : Sierra Leone Development Corp.
SLST : Sierra Leone Selection Trust.
SMGI : Société Minière de la Grande Ile, Malagasy Rep.
SMI : Société Minière d'Ity, Ivory Coast.
SMI : Société Métallurgique d'Imiter, Morocco.
Smmpc : Société Marocaine des Mines et Produits Chimiques.
SMN : Société Minière du Niger.
SMS : Société Minière du Sénégal.
SMTF : Soc. Minière de Tenke-Fungurume, Zaïre.
SNCE : Soc. Nouvelle des Conduites d'Eau, France.
SNDP : Soc. Nationale de Distribution des Produits Pétroliers, Tunisia.

SNEA : Société Nationale Elf-Aquitaine, France.
SNH : Société Nationale des Hydrocarbures, Cameroon.
SNI : Société Nationale d'Investissement, Cameroon.
SNIM : Société Nationale Industrielle et Minière, Mauritania.
SNS : Société Nationale de Sidérurgie, Algérie.
SNSS : Soc. Nationale des Salines de Siné Saloum, Senegal.
Soc.: Société.
Sodimiza : Soc. de Développement Industriel et Minier du Zaïre.
Somair : Soc. des Mines de l'Aïr, Niger.
SOAC : Société anonyme des Ciments, Senegal.
Soarmico : Somali Arab Mining Co.
Socal : Standard Oil of California, U.S.A.
Socim : Société des Ciments du Sud-Ouest, Ivory Coast.
Sococim : Société Ouest-Africaine de Ciments, Senegal.
Socomine : Société de Coopération Industrielle et Minière, France.
Sodecat : Soc. du Développement du Cuivre de l'Anti-Atlas, Morocco.
Sodemi : Société pour le Développement Minier de la Côte d'Ivoire.
Soekor : Southern Oil Exploration Co., South Africa.
Sofratep : Soc. Franco-tunisienne d'Exploitation Pétrolière (Elf).
Sogara : Société Gabonaise de Raffinage.
Sogazcam : Société du Gas du Cameroun.
Sogemak : Société de gestion et d'exploitation des mines de Kalana, Mali.
Somasid : Société Malienne de Sidérurgie.
Somifer : Société des Mines de Fer de Mékambo, Gabon.
Somifer : Société Minière de Bou Saffer, Morocco.
Sominki : Société Minière et Industrielle du Kivu, Zaïre.
Somisa : Société des Mines du Sahel, Burkina Faso.
SONACI : Société Nationale des Ciments, Bénin.
Sonamif : Société Nationale des Mines de M'Fouati, Congo.
Sonamines : Société Nationale des Recherches et d'Exploitations minières, Congo.
Sonangol : Sociedade Nacional de Combustiveís de Angola.
Sonara : Société Nationale de Raffinage, Cameroon.
Sonatrach : Société Nationale pour la Recherche, la Production, le Transport, la Transformation, et la Commercialisation des Hydrocarbures, Algeria
Sonelec : Société Nationale d'Electricité, Senegal.
Sonichar : Société Nigérienne du Charbon d'Anou Araren.
Soremi : Société de Recherches Minières, Burkina Faso.
Soprociment : Société d'Economie Miste Guinéo-Arabe.
Soremi : Soc. de Recherches Minières, Burkina Faso.
Sorepca : Société de Recherches et d'Exploitation des Pétroles du Cameroun.
Sotemi : Société Tunisienne d'Expansion Minière.
Sozir : Société Zaïro-Italienne de Raffinage.
SSPT : Société Sénégalaise des Phosphates de Taïba.
Stamico : State Mining Corp., Tanzania.
STEG : Soc. Tunisienne d'Electricité et de Gaz.
STMM : Société Togolaise de Marbrerie et Matériaux.
Stsid : Société Tuniso-Séoudienne d'Investissement et de Développement.
Suco : Suez Oil Company, Egypt.

TANKS : Tanganyika Concessions Ltd, London.
TCC : Turkish Cement Company.
TCL : Tsumeb Corporation Ltd, Namibia.
TECO : Tsumeb Exploration Co., Namibia.
TFC : Tanzania Fertilizer Co.
Ticaf : Tororo Industrial Chemicals and Fertilizers Ltd, Uganda.
TMFC : Transvaal Mining and Finance Co, South Africa.
TOC : Tanzanian Oil Co.
Tocages : Total Centrafricaine de Gestion.
TPDC : Tanzania Petroleum Development Corporation.
TNSK : Trans Natal Steenkool Korporasie, South Africa.

Ucemco : Uganda Cement Corp.
UDC : Uganda Development Corp.
UMHK : Union Minière du Haut Katanga
Unido : United Nations Industrial Development Organization, Vienna.
UNDP : United Nations Development Programme, New York.
URCA : Compagnie des Mines d'Uranium de Bakouma, Central African Republic.

Valco : Volta Aluminum Co., Ghana.
VRA : Volta River Authority.

WCC : Wankie Colliery Company, Zimbabwe.
WNPC : White Nile Petroleum Corporation, Sudan.

Yskor : Yser en staalkorporasie, South Africa.

ZCCM : Zambia Consolidated Copper Mines Ltd.
ZCI : Zambia Copper Investments (AAC).
Zimalloy : Zimbabwe Alloys Ltd.
Zimasco : Zimbabwe Mining and Smelting Co.
Zimco : Zambia Industrial and Mining Corp.
Zisco : Zimbabwe Iron and Steel Corp.
ZMS : Zimbabwe Mining and Steel Co.

FURTHER READING

1. Africa, London.
2. Africa business, London.
3. Africa economic digest.
4. Africa now, London.
5. African research bulletin, London.
6. Afrique contemporaine.
7. Afrique Industrie, Paris.
8. American Association of Petroleum Geologists, bulletins.
9. Annales des Mines, Bretigny.
10. Annual reports of corporations, departments of energy, water and mines and geological surveys.
11. Backer, H., Schoell, M., 1972 - New deeps with brines and metalliferous sediments in the Red Sea, Nature, Phy. Sci. London 240, pp 153-8.
12. Balek, J., 1977 - Hydrology & Water Resources in South Africa, Elsevier, pp 258 &.
13. Benylague , M., 1972 - La mine d'uranium d'Arlit, Ann. Mines, Mars 1-14.
14. Bigotte, G., Obelliane, J.M., 1968 - Découverte de minéralisations uranifères au Niger, Min. Dep. 317-333.
15. British Geological Survey bulletins, memoirs, statistics.
16. Bulletin de l'Afrique noire, Paris.
17. Bundesanstalt für Geowissenschaften und Rohstoffe, 1983 - Untersuchungen über Angebot und Nachfrage mineralischer Rohstoffe, XVIII, Phosphat, pp 408 + 50, H. Schmidt et al.
18. Bureau de Recherches Géologiques et Minières, bulletins, reports, memoirs, Orléans.
19. Callot, F. - Production et consommation mondiales des minerais, Annales Mines, Paris Nov.-Déc. 1980, pp 13-142.
20. Cembureau, Paris, Statistical Yearbook.
21. Chemical Geology, Elsevier, Amsterdam.
22. Codigaz - Le gaz naturel dans le monde, Paris, annual.
23. Commission Economique pour l'Afrique (U.N.), 1981, Actes de la première Conférence régionale sur la mise en valeur et l'utilisation des ressources minérales en Afrique, pp 328.
24. De Jager, F.S., 1983 - Steenkoolreserwes van die Republiek van Suid-Afrika, Geologiese Opname, Bull. 74 pp 18.
25. de Kun, N., 1964 - The Mineral resources of Africa, pp 740, Elsevier, Amsterdam.
26. de Kun, N., Cavadias, G., Gegan, A., Brewer, W., 1961 - Report on Industrial Development for the Malagasy Republic, Laramore, Douglass and Popham,

New York, pp 257 + 175 + 116.

27. Economic Commission for Africa (U.N.), 1976 - Present status and main trends in the development of mineral resources in the West-African sub-region, Addis Abeba, pp 179.

28. Economic Geology, el Paso.

29. Engineering and Mining Journal.

30. Industrial Minerals, London.

31. Geological Survey of South Africa, 1976, Handbook.

32. Geologie en Mijnbouw.

33. Goudarzi, G. H., 1970 - Geology and mineral resources of Libya, U.S.G.S. Profes. paper 660, pp 104.

34. Halasi-Kun, G., editor, Columbi University, Proceedings of the University Seminar on Water Resources, yearly.

35. International Fertilizer Industry Association, Phosphoric Acid Plants, yearly, Paris, pp 22.

36. Jelenc, D.A., 1966- Mineral occurrences of Ethiopia, Addis Abeba, pp 720.

37. Jeune Afrique économique.

38. Marchés tropicaux et méditerranéens, Paris.

39. Meister, U., 1983 - Quo Vadis südliches Afrika, Vontobel, Zurich, pp 39.

40. Minerals Bureau - South Africa's Mineral Industry (yearly), Braamfontein.

41. Mining Annual Review, London.

42. Mining Journal, London.

43. Mining Magazine, London.

44. Mining in Zimbabwe, Thomson's Publication, Harare, yearly.

45. Moçambique, Serviços de Geologia e Minas, 1974, Carta de jazigos e ocorrências minerais, Noticia explicativa pp 35.

46. OECD, 1982, Reserves production and demand for uranium, 240 pp.

47. Oil and Gas Journal.

48. Pallas, P., 1978, Water resources of the Socialist Peoples Jamahirya of Libya, Tripoli, pp 80.

49. The Petroleum Economist.

50. Revue du Tiers monde, Paris.

51. Schmid, F., 1984, Geology of recent/potential bauxite producing areas in Sierra Leone and ... Guinea, Soc. Min. Eng. AIME. Proc. of the Bauxite symposium, pp 468-499.

52. South, London.

53. Tiratsoo, E.N., 1984, Oilfields of the World, Scientific Press pp 392.

54. Unesco, 1978, World Water Balance and Water Resources of the Earth, 663 pp.

55. U.S. Bureau of Mines Mineral Commodity Profiles.
Mineral Commodity Summaries.
Mineral Industries of Africa, 1984, pp 153.
Minerals yearbook.
Minerals Facts and Problems.
Namibia, 1983, Coakley, G. et al, pp 57.
Zimbabwe, 1981, Shekarchi, E. et al., pp 62.

56. U.S. Geological Survey
- van Alstine, R.E., Schruben, P.G, 1980 - Fluorspar resources of Africa Bulletin 1487, pp 25.
- International Strategic Minerals Inventory Summary Reports.
- De Young, J.H., Sutphin, D.M., Caunou, W.F., 1984, Manganese 930-A, pp 22.
- De Young, J.H., Lee, M.P., Lipin, B.R., 1984, Chromium, 930-B pp 41.
- Krauss, U.H., Saam, H.G., Schmidt, H.W., 1984, Phosphate, 930-C, pp 41.

57. Water Power, Newport Beach.

58. West Africa, London.

59. World Bank Energy Assessment Reports, 1984 - Towards Sustained Development in sub-Saharan Africa, pp 102.

60. World Energy Conference, London.

61. World Mining, San Francisco.

62. World Oil.

63. World Water, Liverpool.

INDEX

Algeria : 38, 39, 44, 48-50, 54
55, 56, 59, 60, 61, 66, 67, 79-88.
Aluminum : 20, 27, 33, 40, 55, 66,
68, 74, 75, 95, 100, 102, 163, 165,
183, 309.
Angola : 38, 39, 41, 44, 46, 48,
58, 61, 62, 64, 197-203.
Andalusite : see industrials.
Antimony and arsenic : 19, 21, 30,
31, 32, 40, 46, 48, 56, 68, 87, 88,
92, 139, 164, 251, 257, 266, 271,
301, 309.
Arsenic : see antimony.
Asbestos : see building materials.
Ascension : 38, 324.
Attapulgite : see industrials.
Barite : see industrials.
<u>Base metals and silver</u> : 19, 21, 26,
27, 29, 31-35, 37, 40, 48-50.
- Algeria : 50, 79, 87, 88.
- Angola : 197, 203.
- Benin : 171.
- Botswana : 50, 259, 264.
- Burkina Faso : 48, 138-140.
- Burundi : 233-4
- Cameroon : 185.
- Central African Rep. : 187.
- Congo : 48, 195-7
- Egypt : 48, 102, 108.
- Ethiopia : 115-119.
- Gabon : 188, 193.
- Ghana : 163.
- Guinea : 153.
- Ivory Coast : 162.
- Kenya : 50, 232.
- Libya : 48.
- Malagasy Rep. : 318.
- Malaŵi : 243.
- Mali : 133,137.
- Mauritania : 50, 130, 133.
- Moçambique : 244, 247.
- Morocco : 48-50, 68, 73-75.
- Namibia : 50, 264, 268-270.
- Niger : 143, 144.
- Nigeria : 48, 172.
- Sea bed : 325-327.
- Senegal : 125-129.
- Somalia : 119.
- South Africa : 50, 56, 277-282, 300.
- Sudan : 48, 109, 114.
- Tanzania : 241.
- Togo : 169.
- Tunisia : 50, 88, 92, 93.
- Uganda : 229-231.
- Zaïre : 48-50, 56, 203, 206-210.
- Zambia : 48-50, 56, 214-8.
- Zimbabwe : 50, 251, 257.

Bauxite Iron : bauxite group.
Benin : 38, 66, 170-1.
Bentonite : see industrials.
Beryl : see tin group.
Bismuth : 30, 31, 78, 185, 229, 231,
246, 318.
Bituminous shales : see energy.
Bophutatswana : see South Africa.
Boron : 40, 66.
Botswana : 38, 40, 44, 46, 50, 53, 61,
62, 259-266.
Bromine : 40, 76.
<u>Building materials</u> : 31, 32, 36, 37,
40, 48, 53-55.
- Algeria : 53, 79, 83, 86.
- Angola : 197, 201, 203.
- Benin : 170-1.
- Botswana : 265-6.
- Burkina Faso : 138-9.
- Burundi : 223-4.
- Cameroon : 54, 55, 182, 184, 185.
- Canary : 319.
- Cape Verde : 319.
- Central African Rep. : 186-7.
- Chad : 144-5.
- Comoro : 319.
- Congo : 201.
- Djibouti : 120.
- Egypt : 53, 102, 107.
- Ethiopia : 115, 118.
- Gabon : 188, 193.
- Ghana : 55, 163, 166.
- Guinea : 148, 153.
- Ivory Coast : 160-1.
- Kenya : 53, 232-3.
- Lesotho : 274.
- Liberia : 158-9.
- Libya : 53, 95, 99.
- Madeira : 319.
- Malagasy Rep. : 318

- o Malaŵi : 241-3.
- o Mali : 133, 135.
- o Mauritania : 54, 130.
- o Mayotte : 319.
- o Moçambique : 244-6.
- o Morocco : 53, 68, 76.
- o Namibia : 266, 272.
- o Niger : 140, 143, 144.
- o Nigeria : 53, 172, 179.
- o Réunion : 319.
- o Rwanda : 226-9.
- o Senegal : 53, 125, 128.
- o Sierra Leone: 154, 157.
- o Somalia : 54, 120, 121, 122.
- o South Africa : 54, 55, 277-282, 302-4, 309.
- o Sudan : 109, 112, 114.
- o Swaziland : 274.
- o Tanzania : 55, 207, 239.
- o Togo : 167, 169, 170.
- o Tunisia : 53, 88, 92.
- o Uganda : 229, 201.
- o Zaïre : 55, 204, 213-4.
- o Zambia : 55, 215, 220.
- o Zimbabwe : 53, 55, 249-251, 253.

Burkina Faso : 38, 48, 138-140.
Burundi : 38, 44, 48, 64, 223-5.
Cadmium : see base and minor metals.
Cameroon : 39, 44, 48, 55, 61, 62, 63, 182-5.
Canary Islands : 38, 40, 51, 53, 55, 64, 319, 322-3.
Cape Verde Republic : 38, 54, 319, 323.
Carbon dioxide : 232, 234.
Cement : see building materials.
Central African Republic (RCA) : 38, 40, 46, 186-7.
Cesium and rubidium : 28, 29, 30, 31, 40, 51, 66, 251, 257, 266, 271.
Chad : 38, 46, 53, 144-5.
Chromium : see ultrabasics.
Coal : see energy.
Cobalt : 15, 22, 23, 36, 37, 40, 50, 66, 68, 77, 139, 162, 203-4, 210, 218, 223-4, 229, 231, 233, 241, 251, 257, 259, 264, 294, 311, 327.
Coke : 44, 251, 291.
Columbium : see tin group.
Comoros : 38, 319, 321.
Congo : 38, 39, 44, 48, 53, 61, 193-7.
Copper : see base metals.
Corundum : see industrials.
Diamonds : see the precious.
Diatomite : see industrial minerals.
Djibouti : 38, 46, 119-120.
Egypt : 38, 39, 44, 48, 51, 53, 55, 56, 57, 60, 62, 64, 66, 101-109.
Emeralds : see the precious.
Energy : 5, 19, 20, 32-35, 37, 40, 44-46, 55, 66-67.

- o Algeria : 44, 55, 79-83, 88.
- o Angola : 44, 198-201.
- o Benin : 170-1.
- o Botswana : 44, 46, 259, 262-4.
- o Burkina Faso : 138-9.
- o Burundi : 223, 225.
- o Cameroon : 44, 55, 182-3, 185.
- o Canary : 319, 322.
- o Cape Verde : 319, 323.
- o Central African Rep. : 46, 186-7.
- o Chad : 46, 144-5.
- o Comoro : 319,321.
- o Congo : 44, 196-7.
- o Djibouti : 46, 120.
- o Egypt : 44, 55, 101-105.
- o Equatorial Guinea : 319, 324.
- o Ethiopia : 44, 46, 115, 117-118.
- o Gabon : 44, 188-192.
- o Gambia : 125.
- o Ghana : 55, 163, 165.
- o Guinea : 46, 148, 153.
- o Guinea-Bissau : 146-8.
- o Ivory Coast : 46, 160-1.
- o Kenya : 55, 232, 234-6.
- o Lesotho : 274-5.
- o Liberia : 158-160.
- o Libya : 44, 55, 95-98.
- o Malagasy Rep. : 311-5.
- o Malaŵi : 241-2.
- o Mali : 45, 46, 133-5.
- o Mauritania : 130, 132.
- o Mauritius : 319.
- o Moçambique : 44, 243-5.
- o Morocco : 44, 45, 68, 78.
- o Namibia : 44, 266, 268-9.
- o Niger : 44, 45, 140.
- o Nigeria : 44, 55, 172-8.
- o Sea bed : 324.
- o Réunion : 319.
- o Rwanda : 44, 226.
- o Sao Thomé : 319, 324.
- o Senegal : 46, 125, 127-8.
- o Seychelles : 319-321.
- o Sierra Leone : 46, 154, 157.
- o Somalia : 120, 121.
- o South Africa : 44, 55, 276-283, 288-293, 308.
- o Sudan : 46, 109-112.
- o Swaziland : 44, 272-3.
- o Tanzania : 44, 236-8.
- o Togo : 167.
- o Tunisia : 44, 88-89.
- o Uganda : 46, 229-231.
- o Zaïre : 44, 45, 46, 55, 203-6, 210.
- o Zambia : 44, 55, 214-5, 219, 220.
- o Zimbabwe : 44, 55, 251, 254-5.

Equatorial Guinea : 38, 319, 324.
Ethiopia : 38, 40, 46, 53, 59, 115-119
Feldspar : see tin group.
Ferroalloys : 19, 23, 33, 34, 36, 37,

56, 102, 188, 249-251, 256, 309-310.
Fertilizers : see food and water.
Flint clay : see industrials.
Food and water : 5, 16, 25, 30, 32, 34, 35, 37, 40-44, 52, 54, 55.
- Algeria : 55, 79, 83 - 85.
- Angola : 43 197 , 201-3.
- Ascension : 324.
- Benin : 170-1.
- Botswana : 43, 53, 265.
- Burkina Faso : 138-9.
- Burundi : 233-4.
- Cameroon : 42, 184.
- Canary : 319, 322-3.
- Cape Verde : 319, 323.
- Central African Rep. : 186-7.
- Chad : 53, 144-5.
- Congo : 53, 197.
- Djibouti : 119, 120.
- Egypt : 53, 55, 102, 106, 107.
- Ethiopia : 42, 53, 115-117.
- Gabon : 42, 53, 193.
- Gambia : 123.
- Ghana : 163, 165.
- Gough : 319, 324.
- Guinea : 153.
- Guinea-Bissau : 146.
- Ivory Coast : 160-1.
- Juan de Nova : 320.
- Kenya : 42, 53, 202-4.
- Lesotho : 42, 274.
- Liberia : 159.
- Libya : 53, 55, 95, 98-99.
- Madeira : 322.
- Malagasy Rep. : 44, 317-8.
- Malaŵi : 241-2.
- Mali : 133,134.
- Mauritania : 130-2.
- Mauritius : 319.
- Moçambique : 244-5.
- Morocco : 52, 55, 68, 70-72.
- Namibia : 41, 43, 53, 266, 270-1.
- Niger : 140, 142-3.
- Nigeria : 55, 172, 178-9.
- Sea bed : 324.
- Prince Edward Isl. : 320.
- Rwanda : 228-9.
- St. Helena : 324.
- Sao Thomé : 319, 324.
- Senegal : 53, 55, 125-128.
- Seychelles : 319, 321.
- Sierra Leone : 53.
- Somalia : 120-121.
- South Africa : 41, 52, 53, 55, 277-282, 307-8.
- Sudan : 109-110, 113.
- Suqutra : 319, 321.
- Tanzania : 43, 53, 256-7.
- Togo : 53, 167-9.
- Tunisia : 52, 55, 88-92.
- Uganda : 53, 229, 230.
- Zaīre : 42, 204, 210, 213-4.
- Zambia : 215, 220-1.
- Zimbabwe : 53, 251, 258.

Fuller's earth : see industrials.
Gabon : 38, 39, 40, 44, 48, 51, 53, 59, 61, 62, 64, 65, 188-193.
Gallium : see minor metals.
Gambia : 38, 123-5.
Garnet , gem : see the precious minerals.
Garnet, industrial : see industrials.
Gas : see energy.
Geothermal : see energy.
Germanium : see minor metals.
Ghana : 38. 40, 46, 48, 55, 51-59, 163-166.
Gough and Tristan de Cunha : 38, 319, 324.
Graphite : see industrials.
Groundwater : see food and water.
Guinea : 38, 40, 46, 48, 55, 62, 63, 148-153.
Guinea Bissau : 38, 48, 146-8.
Gypsum : see building materials.
Hafnium : see tin-titanium group (zircon).
Heavy oil : see energy.
Helium : 19, 31, 40, 66, 79, 83, 239.
Hydropower: see energy.
Ilmenite : see tin-titanium group.
Industrial minerals : 19, 21, 25, 27-29 30-32, 35-37, 40, 50, 52, 53, 66.
- Algeria : 50, 79, 85, 86.
- Angola : 197, 203.
- Benin : 171.
- Botswana : 259, 265-6.
- Burkina Faso : 138, 140.
- Burundi : 223-4.
- Cameroon : 182, 184-5.
- Chad : 144-5.
- Egypt : 53, 102, 107.
- Ethiopia : 115, 118.
- Gabon : 188.
- Gambia : 123.
- Ghana : 166.
- Guinea : 153.
- Ivory Coast : 160-1.
- Kenya : 50, 202-4.
- Liberia : 158, 160.
- Libya : 95, 99.
- Malagasy Rep. : 52, 311-2, 315, 318.
- Malaŵi : 241-3.
- Moçambique : 244, 246.
- Morocco : 26, 50, 68, 75-76.
- Namibia : 50, 266, 271.
- Nigeria : 172, 179, 180.
- Rwanda : 226.
- Sao Thomé : 324.
- Senegal : 53, 125, 128-9.

o Sierra Leone : 155, 157.
South Africa : 50, 54, 277-282, 300,
o 305-7, 309.
o Sudan : 109, 114.
o Swaziland : 272-4.
o Tanzania : 237, 239, 240.
o Togo : 53, 167, 169.
o Tunisia : 50, 88.
o Uganda : 229, 231-233.
o Zaïre : 210-214.
o Zambia : 215, 220-1.
o Zimbabwe : 52, 53, 251, 258.

Iodine : 40.
Iridium : see ultrabasic group.
Iron bauxite group : 15,19, 20-23, 25, 26, 32-37, 40, 47-48, 66.
o Algeria : 48, 79, 83, 87.
o Angola : 48, 197, 202.
o Benin : 170-1.
o Botswana : 259, 265.
o Burkina Faso : 138-140.
o Cameroon : 48, 182-4.
o Central African Rep.: 186-7.
o Chad : 144-5.
o Congo : 197.
o Egypt : 48, 102, 108.
o Ethiopia : 115, 119.
o Gabon : 48, 188, 192-3.
o Ghana : 48, 163, 165-6.
o Guinea : 48, 148-152.
o Guinea-Bissau : 146.
o Ivory Coast : 48, 160, 162.
o Kenya : 232.
o Liberia : 48, 158-160.
o Libya : 95, 100.
o Malagasy Rep.: 48, 317.
o Malaŵi : 48, 241-2.
o Mali : 48, 133, 136-7.
o Mauritania : 48, 130-1.
o Moçambique : 244, 247.
o Morocco : 48, 77, 78.
o Namibia : 266, 271-2.
o Niger : 144.
o Nigeria : 172, 180.
o Sea bed : 311, 326-7.
o St. Helena : 319, 324.
o Senegal : 48, 125, 138-9.
o Sierra Leone : 48, 154-5.
o Somalia : 120, 122.
o South Africa : 48, 277-282, 297-300, 301.
o Sudan : 109, 113, 114.
o Swaziland : 272-3.
o Tanzania : 237-240.
o Togo : 167-169.
o Tunisia : 88, 93.
o Uganda : 229, 231.
o Zaïre : 48, 204, 214.
o Zambia : 215, 220.
o Zimbabwe : 48, 251, 257.

Ivory Coast : 38, 46, 48, 61, 62, 64, 160-2.
Juan de Nova (French) : 320.
Kaolin : see industrials.
Kenya : 38, 40, 50, 51, 53, 55, 61, 232-6.
Kyanite : see industrials.
Lead : see base metals.
Lesotho : 38, 40, 274-5.
Liberia : 38, 40, 48, 61, 62, 151-160.
Libya : 38, 39, 40, 44, 48, 53, 55, 56, 60, 63, 64, 67, 93-100.
Lignite : see energy.
Limestone : see building materials.
Lithium : see tin group.
Madeira : 38, 319, 321-2.
Magnesite : see ultrabasic group.
Magnesium salts : see industrials.
Malagasy Republic : 38, 40, 44, 48, 51, 59, 314-8.
Malaŵi : 38, 48, 241-3.
Mali : 38, 45, 46, 48, 51, 133-8.
Manganese ore : see Iron bauxite group.
Manganese metal : 19, 32, 33, 36, 56, 310.
Marble : see building materials.
Mauritania : 38, 40, 48-50, 54, 129-133.
Mauritius : 38, 319-320.
Mayotte : 38, 319, 320-1.
Mercury : 26, 27, 40, 48, 79, 80, 87, 88, 93.
Meerschaum : 29, 40, 120, 121, 239.
Mica : see tin group.
Moçambique : 38, 41, 44, 48, 51, 61, 63, 64, 243-7.
Molybdenum : 23, 40, 51, 77, 78, 108, 119, 122, 140, 143, 144, 157, 162, 184, 230, 233, 257, 272-3, 318.
Minor metals : 19, 20, 23, 24, 26, 27, 28, 29, 36, 40, 50, 66, 78, 148, 153, 204, 210, 215, 218, 266, 270.
Monazite : see tin-titanium group.
Morocco (and Western Sahara) : 38, 39, 44, 45, 46, 48-50, 52, 53, 55, 59, 61, 65, 68-79.
Namibia : 38, 40, 41, 44, 46, 50, 51, 53, 58, 59, 62, 266-272.
Nickel : see ultrabasic group.
Niger : 38, 40, 44, 45, 51, 54, 62, 65, 140-4.
Nigeria : 38, 39, 44, 48, 51, 53, 55, 56, 57, 62, 64, 67, 171-182.
Niobium : see tin group.
Oil : see energy.
Palladium : see ultrabasic group.
Pantelleria and Pelagians : 38, 319, 321.

Peat : see energy.
Perlite : see building materials.
Petroleum (products) : see energy.
Phosphates : see food and water.
Platinum group metals : see ultrabasic group.
Potash : see food and water.
Pozzuolana and
Pumice : see building materials.
Precious metals and minerals : 19, 21, 22, 25, 26, 32-34, 37, 40, 51, 66.
- Angola : 197, 201-3.
- Botswana : 46, 259-262.
- Burkina Faso : 138-140.
- Burundi : 223-4.
- Cameroon : 182, 184.
- Central African Rep. 46, 186-7.
- Chad : 144-5.
- Congo : 195-6.
- Egypt : 107.
- Ethiopia : 115, 118-119.
- Gabon : 188, 193.
- Ghana : 46, 163-5.
- Guinea : 46, 148, 153.
- Ivory Coast : 160-2.
- Kenya : 51, 232-4.
- Lesotho : 274.
- Liberia : 159.
- Malagasy Rep.: 51, 318.
- Malaŵi : 51.
- Mali : 46, 133-5.
- Mauritania : 130, 133.
- Moçambique : 244, 246.
- Morocco : 68.
- Namibia : 46, 51, 266-8, 271-2.
- Niger : 144.
- Nigeria : 172, 180-1.
- Sea bed : 325-6.
- Rwanda : 226.
- Senegal : 125, 128, 129.
- Sierra Leone : 46, 154-155, 157.
- South Africa : 46, 51, 56, 276-288, 308.
- Sudan : 109, 113, 114.
- Swaziland : 273.
- Tanzania : 46, 51, 236-7, 239-241.
- Uganda : 229, 231.
- Zaïre : 46, 204, 211, 214.
- Zambia : 51, 214-5, 218, 220.
- Zimbabwe : 46, 51, 249-253, 258.

Presidios, & : 38, 319, 321.
Prince Edward Island : 320.
Pyrite : see base metals and gold.
Quartz crystal, gem and industrial : see semiprecious stones and tin group.
Quartzite : see industrials.
Rare earth metals : see tin-titanium group.
Réunion : 38, 319, 321.
Rubidium : see cesium.
Rutile : 19, 36, 40, 51, 52, 108, 123, 154-5, 162, 171, 185, 244, 301.
Rwanda : 38, 40, 44, 51, 64, 225-9.
Salt : see food and water.
Sand : see industrials.
Saõ Thomé e Principe : 38, 319, 324.
Scandium : 40, 140, 143, 188, 192, 266, 318.
Sea bed and marine resources : 311, 324-8.
Selenium : see minor metals.
Semi-precious stones : see the precious (and tin group).
Senegal : 38, 40, 44, 46, 48, 53, 57, 59, 62, 63, 64, 125-129.
Seychelles : 38, 63, 319, 321.
Sierra Leone : 38, 40, 46, 48, 51, 53, 62, 154-7.
Silica : see industrials.
Silicon : 32, 310.
Sillimanite : see industrials.
Silver : see gold and base metals.
Soapstone : see ultrabasics.
Somalia : 38, 54, 120-122.
South Africa : 38, 39, 40, 44, 46, 48-52, 55, 56, 58, 66, 276-313.
South-West Africa : see Namibia.
Staurolite : 123, 166.
Steel : 23, 33, 56, 68, 79, 80, 87,88, 93, 95, 100, 102, 163, 167, 169, 172, 180, 232-3, 244, 247, 251, 309.
Stone : see building materials.
Strontium : 29, 30, 48, 76, 79, 80, 85, 92, 230, 243.
Sudan : 38, 46, 48, 52, 58, 64, 109-114.
Suqutra (South Yemen) : 38, 319, 34.
Sulfur : byproduct : see (oil) energy.
native : see industrials.
pyrite : see base metals and gold.
Swaziland : 38, 40, 44, 51, 272-4.
Talc : see ultrabasics.
Tantalum : see tin group.
Tanzania : 38, 40, 41, 44, 46, 51, 53, 55, 64, 236-241.
Tellurium : see minor metals.
Thorium : see tin-titanium group (monazite).
Tiger's eye : see the precious.
Tin and titanium group : 19, 20, 21, 23-30, 32, 33, 36, 37, 40, 50-52, 66.
- Algeria : 83, 88.
- Angola : 203.
- Botswana : 264-5.
- Burkina Faso : 140.
- Burundi : 223-4.
- Cameroon : 182, 184-5.
- Central African Rep. : 187.

- Chad : 144-5.
- Congo : 196.
- Egypt : 55, 102, 107, 108.
- Ethiopia : 115, 119.
- Gabon : 193.
- Gambia : 123.
- Ghana : 166.
- Guinea Bissau : 146.
- Ivory Coast : 160, 162.
- Kenya : 252-3.
- Liberia : 160.
- Malagasy Rep. : 311, 315.
- Malaŵi : 241-3.
- Mali : 51, 133, 137.
- Mauritania : 132, 133.
- Moçambique : 51, 244, 246-7.
- Morocco : 68, 75, 77, 78.
- Namibia : 51, 266, 271.
- Niger : 140, 143, 144.
- Nigeria : 51, 172, 180-1.
- Rwanda : 51, 225-7.
- Senegal : 128, 129.
- Sierra Leone : 51, 155.
- Somalia : 120, 122.
- South Africa : 51, 55, 277-282, 297, 301, 304-5, 308.
- Sudan : 113, 114.
- Swaziland : 273.
- Tanzania : 236-7, 241.
- Uganda : 229, 230, 231.
- Zaīre : 55, 204, 211-4.
- Zambia : 215, 220.
- Zimbabwe : 51, 251, 254, 257.

Titanium : see tin group.
Togo : 38, 40, 54, 167-170.
Trona : see industrials.
Tungsten : see tin group.
Tunisia : 38, 39, 40, 44, 48-50, 52, 53, 55, 88-93.
Uganda : 38, 46, 53, 64, 229-231.
Ultrabasic group : 15, 19, 20, 21, 23, 26, 27, 29, 30-37, 40, 46, 66.
- Algeria : 83.
- Benin : 171.
- Botswana : 46, 259, 264, 266.
- Burkina Faso : 138-140.
- Burundi : 46, 223-4.
- Cameroon : 185.
- Central African Rep. : 187.
- Congo : 197.
- Egypt : 46, 107, 108.
- Ethiopia : 118, 119.
- Gabon : 188, 193.
- Ghana : 166.
- Guinea : 153.
- Ivory Coast : 46, 160, 162.
- Kenya : 46, 202.
- Liberia : 160.
- Malagasy Rep. : 46, 311, 315-8.
- Malaŵi : 243.
- Mauritania : 133.
- Moçambique : 244, 246-7.
- Morocco : 68, 76.
- Sea bed : 311, 326-7.
- Sierra Leone : 157.
- South Africa : 46, 56, 276-282, 293-7, 308.
- Sudan : 46, 109, 112, 113.
- Swaziland : 46, 273.
- Tanzania : 46, 207, 239-241.
- Togo : 169.
- Uganda : 231.
- Zaīre : 204, 210.
- Zambia : 215, 218, 221.
- Zimbabwe : 46, 55, 249-251, 254-6.

Uranium : see energy.
Vanadium : 19, 32, 33, 34, 36, 37, 48, 56, 66, 78, 120, 121, 133, 138-9, 203, 215, 223-4, 230, 244, 247, 266, 295-6, 309.
Vermiculite : 19, 25, 31, 32, 46, 107, 118, 122, 229, 230, 232, 239, 241-2, 266, 304.
Western Sahara : see Morocco.
Wollastonite : 28, 30, 40, 52, 76, 109, 113, 114, 232, 234, 271, 307, 318.
Wonderstone : see the precious and building materials.
Yttrium : see tin-titanium group.
Zaīre : 38, 39, 44, 45, 46, 48-51, 55, 56, 58, 59, 60, 61, 62, 64, 203-214.
Zambia : 38, 39, 44, 48-51, 55, 56, 63, 64, 214-221.
Zimbabwe : 38, 40, 44, 46, 48, 50, 51, 53, 55, 56, 59, 61, 62, 64, 249-258.
Zinc :see base metals.
Zirconium : see tin-titanium group.